blackout.

Principles of Fracture Mechanics

R. J. Sanford

Professor Emeritus
Mechanical Engineering
University of Maryland
College Park

Pearson Education, Inc.
Upper Saddle River, NJ 07458

Dedicated to the memory of Leona K. Sanford (1912–1969) and Paul H. Sanford (1910–1993), parents in the grandest sense that parenting implies.

Preface

OVERVIEW

Fracture mechanics as an engineering discipline was introduced in the 1950s under the leadership of George R. Irwin at the Naval Research Laboratory (NRL). The concepts of fracture mechanics were further developed and refined throughout the 1960s by a collaboration of researchers in universities, government laboratories, and the commercial aircraft and aerospace industries. Definable within the context of the linear theory of elasticity, the fundamentals of fracture mechanics have a wide range of engineering design applications, including the analysis of brittle fracture of low-toughness structural materials and many nonmetallics, and the quantitative prediction of fatigue crack growth in a wide range of engineering materials. This latter application is of major importance in contemporary engineering design since over 80 percent of all brittle fractures have their origins in fatigue crack growth. In its current state of development, the discipline of Linear Elastic Fracture Mechanics (LEFMs) is a mature science that can be and, indeed, is being introduced into the basic programs of instruction in mechanical, civil, aerospace, and engineering mechanics departments.

The focus of this book, intended for a first course in the mechanics of fracture at the graduate level (or for senior undergraduates with a background in engineering mechanics), is on the mathematical principles of linear elastic fracture mechanics and their application to engineering design. The selection of topics and order of presentation in the book evolved from a graduate course in fracture mechanics developed by the author over the last two decades. The material has been tested on several hundred students over that time, and the level of treatment and extent of mathematical development presented in the text are the result of feedback provided by the students. Many of the chapter exercises and comprehensive design problems are taken from examinations given in the course. A Web site at the University of Maryland (www.wam.umd.edu/~sanford) contains supplemental material, including detailed graphics, data sets, and relevant links to supporting material. The site also provides a convenient way to communicate your suggestions and comments to the author.

The material is presented in a conversational, yet rigorous, manner, with the focus on the general formulation of the theory. In this way the origins and limitations of the simplified results presented in other introductory texts are apparent. Throughout the text, key historical results are emphasized to provide a sense of the history of fracture mechanics. This feature makes the book of interest to practicing engineers and researchers interested in a broad overview of the field.

Ideally, the study of LEFM should have as a prerequisite a thorough understanding of the linear theory of elasticity: however, the practicalities of scheduling graduate instruction often result in the student having no choice but to study both topics concurrently. The organization of the book anticipates this possibility by including, early on, a chapter on those elements of solid mechanics necessary for understanding the remainder of the text.

The minimum mathematical background required to gain a full appreciation of the material is a one-semester introductory course in partial differential equations (often taught at the undergraduate junior level) or its equivalent. Some familiarity with complex variables and functions, but not necessarily a comprehensive knowledge, is also required. Also, some of the exercises assume proficiency in computer-based problem solving at the PC or workstation level.

KEY FEATURES OF THE BOOK

The book is a self-contained manual on the mechanics aspects of the theory of brittle fracture and fatigue and is suitable for either self-study or classroom instruction. It includes a guided introduction to the linear theory of elasticity with pivotal results for the circular hole, the elliptical hole, and the wedge, leading up to the general problem of bodies containing cracks.

The book draws upon extensive original material on the mathematical formulation of the stress field around crack tips, the numerical analysis of bodies with finite dimensions, and the use of experimental methods to determine the stress intensity factor, based on the author's nearly 40 years experience in the field.

Designed around pedagogical needs, the exercises at the end of each chapter provide a mixture of problematic approaches that can be emphasized, depending on the instructor's focus. A typical chapter includes problems that extend the mathematical developments presented in the chapter, applications problems that require numerical and/or graphical responses, and essay/literature study questions. In addition, more comprehensive exercises requiring integration of the knowledge presented throughout the text of the book are included as an appendix. The majority of these comprehensive exercises were adapted from take-home final exams given by the author over the years and are suitable for that purpose. To protect the integrity of these questions they will not be included in the solutions manual.

Curiously missing from all of the existing introductory textbooks on fracture mechanics is even a modest tabulation of elastic and fracture properties of engineering materials. Included in this text are two appendices listing (a) strength and fracture properties and (b) fatigue data for a wide variety of metallic materials, adapted from the NASA/NASGRO database.

All of the above features of the book notwithstanding, the single most unique feature of the book is the use throughout of a unified mathematical treatment based on the generalized Westergaard formulation of the elastic problem of stresses in bodies containing cracks. This mathematically complete formulation, developed by the author in 1978 and extended to multiple categories of problems in the years since, provides all of the rigor needed to carry out the theme of the book without demanding mathematical proficiency beyond that of most potential readers (and instructors). An appendix on the use of complex variables in elasticity describes the simple complex-variable operations required with this method. The Westergaard method presented in this book differs from that presented in every other book on fracture mechanics, including even the most recent topical monographs, in that the formulation presented here is mathematically equivalent to any of the alternative complex-variable formulations but is significantly less difficult to manipulate. As a consequence, readers of this book will be able to derive very general mathematical expressions and solve complex problems, even if they have had only nominal mathematical training. In addition, this formulation is highly compatible with symbolic manipulation languages such as Mathematica™ and MathCAD™. Consistent with the self-contained philosophy of the book, an appendix contains an extensive tabulation of Westergaard stress functions and the corresponding K solutions.

OUTLINE OF THE BOOK

The book consists of 11 chapters and 5 appendices. The main body of the book divides naturally into two roughly equal parts that focus on the two complementing concepts of fracture mechanics; namely, the stress state at the crack tip and the material's resistance to fracture. The appendices provide supplemental material necessary for completeness and self-sufficiency as a textbook.

The first focal concept presented in the book, discussed in Chapters 1 through 5, is the development of the mathematical theory of the state of stress at the crack tip. By the middle of Chapter 3 the reader is introduced to the universal nature of this stress state and its characterizing parameter, the geometric stress intensity factor, K. The remainder of the chapter demonstrates analytical methods (exact and approximate) to determine K without formally solving the complete elasticity problem for each new geometry. In Chapters 4 and 5 the theoretical basis for various numerical (Chapter 4) and experimental (Chapter 5) methods to compute K is developed and demonstrated by practical examples. The goal of these two chapters is to encourage creativity in extracting the stress intensity factor and other key variables for realistic geometries.

The second focal concept, encompassing Chapters 6 through 11, is the development of appropriate theories of failure based principles of elastic mechanics. In

Chapter 6 the concept of a critical stress intensity factor as a material property, distinct from the geometric stress intensity factor, is introduced, and the consequences explored. Chapter 7 follows a similar development for the strain energy release rate. Having defined new material properties in the preceding chapters, Chapter 8 describes established and novel ways to determine them.

In Chapter 9 the focus shifts from brittle, sudden fracture to its frequent precursor, sub-critical fatigue crack growth. The emphasis in this chapter necessarily is on the origin and rationale behind the various empirical fatigue crack growth laws. Starting from the Paris law, the historical progression of fatigue laws is presented, including the latest variant of the Foreman–Neuman–de Konig form used in the NASA/NASGRO 3.0 computer program for fatigue life prediction, which is treated in detail.

The incorporation of a fracture-resistant mentality into the design process by adding the fracture failure analysis scenario into the design calculation sequence is described in Chapter 10 and demonstrated with an example taken from the literature. The philosophy and requirements of the U.S. Air Force damage-tolerant design model are presented. The role of nondestructive evaluation (NDE) in the design process is also introduced in Chapter 10. Finally, the lessons to be learned from reviewing case studies in fracture are discussed, and several examples are presented.

The final chapter, Chapter 11, is intended as an introduction to the intermediate mechanism of failure lying between brittle fracture and general yielding—specifically—elastoplastic fracture or ductile tearing. The intent of this chapter is to introduce the reader to the basic concepts and the terminology of elastoplastic fracture mechanics (EPFM).

ACKNOWLEDGMENTS

No textbook in an established field, such as fracture mechanics, can purport to be entirely original, and this one is no exception. While I do lay claim to some original material, mostly in Chapters 3, 4, and 5, the remainder of the subject matter is based on the published works of pioneers in fracture mechanics over the last 40 years. I have placed a special emphasis on citing the original works whenever a new topic is first introduced, both to maintain historical accuracy and to acknowledge the contributor of that particular idea. Rather than repeat all of those names here, I refer the reader to the list of references at the end of each chapter. Included among them are the contributions of many of the talented graduate students whom it has been my privilege to direct over the years. There is, however, one individual whose impact on fracture mechanics (and my personal involvement) is so profound that he deserves special recognition, and that is George R. Irwin. It was Dr. Irwin who hired me into the Ballistics Branch, Mechanics Division, of the Naval Research Laboratory in 1960 while I was still an undergraduate student at a local university. That generous act placed me among the likes of Joe Kies, Irvin Wolock, J. M. Krafft, A. M. Sullivan, and H. L. Smith, as well as the many leaders in the fracture mechanics field who came to NRL to visit and lecture. Following Dr. Irwin's retirement in 1967, we went our different paths until 1978, when I first became involved with the University of Maryland, where Dr. Irwin was a Visiting Professor. During my early years at the university, Dr. Irwin graciously provided me with research support on his research contracts, and we were once again in daily contact. Our offices were two doors apart, and I used the proximity to question him regularly on the fine points of fracture mechanics as only he knew them. He was instrumental in the selection of topics to be included in this book and took an active role in reviewing the early chapters. Later, as his eyesight failed, we conversed regularly on material to be included. There is no question that the topics and flow of ideas presented here have been greatly influenced by his presence. However, of all the things for which I have to thank Dr. Irwin, the most important is the atmosphere of "research for self-enlightenment" that he fostered at NRL. That sense of research solely for the joy of understanding has been his greatest gift to me and my contemporaries. The large number of publications coming out of the Ballistics Branch during those early years of fracture mechanics comprises only a small sampling of ideas spawned in the laboratories up and down that hall.

I wish to express my appreciation to the professional societies and technical journals that have given me permission to use copyrighted material in the preparation of this textbook. The vast library of material on fracture mechanics published by the American Society for Testing and Materials (ASTM) has been especially helpful. Appendix C was made possible through access granted by the American Society of Mechanical Engineers (ASME). My thanks also go to the Institute of Mechanical Engineers (IMechE) for granting me its permission to reproduce Figures 6.39 and 6.43 from the International Congress on Fatigue of Metals in 1956. For further information on IMechE publications, please visit www.imeche.org.uk or www.pepublishing.com.

I want to express my thanks to the doctors and staff of the Comprehensive Transplant Center and related departments of the Johns Hopkins Hospital in Baltimore whose skills and continued vigilance have made this work possible. I also want to thank that unknown family somewhere in central Florida for its most generous gift, and I urge every reader of this book to consider being an organ donor. Please, do not bury or burn useful parts for the body human. Moreover, make sure that you tell your family of your wishes. Somewhere, there is a book unwritten, a song uncomposed, or a work of art uncreated for want of an organ transplant. Recycle.

Finally, I would like to give a special acknowledgment to Tina, my wife and life partner. She was by my side through my darkest hours. It is difficult for me to imagine how I would have fared without her continued encouragement and support. This book, now completed at last, is my gift to her to mark our 40th wedding anniversary (June 11, 2001). Happy anniversary—with all my love.

R. J. SANFORD

CHAPTER 1

Introduction to Fracture Mechanics

1.1 HISTORICAL OVERVIEW OF BRITTLE FRACTURE

The field of fracture mechanics is focused on the prevention of brittle fracture and, as a scientific discipline in its own right, is less than 40 years old. However, the concern over brittle fracture is not new. The ancients were aware of the problem and designed against fracture by ensuring (unknowingly) that the structures were always in compression. Several classic examples come to mind. The Roman arch, which played an important role in the development of the Roman Empire, accomplished its function by incorporating a massive capstone to transfer compressive stress throughout the arch. In contrast to the simple arch are the elaborate and beautiful buttresses of the European cathedrals. We now understand that the function of the buttress was to add weight to the cathedral wall to ensure that the combined effects of the membrane and bending stress remained in compression at all times. Based on accounts of the numerous collapses of cathedrals under construction, it is clear that the architects of the time were unaware of the critical role that eliminating tensile stresses plays in preventing failure in brittle materials.

One of the earliest recorded attempts to systematically study strength of materials is found in the sketches and notes by Leonardo da Vinci on the strength of beams and wires. In comments adjoining a sketch of an apparatus for studying the strength of iron wires (see Figure 1.1), da Vinci describes an experiment intended to establish a "law" for the influence of length on the strength of all types of materials, including "metals and woods, stones, ropes, and anything that will support other objects" [Uccelli, 1956]. It is not known whether such experiments were ever performed. Perhaps, if they had been, the concept of the size effect would have been discovered long before the 20th century.

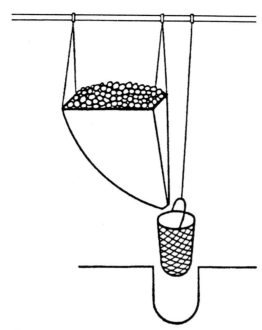

Figure 1.1   Da Vinci's apparatus for measuring the strength of wire.

Despite the earlier works of da Vinci, the science of the strength of materials is generally attributed to Galileo. During the period of his seclusion following the Inquisition, Galileo wrote *Two New Sciences* [1638], in which he describes the results of his earlier studies of as long as 40 years on the strength of materials. He introduced the concept of tensile strength in simple tension, which he called the "absolute resistance to fracture," by observing that the strength of a bar is proportional to the cross-sectional area and is independent of the length. (See Figure 1.2.)

A major shift in emphasis in the theory of strength of materials occurred in the 19th century with the introduction of malleable iron to replace wood, stone and cast iron. This change in primary construction material permitted rapid advancement in both the bridge and railroad industries and brought with it a new type of failure behavior to be accounted for—fatigue. The accepted failure theory of the day was that, under the action of cyclic stresses, the tough, fibrous character of the malleable iron was transformed into a brittle, crystalline form—a perception, no doubt, reinforced by casual observation of a typical fatigue fracture surface, with its characteristic smooth, flat fatigue crack surface and rough, irregular unstable growth region. It was also observed that thin cracks often appeared in the brittle material.

The proper role of slow crack growth in fatigue failure appears to have been first discussed by Rankine [1843] in relation to the failure of railroad axles. He discounts the recrystalization theory and states that "the fractures appear to have commenced with a smooth, regularly-formed, minute fissure, extending all round the

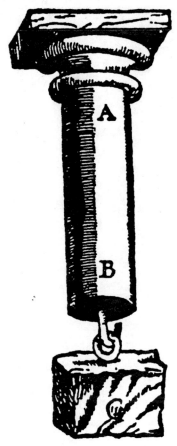

Figure 1.2 Galileo's illustration of
tensile strength in a column.

neck of the journal and penetrating on an average to a depth of half an inch. They
would appear to have gradually penetrated from the surface toward the centre...."
Rankine's description of the fracture surface of malleable iron is characteristic of
most fatigue crack patterns in rotating metal shafts, such as the one illustrated in
Figure 1.3.

The influence of a crack on the fracture strength was widely appreciated by the
end of the 19th century, but the exact nature of its influence was not. The milestone
work of Inglis [1913], which we will treat in detail in Chapter 2, was inspired by a
remark attributed to B. Hopkinson before the Sheffield Society of Engineers and
Metallurgists concerning the desirability of computing the intensities of the stresses
around a crack. To address this issue, Inglis solved the elasticity problem of an
elliptical hole in an otherwise uniformly loaded plate. By letting the ellipticity ratio
approach zero, he was able to deduce some important observations about the stress
state around the tip of a crack-like defect.

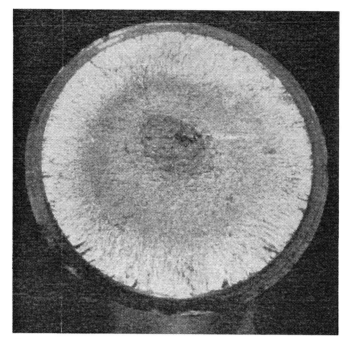

Figure 1.3 Appearance of a fatigue fracture surface in
a rotating steel shaft [courtesy Naval Surface Warfare
Center, Carderock Division].

Building on the mathematical foundation laid by Inglis, A. A. Griffith [1921]
attempted to explain the large discrepancies between the theoretical and measured
tensile strengths of glass (one million psi versus one thousand psi). He attributed the
lower value of the measured tensile strength to the high stresses in the neighborhood
of microscopic cracks and developed a theory of brittle fracture based on energy
concepts. We will treat the Griffith theory in full detail in Chapter 7.

With the onset of full-scale industrialization throughout the first half of the 20th
century, the problem of brittle fracture was seemingly overcome by the introduction
of ductile metals to replace the more brittle irons. Nonetheless, failures due to
the growth of cracks were still common, although most often these failures were
attributed to design flaws rather than real flaws. The solution always seemed to be
to increase the amount of material used. A. E. H. Love [1926], in his widely read
Treatise on the Mathematical Theory of Elasticity, describes a factor of safety of 6 for
boilers and axles, 6 to 10 for railway bridges, and 12 for propeller shafts, relative to the
tensile strength. However, because of the need to manufacture large quantities of
military hardware with limited natural resources during World War II, and due to the
increased use of aircraft with their increased strength-to-weight ratio requirements,
old design methods based on massive structures with large safety factors were forced

to give way to more efficient designs based on a more realistic theory of failure. By 1925 the concepts of both stress analysis (Inglis) and energy (Griffith) had been introduced to explain material behavior in the presence of a crack. Griffith had even introduced the idea of progressive crack extension by proposing a criterion for sustaining crack propagation. With these concepts firmly in place, it seems plausible that the development of a theory of fracture would ultimately emerge in some form. However, there were a series of major events in the period from 1940 to 1960 that played a pivotal role in shaping the discipline that we now call Linear Elastic Fracture Mechanics (LEFM).

In order to replace the large number of cargo ships lost to German U-boats, England and the United States embarked on a radically new approach to shipbuilding. Using techniques of mass production and heavy reliance on welding rather than riveting to assemble ships, U.S. shipyards produced the Liberty ship (formally designated the EC–2). Liberty ships were basic cargo ships, 441 feet long, with cargo-carrying capacity of 9,000 tons. In the four years of production, 1940–1944, a total of 2,708 ships were constructed. At the peak of the effort, one of these ships, the Robert E. Peary, was launched only 63 hours after her keel was laid! The key to this rate of production was the extensive use of welding, but along with this new technology came a new problem: Liberty ships exhibited a tendency to crack, usually in cold weather and rough seas. An eyewitness report of one such failure to a ship in the North Atlantic in March 1944 recounts that "a loud report, followed by two smaller ones, was heard... Immediately afterwards, the forward end of the ship separated from the after end and floated away" [Chiles, 1988]. While this account represents an extreme example, hundreds of ships did experience large cracks in their steel plates, such as the one shown in Figure 1.4.

In retrospect, the causes of these failures can be easily explained with modern-day principles of fracture mechanics. First, the composition of the steel was such that the transition from ductile to brittle behavior occurred at temperatures that the ships experienced while in service, particularly in the North Atlantic, where many of them were dispatched. Second, the design of the Liberty ship called for hatch openings with square corners. These sharp corners acted much like starter cracks. In fact, nearly half of the serious fractures suffered by these ships started at hatch covers. In the example show in Figure 1.4, the two key elements of fracture-resistant design—material selection and stress-state control—are readily apparent. There is no question that an elementary understanding of contemporary fracture mechanics (were it available at the time) would have prevented these failures.

Although the application was dramatically different, the failures of the de Havilland "Comet" commercial aircraft experienced features very similar to those found in the Liberty ship failures. The Comet, first manufactured in 1952, was the first

Figure 1.4   Failure of a Liberty ship, SS Schenectady, at dockside, January 1943 [courtesy B. B. Rath, Naval Research Laboratory].

twin-jet-engine passenger aircraft to fly at 40,000 feet with a pressurized cabin. After about a year in service, three aircraft failed, with considerable loss of life. From the study of recovered fuselage segments from one of these aircraft, the origin of the failure was identified as a short fatigue crack that started from an overhead observation window and caused the fuselage to burst [Wells, 1975]. Fracture tests in the United Kingdom on panels of the same aluminum alloy with fatigue cracks comparable with those found on the recovered aircraft failed to correlate with the stress levels experienced in service, even if the stress-raiser effect of the window was included. However, Irwin and others at the Naval Research Laboratory in Washington, DC, argued that the effective crack length should include the diameter of the window. Using this larger value of "effective crack length" resulted in critical stress levels that accounted for the failures.

One of the first applications of the infant science of fracture mechanics was the analysis of crack propagation in high-strength steel (with a yield stress of 190–215 ksi) solid propellant rocket motor casings for the Polaris and Minuteman missile programs. Beginning in 1958 extensive failure analyses and material characterization tests were conducted to determine the cause of occasional failures of the rocket motors (usually during hydro-testing). In nearly all of the cases, the cause of the failure was linked to fabrication cracks in welds or along weld borders. The problem was ultimately solved by a combination of smooth finishing of the welds, greatly improved inspection techniques, and material modification to improve fracture toughness. It was during this period that a major cooperative effort was undertaken to develop test methods and specimen geometries to measure the newly identified material property, the fracture toughness. In 1959 the American Society for Testing and Materials (ASTM) established a special committee for fracture testing of high-strength materials to assess the significance of these tests. In January of the following year the committee issued its first report which concluded that "the validity of the analytical methods of fracture mechanics is sufficiently well established" to permit their use in determining whether a fracture test "is measuring the significant quantities governing performance" and the degree to which the results of the fracture test "may be generalized to the more complex structure existing in service"[ASTM, 1960]. This committee was the forerunner of the ASTM E–24 Committee on Fracture Testing,[1] which continues to play a leadership role in the development and acceptance of fracture mechanics as an engineering discipline.

1.2 ELEMENTARY BRITTLE-FRACTURE THEORIES

Prior to the advent of fracture mechanics, failure theories for brittle fracture were based on the results of uniaxial strength tests. In many cases, these theories of failure apply to failure either by yielding or by brittle fracture, depending on whether the uniaxial yield stress, σ_{ys}, or the fracture stress, σ_f, is used as the critical parameter. Because brittle fracture failures tend to occur along planes normal to the principal planes, we will confine the discussion here to theories that do not depend on the shear stress. One of the earliest failure theories to be proposed was the maximum normal strain theory (also referred to as the St. Venant theory). According to this theory, failure will occur when the magnitude of the largest normal strain exceeds the uniaxial fracture strain, ϵ_f. With the introduction of Hooke's law, this failure theory can be expressed as

$$|\sigma_1 - \nu(\sigma_2 + \sigma_3)| = \sigma_f \quad \text{or} \quad |\sigma_3 - \nu(\sigma_1 + \sigma_2)| = \sigma_f \tag{1.1}$$

where, as is customary, $\sigma_1 > \sigma_2 > \sigma_3$ are the principal stresses [Ugural and Fenster, 1987]. For the case of plane stress, these conditions can be rewritten in the form

$$\frac{\sigma_1}{\sigma_f} - \nu\frac{\sigma_2}{\sigma_f} = \pm 1 \quad \text{or} \quad \frac{\sigma_2}{\sigma_f} - \nu\frac{\sigma_1}{\sigma_f} = \pm 1 \tag{1.2}$$

These equations represent the boundary lines on a two-dimensional failure locus plot. For the case of $\nu = 0.25$, the failure locus as determined by the maximum normal strain theory is illustrated in Figure 1.5a. As illustrated by the figure, failure will occur whenever the normalized stresses are outside of the shaded region.

---

[1] In 1992, ASTM Committees E–24 and E–09 were merged and renamed "Committee E–08 on Fatigue and Fracture."

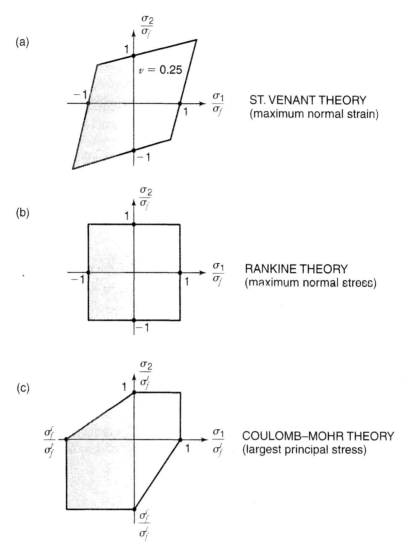

Figure 1.5 Failure loci in plane stress for three common brittle-failure theories.

A similar theory, based on maximum principal stresses, was proposed somewhat later by W. J. M. Rankine. The Rankine theory states that failure will occur when the magnitude of the largest principal stress exceeds the uniaxial fracture stress. Expressed mathematically, the Rankine theory is

$$|\sigma_1| = \sigma_f \quad \text{or} \quad |\sigma_3| = \sigma_f \tag{1.3}$$

[Ugural and Fenster, 1987]. For plane stress conditions, these equations can be expressed as

$$\frac{\sigma_1}{\sigma_f} = \pm 1 \quad \text{or} \quad \frac{\sigma_2}{\sigma_f} = \pm 1 \tag{1.4}$$

Using these equations, the failure locus in two dimensions is shown in Figure 1.5b for the maximum normal stress theory.

It has been widely observed that the fracture stress in uniaxial compression tests, σ_f^c, greatly exceeds the corresponding result in tension, σ_f^t, in typical brittle materials, such as cast iron, concrete and ceramics. To address this condition, Otto Mohr proposed a failure theory based on a graphical representation of the two extreme cases with his Mohr's circle. Independently, Coulomb proposed a failure theory based on an internal-friction argument. Since both of these theories reduce to the same mathematical form, they have been combined into one as the Coulomb–Mohr theory. To develop the locus equations it is convenient to treat three separate cases [Dally and Riley, 1978]. For plane stress, the cases are:

Case 1: σ_1 and $\sigma_2 \geq 0$

Case 2: σ_1 and $\sigma_2 \leq 0$

Case 3: $\sigma_1 \geq 0 \geq \sigma_2$ or $\sigma_2 \geq 0 \geq \sigma_1$

For Case 1, the Coulomb–Mohr theory coincides with the normal stress theory, and the locus equations are

$$\frac{\sigma_1}{\sigma_f^t} = 1 \quad \text{or} \quad \frac{\sigma_2}{\sigma_f^t} = 1 \tag{1.5a}$$

where σ_f^t is the failure stress in uniaxial tension. Case 2 is similar to Case 1 (after compensating for the different failure stress, σ_f^c, in uniaxial compression) when both stresses are compressive:

$$\frac{\sigma_1}{|\sigma_f^c|} = -1 \quad \text{or} \quad \frac{\sigma_2}{|\sigma_f^c|} = -1 \tag{1.5b}$$

In order to construct the yield locus equation for the regions in which the two principal stresses are of opposite sign (Case 3), we will assume that the governing equation is a straight line. Since the yield surface must be closed, the boundary points from Cases 1 and 2 provide the necessary information to solve for the governing equations for Case 3. Hence, in the region in which $\sigma_1 > 0$ and $\sigma_2 < 0$, the conditions

$$\sigma_1 = \sigma_f^t \quad \text{when} \quad \sigma_2 = 0$$

and

$$\sigma_2 = \sigma_f^c \quad \text{when} \quad \sigma_1 = 0 \tag{1.5c}$$

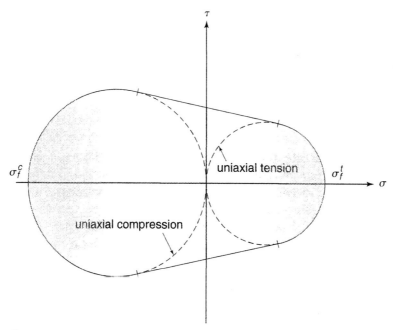

Figure 1.6 Mohr's-circle representation of the Coulomb–Mohr failure theory.

result in an equation of the following form for the yield locus in the fourth quadrant:

$$\frac{\sigma_1}{\sigma_f^t} - \frac{\sigma_2}{\sigma_f^c} = 1 \qquad (1.5d)$$

By a similar argument the yield locus equation for the second quadrant can be constructed. The resulting yield locus for plane stress is shown in Figure 1.5c. As with the previously described failure theories, failure is assumed to occur when the normalized stresses are outside of the shaded region.

An alternative graphical description of the Coulomb–Mohr theory can be constructed by considering the limiting uniaxial cases on a Mohr circle plot. According to the Mohr theory, all possible states of allowable stress can be represented by Mohr's circles bounded by the limiting circles of uniaxial tension and uniaxial compression and the straight lines corresponding to points connecting the circles' common tangent. The Mohr's circle representation of the Coulomb–Mohr failure theory is shown in Figure 1.6.

1.3 CRACK EXTENSION BEHAVIOR

The classical theories of fracture described in the previous section implicitly assume that failure occurs by separation of the atomic lattice across the entire plane of fracture as a single event. However, from theoretical considerations, the forces required to accomplish this feat are very large—much larger than the measured tensile strength of brittle materials. More realistically, the measured forces are more closely related to the force required to separate a few lines of atoms at a time. This perspective of fracture by progressive separation of the fracture plane is fundamental to the development of a modern fracture theory.

A series of experiments conducted at the Naval Research Laboratory in Washington, DC, in the late 1940s on crack propagation behavior in zinc foils has provided important insight into the mechanism of progressive fracture in metals. In these studies, a razor blade was used to introduce an initial crack into a thin sheet of zinc foil loaded in tension. The load was slowly increased until slow extension of the crack occurred. By using back lighting, it was observed that small holes formed ahead of the crack tip, and the fracture proceeded by successive joining of these advanced voids. This process, now called "void coalescence," is characteristic of crack propagation in most metallic materials. Because of the discontinuous nature of this type of crack propagation process—that is, the joining of discrete holes by isolated events—the fracture surface has a rough appearance when viewed on a fine scale. This type of fracture is also called "fibrous" because of the stretching of the ligaments between holes prior to their linking up. These "fibers" are readily observed under an electron microscope, as illustrated in Figure 1.7. This fracture mechanism has associated with it a relatively large amount of dissipated energy related to the plastic work done as the crack progresses.

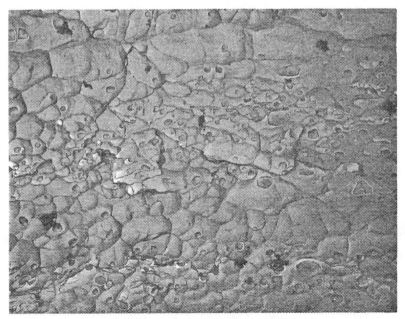

Figure 1.7 Electron micrograph of fibrous fracture in tough pitch copper [courtesy C. D. Beachem, P.E.].

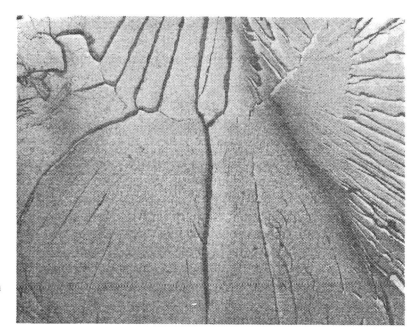

Figure 1.8
Cleavage fracture
surface in tungsten
[courtesy C. D.
Beachem, P.E.].

In contrast to the fibrous appearance associated with ductile fractures are the smooth, flat fracture surfaces characteristic of crack propagation associated with brittle or semibrittle materials. This mechanism, called "cleavage" fracture, is characterized by the progressive separation of bonds between atoms or molecules as a result of the high strain levels that exist at the crack tip. The fracture surface associated with this type of fracture is often mirrorlike in brittle, nonmetallic materials. Although characteristic of fractures in nonmetals, cleavage fractures are also observed in otherwise ductile metals in the presence of lower temperatures and weak initiation sites, as shown in Figure 1.8. In marked contrast to the large energy requirements of fibrous fracture, cleavage fractures proceed with only small amounts of energy required for each increment of crack extension.

From the preceding discussion it would appear natural (in hindsight) to describe fracture in terms of the energy dissipated per unit extension of the crack. This description was the view of the mechanics of fracture first described by Irwin [1948] and developed in simplistic form by Irwin and Kies [1952, 1954]. These papers developed the concept of energy release during progressive fracture and provided a means to measure the material's resistance to fracture using the compliance approach. We can gain an understanding of this concept by considering a hypothetical experiment. Imagine a body containing an initial crack of area, A, that is subjected to a slowly increasing load. If the body is linearly elastic, the load-deflection behavior is given by the line OB shown in Figure 1.9.

After the cracked body has been loaded to point B, the body is fixed so that no deflection of the load points is possible. This condition, called "system isolated," is

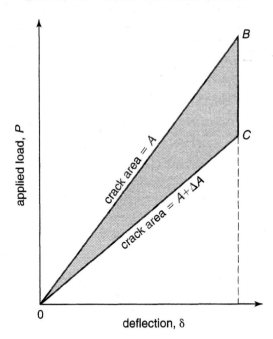

Figure 1.9 Load-deflection diagram for a "system-isolated" fracture event.

necessary in the present discussion to ensure that no energy is lost due to potential energy changes through the load points. (We will relax this requirement in the more general discussion of energy concepts in Chapter 7.) Imagine now that some process results in a change in crack area to $A + \Delta A$. Since the body is now less stiff, the load must drop to point C in Figure 1.9. If we now release the constraint at the load point and remove the applied load, the load-deflection diagram must return to zero along path OC. The triangle OBC represents the energy lost during the load–unload process; however, because the "system isolated" condition prevented any loss of energy through load points, all of this energy must be related to the crack extension area, ΔA.

Following Irwin, we will define a quantity called the strain energy release rate, $\mathcal{G}$, as the spatial rate of change of stored strain energy under system isolated conditions, i.e.,

$$\mathcal{G} = -\left.\frac{\partial U}{\partial A}\right|_{\delta} \tag{1.6}$$

where U is the strain energy of the system, and the negative sign is introduced to make $\mathcal{G}$ a positive quantity.

For the linear, elastic conditions of our experiment,

$$U = \frac{1}{2}P \cdot \delta \tag{1.7}$$

but

$$\delta = C \cdot P \tag{1.8}$$

where C, called the compliance, is the reciprocal of the slope of the load-deflection line. Since C is a measure of the stiffness of the body, it is a function of the crack length. Substituting Eq. (1.8) into Eq. (1.7) and differentiating with respect to A, we obtain the result

$$G = \frac{P^2}{2} \frac{\partial C}{\partial A} \tag{1.9}$$

In deriving equation (1.9), no restrictions were placed on the form of the specimen, and, as a result, the equation can be applied to any geometrical shape. Accordingly, the compliance calibration method can be used to develop a fracture mechanics measure for any specimen geometry. This was the approach taken by Irwin and Kies to develop suitable specimens and to compute the fracture mechanics parameter, G_c, that formed the basis for the early confirmation of linear elastic fracture mechanics.

1.4 SUMMARY

Spurred on by several major failure events and by an increasing awareness that classical failure theories did not predict observed phenomena (such as the effect of size on failure strength), the development of a new theory of fracture was inevitable. Building on ideas developed by G. R. Irwin and his coworkers at the Naval Research Laboratory, the new engineering discipline of linear elastic fracture mechanics quickly spread throughout the 1950s to interested researchers, particularly in the aircraft, aerospace, and electric utility industries, in both the United States and England.

Irwin's generalization of the Griffith argument of an energy-related fracture criterion provided a convenient starting point for the development of a much more comprehensive theory of fracture. However, calculation of energy changes is not always practical and provides no insight into the behavior of the material near the crack tip. Further, although phenomenologically sound, the energy approach has an intangible quality that makes it hard to visualize. For these and other reasons, the focus of attention in the theory of fracture mechanics in the 1960s was on the stress field approach to LEFM. From this viewpoint, the theory of fracture is treated as a subset of the engineering discipline of solid mechanics. In particular, fracture mechanics is a study of the mechanics of deformable bodies containing crack-like singularities. In order to develop the concept of fracture mechanics from this perspective, it is first necessary to master a working knowledge of linear elasticity. Accordingly, in the next few chapters, we will first treat those aspects of solid mechanics necessary to develop a theory of fracture based on the stress analysis approach and then go on to formulate a suitable theory and study its consequences.

REFERENCES

ASTM, 1960, "Fracture Testing of High-Strength Sheet Materials (A Report of a Special ASTM Committee)," *ASTM Bulletin*, No. 243, January 1960, pp. 29–40, and No. 244, February 1960, pp. 18–28.

Chiles, J. R., 1988, "The Ships That Broke Hitler's Blockade," *Amer. Heritage of Invention and Technology*, Vol. 3, No. 3, pp. 26–32, 41.

Dally, J. W., and Riley, W. F., 1978, *Experimental Stress Analysis*, McGraw-Hill, New York, pp. 93–95.

Galileo, 1638, *Two New Sciences*, H. Crew and A. de Salvio, Macmillan Co., New York, 1933.

Griffith, A. A., 1921, "The Phenomena of Rupture and Flow in Solids," *Phil. Trans. Royal Society*, Series A, Vol. 221, pp. 163–198.

Inglis, C. E., 1913, "Stresses in a Plate Due to the Presence of Cracks and Sharp Corners," *Trans. Inst. Naval Architects*, Vol. 55, pp. 219–230.

Irwin, G. R., 1948, "Fracture Dynamics," *Fracturing of Metals*, Amer. Soc. Metals, Cleveland.

Irwin, G. R., and Kies, J. A., 1952, "Fracturing and Fracture Dynamics," *Welding Jnl*, Vol. 31, Research Supplement, pp. 95s–100s.

Irwin, G. R., and Kies, J. A., 1954, "Critical Energy Rate Analysis of Fracture Strength of Large Welded Structures," *Welding Jnl*, Vol. 33, Research Supplement, pp. 193s–198s.

Love, A. E. H., 1926, *A Treatise on the Mathematical Theory of Elasticity*, Cambridge University Press, London.

Rankine, W. J. M., 1843, *Proc. Inst. Civil Engineers (London)*, Vol. 2, p. 105.

Uccelli, A., 1956, *Leonardo da Vinci*, Reynal and Company, Inc., New York, p. 274.

Ugural, A. C., and Fenster, S. K., 1987, *Advanced Strength and Applied Elasticity*, Elsevier Science Pub. Co., New York, pp. 104–120.

Wells, A. A., 1975, "The Interaction of American and British Work on Fracture Mechanics in the Period 1953/69," *Linear Fracture Mechanics*, Envo Publishing Co., Inc., Lehigh Valley, PA, pp. 29–40.

EXERCISES

1.1 Identify at least three scholarly references that discuss the Boston molasses tank failure of 1919, including one written within three years of the event. Write a brief essay describing the circumstances of the failure, its consequences and speculate on ways the failure might have been prevented.

1.2 Repeat Exercise 1.1 for the Point Pleasant Bridge, West Virginia, failure of 1967.

1.3 Repeat Exercise 1.1 for the Kings Bridge, Melbourne, Australia, failure of 1962.

1.4 Repeat Exercise 1.1 for the Aloha 737 aircraft failure of 1988.

CHAPTER 2

Elements of Solid Mechanics

2.1 CONCEPTS OF STRESS AND STRAIN

In this chapter we will focus our attention on the internal response of a continuous, homogeneous body to the action of external forces. This hypothetical body, illustrated in Figure 2.1a, is assumed to be in static equilibrium. Using Newton's principle of action and reaction, we can imagine that the body is cut by a fictitious plane passing through a point, P, within the body and replace the removed portion by an equivalent force, $\mathbf{F}$, acting on the cross-sectional area, A, as shown in Figure 2.1b. This force has the magnitude and direction needed to restore static equilibrium to the remaining portion of the body. Further, we can resolve $\mathbf{F}$ into components normal to the plane, $\mathbf{F}_n$, and tangential to the plane, $\mathbf{F}_s$. Since the magnitudes of these force components depend on the area, it is reasonable to define "normalized" values of these forces as the average force per unit area, i.e., $\mathbf{F}_n/A$ and $\mathbf{F}_s/A$ respectively. The concept of stress at a point is obtained by shrinking the area, A, to infinitesimal dimensions. Formally, the normal stress, σ, and the shear stress, τ, are defined as

$$\sigma = \lim_{A \to 0} \frac{|\mathbf{F}_n|}{A} \quad \text{and} \quad \tau = \lim_{A \to 0} \frac{|\mathbf{F}_s|}{A} \tag{2.1}$$

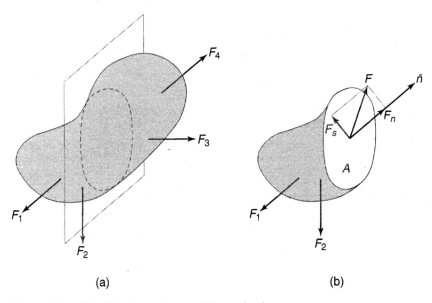

(a)   (b)

Figure 2.1   Equilibrium of an arbitrary body.

Clearly, σ and τ depend on the orientation of the plane passing through P and will vary from point to point.

A complete description of the stresses acting at a point can be obtaining by constructing a Cartesian coordinate system at the point and examining the average forces per unit area acting on the faces of an infinitesimal cube surrounding the point. Using the definition of normal and shear stresses from Eq. (2.1), the stresses on each face can be represented in Cartesian form, as shown in Figure 2.2. In the figure, positive (tensile) normal stresses are shown by outwardly directed arrows in the direction of the surface normal for that face and are denoted with the corresponding coordinate as a subscript, i.e., σ_x, σ_y, and σ_z. Shear stresses require two subscripts. The first subscript denotes the face on which the shear force acts, and the second subscript indicates the direction in which the resultant shear is being resolved. Thus, the shear stress component, τ_{xy}, corresponds to the y component of the shear force on the x face. By convention a shear stress is considered positive if the resulting component is in the positive coordinate direction on the positive face of the infinitesimal cube. (A face is positive if its outwardly directed normal is in the positive coordinate direction.) On the negative faces, positive shear stresses are in the negative coordinate direction. All of the stresses illustrated in Figure 2.2 are positive as shown.

If the stresses are slowly varying across the infinitesimal cube, moment equilibrium about the centroid of the cube requires that

$$\tau_{xy} = \tau_{yx} \qquad \tau_{xz} = \tau_{zx} \qquad \tau_{yz} = \tau_{zy} \qquad \qquad (2.2)$$

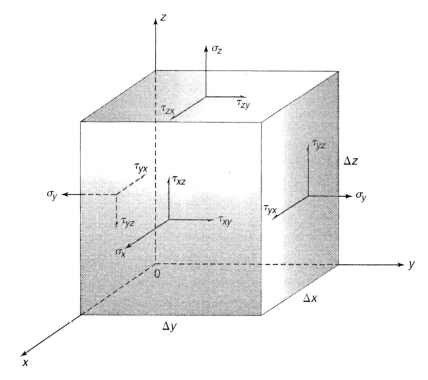

Figure 2.2 Cartesian components of stress in three dimensions.

As a result, the nine stress components depicted in Figure 2.2 reduce to six independent quantities that we can write in array form as

$$\sigma_{ij} = \begin{pmatrix} \sigma_x & \tau_{xy} & \tau_{xz} \\ \tau_{xy} & \sigma_y & \tau_{yz} \\ \tau_{xz} & \tau_{yz} & \sigma_z \end{pmatrix} \tag{2.3}$$

It can easily be shown that this array has the transformation properties of a symmetric second order tensor [e.g., see Borsesi and Chong, 1987] and is called the *stress tensor*. We recognize that the Cartesian coordinate system is not intrinsic to the body. Our choice of this coordinate system was completely arbitrary. In fact, for many problems in solid mechanics, this system is not ideal, and another coordinate system might be a better choice.

One alternative coordinate system that is often used to represent the state of stress is the cylindrical coordinate system (r, θ, z). The stress components in this system are shown in Figure 2.3. In tensor form, the components are expressed as

$$\sigma_{ij} = \begin{pmatrix} \sigma_r & \tau_{r\theta} & \tau_{rz} \\ \tau_{r\theta} & \sigma_\theta & \tau_{\theta z} \\ \tau_{rz} & \tau_{\theta z} & \sigma_z \end{pmatrix} \tag{2.4}$$

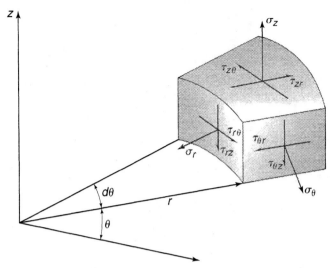

Figure 2.3 Cylindrical components of stress in three dimensions.

where the symmetry of the orthogonal shear stresses has been used. Both the Cartesian and cylindrical coordinate systems are orthogonal and, for arbitrary rotation of the coordinate system, transform according to similar rules. In either case, if we rotate the infinitesimal element and observe the changes in the stress components, we will see that there is one orientation of the element for which all of the shear components vanish. The coordinate directions corresponding to this unique orientation are called the *principal stress directions*, and, in this orientation, the stress tensor becomes

$$\sigma_{ij} = \begin{pmatrix} \sigma_1 & 0 & 0 \\ 0 & \sigma_2 & 0 \\ 0 & 0 & \sigma_3 \end{pmatrix} \tag{2.5}$$

where, by convention, $\sigma_1 > \sigma_2 > \sigma_3$.

From a physics standpoint the principal stress directions would form an ideal coordinate system. Since the principal stresses represent physical quantities, they will always be the same, regardless of the coordinate system initially chosen. In other words the transformation from the Cartesian or cylindrical coordinate system (or any other system) to the principal coordinate system will always result in the final orientation of the element being in the same physical direction. For this reason many theories of brittle fracture, such as those discussed in Chapter 1, and some yield criteria are often represented in terms of principal stress values. Unfortunately, the principal coordinate system is not a convenient system to use for development of the fundamental equations of elasticity since the orientation most often varies from point to point within the body.

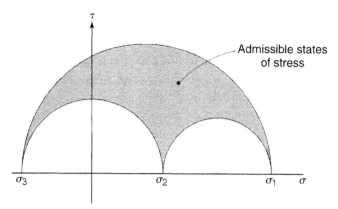

Figure 2.4 Three-dimensional Mohr's-circle representation of the state of stress at a point.

The states of stress at a point for all orientations of the cubical element can be represented on a three-dimensional Mohr's-circle diagram. This diagram is constructed by drawing the Mohr's circle for each pair of principal planes separately on a common $\sigma\tau$ plane. Since the principal planes represent the extreme values of stress, the stress on any arbitrary plane must lie within the area bounded by these limiting circles, as shown by the shaded area in Figure 2.4. The three-dimensional Mohr's circle provides a qualitative overview of the possible normal and shear stresses on an arbitrarily oriented face of the infinitesimal cube at the point. From the diagram various key features of the state of stress can be readily obtained. A parameter of particular interest is the extreme value of the shear stress, denoted τ_{max}. The Tresca yield criterion is based on this parameter. From Figure 2.4, the magnitude of the maximum shearing stress is given by

$$\tau_{max} = \frac{\sigma_1 - \sigma_3}{2} \tag{2.6}$$

and acts on planes at 45° to the σ_1 and σ_3 directions, as illustrated in Figure 2.5. We will see in Chapter 6 that the orientation of these planes of maximum shear stress plays a central role in fracture behavior.

In the preceding paragraphs a systematic procedure for describing the action of forces on an infinitesimal element of material was presented in various coordinate systems. We now turn our attention to the corresponding descriptions of the distortion of the element. These measures of distortion, called the *state of strain*, are described in terms of the displacements of the point from its undistorted position and its derivatives. In the Cartesian system we will define the displacements in the coordinate directions x, y and z as u, v and w, respectively. Then, for small displacements in a continuous body, the customary definitions of the normal (i.e.,

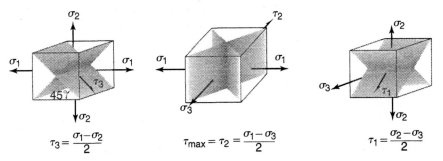

$$\tau_3 = \frac{\sigma_1 - \sigma_2}{2} \qquad \tau_{max} = \tau_2 = \frac{\sigma_1 - \sigma_3}{2} \qquad \tau_1 = \frac{\sigma_2 - \sigma_3}{2}$$

Figure 2.5 Planes of maximum shear stress relative to the principal planes.

dilatational) strains in terms of the displacement gradients are given by

$$\epsilon_x = \frac{\partial u}{\partial x} \qquad \epsilon_y = \frac{\partial v}{\partial y} \qquad \epsilon_z = \frac{\partial w}{\partial z} \tag{2.7}$$

and the engineering shear strains (i.e., the measures of the angular distortion of the element) are

$$\gamma_{xy} = \frac{\partial v}{\partial x} + \frac{\partial u}{\partial y}$$

$$\gamma_{xz} = \frac{\partial w}{\partial x} + \frac{\partial u}{\partial z} \tag{2.8}$$

$$\gamma_{yz} = \frac{\partial w}{\partial y} + \frac{\partial v}{\partial z}$$

As we did for the state of stress, we can write the complete description of the state of strain compactly as an array:

$$\epsilon_{ij} = \begin{pmatrix} \epsilon_x & \frac{\gamma_{xy}}{2} & \frac{\gamma_{xz}}{2} \\ \frac{\gamma_{xy}}{2} & \epsilon_y & \frac{\gamma_{yz}}{2} \\ \frac{\gamma_{xz}}{2} & \frac{\gamma_{yz}}{2} & \epsilon_z \end{pmatrix} \tag{2.9}$$

Note that from their definitions, Eqs. (2.7) and (2.8), the strains are naturally symmetric, and the state of strain is described by six independent quantities. Also, the factor of 2 in the array description of the shear strains is needed in order for the strain array to have the transformation properties of a second-order tensor. When strains are expressed in this form, there is a one-to-one correspondence between the elements of the stress tensor, Eq. (2.3), and the elements of the strain tensor, Eq. (2.9). As a result, the equations of coordinate transformation or other functionals involving strain can be obtained directly from their counterparts for stress by substituting the corresponding array elements into the equation being converted. As a practical matter, this correspondence means that, in order to convert a stress-related equation to its strain counterpart, one must replace a normal stress component by its normal strain equivalent and replace a shear stress component by one half of the corresponding shear strain element in the equation to be converted.

In cylindrical coordinates (r, θ, z) the displacement components are denoted u_r, u_θ and u_z, respectively. Because of the curvature in the coordinate system, the representation of the strains in cylindrical coordinates as derivatives of the displacement components is somewhat more complex. In cylindrical coordinates the normal strain counterparts to Eqs. (2.7) become

$$\epsilon_r = \frac{\partial u_r}{\partial r} \qquad \epsilon_\theta = \frac{u_r}{r} + \frac{1}{r}\frac{\partial u_\theta}{\partial \theta} \qquad \epsilon_z = \frac{\partial u_z}{\partial z} \tag{2.10}$$

and, for the shear strains, Eqs. (2.8) become

$$\gamma_{r\theta} = \frac{\partial u_\theta}{\partial r} + \frac{1}{r}\frac{\partial u_r}{\partial \theta} - \frac{u_\theta}{r}$$

$$\gamma_{rz} = \frac{\partial u_z}{\partial r} + \frac{\partial u_r}{\partial z} \tag{2.11}$$

$$\gamma_{\theta z} = \frac{1}{r}\frac{\partial u_z}{\partial \theta} + \frac{\partial u_\theta}{\partial z}$$

The cylindrical strain components can be written in tensor form as

$$\epsilon_{ij} = \begin{pmatrix} \epsilon_r & \frac{\gamma_{r\theta}}{2} & \frac{\gamma_{rz}}{2} \\ \frac{\gamma_{r\theta}}{2} & \epsilon_\theta & \frac{\gamma_{\theta z}}{2} \\ \frac{\gamma_{rz}}{2} & \frac{\gamma_{\theta z}}{2} & \epsilon_z \end{pmatrix} \tag{2.12}$$

As was the case for the state of stress, there is a unique orientation of the cubical element for which the deformation is purely dilatational. Along these three orthogonal directions, a line, initially drawn in the undeformed material, will be longer (tension) or shorter (compression) after deformation but will not rotate. These directions are the principal strain directions, and in this orientation the strain tensor becomes

$$\epsilon_{ij} = \begin{pmatrix} \epsilon_1 & 0 & 0 \\ 0 & \epsilon_2 & 0 \\ 0 & 0 & \epsilon_3 \end{pmatrix} \tag{2.13}$$

where $\epsilon_1 > \epsilon_2 > \epsilon_3$. The principal stress directions and principal strain directions do not necessarily coincide unless the material is isotropic.

Chapter 1 introduced the concept of stored strain energy in a mechanical system. We can apply the same concept to an infinitesimal element in order to define a local-

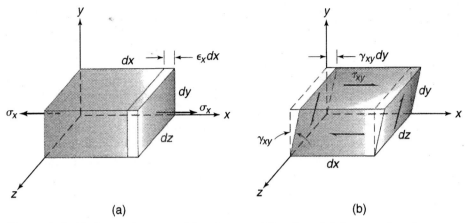

Figure 2.6 Deformations resulting from unidirectional (a) normal and (b) shear stresses acting on a differential element of elastic material.

ized energy parameter. Consider the infinitesimal cube of a linear elastic material subjected to a uniaxial stress in the x-direction illustrated in Figure 2.6a. Then, from Eq. (1.7), we have

$$U = \frac{1}{2}P \cdot \delta = \frac{1}{2}\sigma_x dydz \cdot \epsilon_x dx \qquad (2.14)$$

Since the volume of the cubical element is $dxdydz$, we can define the strain energy density, W, as

$$W = \frac{1}{2}\sigma_x \epsilon_x \qquad (2.15)$$

By similar reasoning, the strain energy density corresponding to the shear stress, τ_{xy}, as in Figure 2.6b, is

$$W = \frac{1}{2}\tau_{xy}\gamma_{xy} \qquad (2.16)$$

From superposition the strain energy density of an infinitesimal element with an arbitrary stress state for a linear elastic body is given by

$$W = \frac{1}{2}\sigma_x\epsilon_x + \frac{1}{2}\sigma_y\epsilon_y + \frac{1}{2}\sigma_z\epsilon_z + \frac{1}{2}\tau_{xy}\gamma_{xy} + \frac{1}{2}\tau_{xz}\gamma_{xz} + \frac{1}{2}\tau_{yz}\gamma_{yz} \qquad (2.17)$$

Since we recognize that Eqs. (2.15) and (2.16) represent the area under their respective stress–strain curves, we can extend the concept of strain energy density to nonlinear elastic bodies by replacing these equations by their integral forms, i.e.,

$$W = \int \sigma_x d\epsilon_x \quad \text{and} \quad W = \int \tau_{xy} d\epsilon_{xy} \qquad (2.18)$$

2.2 EQUATIONS OF ELASTICITY IN CARTESIAN COORDINATES

In the previous section the concepts of stress and strain were introduced without specific mention of a material. Stress is a conceptual quantity related to the force acting at a point. It is assumed that a material medium exists such that the stress acts against some resistance. Similarly, strain is a kinematic quantity relating relative motion between points in a material medium. For our purposes we will require that the medium be continuous, so that derivative definitions of strain, such as Eqs. (2.7) and (2.8), have meaning. Other than these minimal requirements, the definitions of stress and strain are independent of the properties of the material medium in which they act.

The relation between the state of stress and the state of strain is called the *constitutive equation*. The form of this equation is not dictated by the laws of physics, but rather can take any suitable form in order to model the behavior of any material (real or hypothetical). Depending on the form of the constitutive equation that is chosen, various branches of continuum mechanics can be explored (e.g., elasticity, viscoelasticity, plasticity). For our purposes, we will restrict our attention to the class of materials that are nondissipative and exhibit a unique, reversible relation between stress and strain, namely elastic behavior. The influence of localized, small-scale yielding will be taken into account in Chapter 6. To simplify matters further, we will assume that the material is isotropic and homogeneous. As a result, the properties of the material will not be directionally or spatially dependent; in other words, they can be treated as bulk properties. Neither of these assumptions is necessary in order to develop the theory of linear elastic fracture mechanics, but their adoption greatly simplifies the development of the theory and embodies the majority of practical applications. Since the constitutive equations are models for characterizing the behavior of materials, we are free to choose whatever restrictions we wish in order to restrict the behavior to match the model. Whether the model matches the behavior of a real material or not can only be determined by experiment. Accordingly, we will adopt the following restrictions, which tend to be supported to some degree by experimental observations of many materials of practical importance:

(a) The material properties are independent of orientation (i.e., the behavior is assumed to be isotropic);

(b) the stresses and strains are linearly related;

(c) the strains are sufficiently small that squares and products of displacement derivatives can be ignored (i.e., Eqs. (2.7) and (2.8) or their equivalents in other coordinate systems apply);

(d) normal stresses produce only normal strains, and shear stresses produce only shear strains (i.e., the normal and shear modes of deformation are uncoupled).

A set of constitutive equations that conforms to these assumptions and is widely employed to model the behavior of linear, elastic, and isotropic materials is the classical equations of elasticity referred to as "generalized Hooke's law." We recognize, of course, that, from a physics viewpoint, there is no "law" to be followed in formulating this constitutive model. In the three-dimensional Cartesian coordinate system, the equations have the form

$$\epsilon_x = \frac{1}{E}[\sigma_x - \nu(\sigma_y + \sigma_z)]$$

$$\epsilon_y = \frac{1}{E}[\sigma_y - \nu(\sigma_x + \sigma_z)]$$

$$\epsilon_z = \frac{1}{E}[\sigma_z - \nu(\sigma_x + \sigma_y)]$$

$$\gamma_{xy} = \frac{1}{G}\tau_{xy} \qquad\qquad (2.19)$$

$$\gamma_{xz} = \frac{1}{G}\tau_{xz}$$

$$\gamma_{yz} = \frac{1}{G}\tau_{yz}$$

where E is the elastic modulus, ν is Poisson's ratio, and G, the shear modulus, equals $E/[2(1+\nu)]$.

Provided that $\nu \neq \frac{1}{2}$, Eqs. (2.19) can be inverted to express the stress in terms of the strain components

$$\sigma_x = \frac{E}{1+\nu}\epsilon_x + \frac{\nu E}{(1+\nu)(1-2\nu)}(\epsilon_x + \epsilon_y + \epsilon_z)$$

$$\sigma_y = \frac{E}{1+\nu}\epsilon_y + \frac{\nu E}{(1+\nu)(1-2\nu)}(\epsilon_x + \epsilon_y + \epsilon_z)$$

$$\sigma_z = \frac{E}{1+\nu}\epsilon_z + \frac{\nu E}{(1+\nu)(1-2\nu)}(\epsilon_x + \epsilon_y + \epsilon_z) \qquad (2.20)$$

$$\tau_{xy} = G\gamma_{xy}$$

$$\tau_{xz} = G\gamma_{xz}$$

$$\tau_{yz} = G\gamma_{yz}$$

Having now developed the concepts of stress, strain, and Hooke's law in three dimensions, we could go on to develop the general governing differential equations in three dimensions as well. Unfortunately, the solution of elasticity problems in three

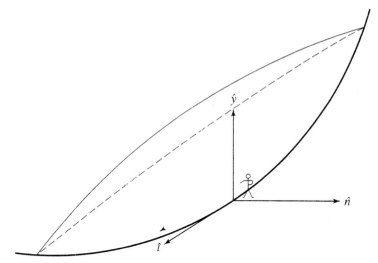

Figure 2.7 Local crack-tip coordinate system along a slowly curving crack.

dimensions is extremely difficult. The number of available solutions reported in the literature is very small, and, of these solutions, only a few are for fracture problems. It would appear then that for practical reasons, we need to restrict our attention to problems that can be formulated with a two-dimensional theory. There is, moreover, a fundamental argument for considering the two-dimensional formulation.

For our purposes every problem to be solved will contain a crack with a very small root radius, and all other dimensions will be large by comparison. If we construct a local Cartesian coordinate system at the crack tip, with one axis tangent to the crack front, as illustrated in Figure 2.7, variations in stress in the tangent direction will be small in comparison to these in the plane perpendicular to the plane of the crack. Imagine yourself as a tiny observer standing at the crack tip in a body containing a gently curving crack. As you look forward into the material, stresses, strains, and displacements change rapidly. Similarly, looking up into the material the same is true. But if you were to look along the direction of the tangent vector to the crack front, these quantities would appear to change little, since, on this scale, the curvature of the crack front is very small in comparison with events occurring in the xy-plane. Consequently, we would expect that, to the first order, the rates of change (i.e., the derivatives) of any of the measures of deformation will be small in the tangent direction and can be neglected. It is for this reason that the two-dimensional formulation of the fracture problem is adequate for most problems of practical interest.

With these considerations in mind, we can examine two special cases in which the antiplane behavior is restricted and the governing equations of elasticity can be formulated in two dimensions. First consider the case of a long prismatic bar with a constant cross-section and loaded uniformly along its length. If the ends of the bar are held rigid, the antiplane displacement will be zero. Under these conditions the body is said to be in a state of plane strain, and all strain components in the z-direction are required to be zero, i.e.,

$$\epsilon_z = 0 \qquad \gamma_{xz} = 0 \qquad \gamma_{yz} = 0 \tag{2.21}$$

Substituting these conditions into Eqs. (2.19) and (2.20) results in the plane-strain form of Hooke's law:

$$\epsilon_x = \frac{1 + \nu}{E}[(1 - \nu)\sigma_x - \nu\sigma_y]$$

$$\epsilon_y = \frac{1 + \nu}{E}[(1 - \nu)\sigma_y - \nu\sigma_x] \tag{2.22}$$

$$\sigma_x = \frac{E}{(1 + \nu)(1 - 2\nu)}[(1 - \nu)\epsilon_x + \nu\epsilon_y]$$

$$\sigma_y = \frac{E}{(1 + \nu)(1 - 2\nu)}[(1 - \nu)\epsilon_y - \nu\epsilon_x] \tag{2.23}$$

$$\tau_{xy} = G\gamma_{xy}$$

From the third of Eqs. (2.19), the only nonzero stress in the z-direction is

$$\sigma_z = \nu(\sigma_x + \sigma_y) \tag{2.24}$$

The second two-dimensional case differs from plane strain in that, instead of requiring the antiplane strains to be zero, it requires that the antiplane stresses be zero, i.e.,

$$\sigma_z = 0 \qquad \tau_{xz} = 0 \qquad \tau_{yz} = 0 \tag{2.25}$$

As a consequence of Eqs. (2.25), the stresses and strains can be written as:

$$\epsilon_x = \frac{1}{E}(\sigma_x - \nu\sigma_y)$$

$$\epsilon_y = \frac{1}{E}(\sigma_y - \nu\sigma_x)$$

$$\sigma_x = \frac{E}{1 - \nu^2}(\epsilon_x + \nu\epsilon_y) \tag{2.26}$$

$$\sigma_y = \frac{E}{1 - \nu^2}(\epsilon_y + \nu\epsilon_x)$$

$$\tau_{xy} = G\gamma_{xy}$$

As in the previous case, the third of Eqs. (2.19) yields

$$\epsilon_z = -\frac{\nu}{E}(\sigma_x + \sigma_y) \tag{2.27}$$

for the only nonzero strain in the z-direction.

The requirements of Eq. (2.25) are called the plane stress conditions, and Eqs. (2.26) and (2.27) are called the plane-stress form of Hooke's law. These conditions can be approximately satisfied by a very thin plate loaded only by forces or displacements in the plane of the plate. In this case the stress-free faces satisfy Eq. (2.25) exactly, and, if the plate is thin enough, there is insufficient thickness to permit these stresses to reach any significant magnitude. To be strictly correct, we should interpret the in-plane stresses and strains of Eqs. (2.26) as quantities averaged through the thickness, and this case is referred to as "generalized" plane stress [Timoshenko, 1951]. The plane stress conditions can also be satisfied on the stress-free surface of any arbitrary three-dimensional body, but not in the interior. In fracture mechanics the plane-stress condition on the lateral faces of a thick test specimen leads to a condition referred to as "lack of constraint" and affects the fracture behavior significantly. For either of these two special cases—plane strain or plane stress—it is important to observe that the state of stress and strain in all three dimensions is completely defined. The key point is that the antiplane quantities are described in terms of in-plane quantities, which are themselves only functions of the in-plane coordinates, x and y. As a consequence, either of these plane problems can be mathematically formulated in two dimensions. However, the deformation field is still three-dimensional, and no generality has been lost if the conditions of Eqs. (2.21) or (2.25) can be assumed.

Throughout the remainder of this chapter, we will assume that the conditions of either plane stress or plane strain apply. As a result, the general problem of linear elasticity in the (x, y) plane can be fully described by the eight unknowns σ_x, σ_y, τ_{xy}, ϵ_x, ϵ_y, γ_{xy}, u, and v. To solve for these eight unknowns, we will need eight algebraic or partial differential equations. Three of the equations are obtained from the Hooke's law relations of Eqs. (2.23) or (2.26), depending on whether we define the problem to be one of plane strain or plane stress. In addition, selecting the in-plane equations for strain from Eqs. (2.7) and (2.8) provides three more equations. To obtain the remaining two equations, let us consider the static equilibrium of an infinitesimal element in a slowly varying stress field, as shown in Figure 2.8. For convenience we will assume that the element is of unit thickness, and we will neglect body forces. Noting that Newton's first law applies to forces, not stresses, summing the forces in the x-coordinate direction yields

$$-\sigma_x dy \cdot 1 + \left(\sigma_x + \frac{\partial \sigma_x}{\partial x}dx\right)dy \cdot 1 + \left(\tau_{xy} + \frac{\partial \tau_{xy}}{\partial y}dy\right)dx \cdot 1 = 0 \tag{2.28}$$

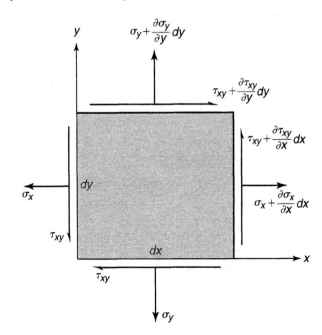

Figure 2.8 Stresses acting on an infinitesimal element in a slowly varying two-dimensional deformation field.

Since the volume of the infinitesimal element is finite (i.e., $dxdy \cdot 1 \neq 0$), we can divide by the volume to obtain

$$\frac{\partial \sigma_x}{\partial x} + \frac{\partial \tau_{xy}}{\partial y} = 0 \qquad (2.29a)$$

Similarly, equilibrium in the y-coordinate direction requires that

$$\frac{\partial \tau_{xy}}{\partial x} + \frac{\partial \sigma_y}{\partial y} = 0 \qquad (2.29b)$$

The equilibrium Eqs. (2.29) provide the two additional equations needed to formulate the problem of linear elasticity in two dimensions. However, there is one issue to be resolved.

From Eqs. (2.7) and (2.8), the in-plane strains and displacements are related by

$$\epsilon_x = \frac{\partial u}{\partial x} \qquad \epsilon_y = \frac{\partial v}{\partial y} \qquad \gamma_{xy} = \frac{\partial u}{\partial y} + \frac{\partial v}{\partial x} \qquad (2.30)$$

These equations can be viewed as a set of partial differential equations for u and v in terms of the in-plane strains ϵ_x, ϵ_y, and γ_{xy}. But, if the strains are arbitrarily prescribed, there will be three equations for the two unknowns, and thus no unique solution for u and v is possible. Accordingly, there must be some relation between the strains to ensure that a unique solution for the displacements exists. Differentiating the last of Eqs. (2.30) with respect to y, and then x yields

$$\frac{\partial^2 \gamma_{xy}}{\partial x \partial y} = \frac{\partial^3 u}{\partial x \partial y \partial y} + \frac{\partial^3 v}{\partial x \partial y \partial x} \qquad (2.31a)$$

$$\frac{\partial^2 \gamma_{xy}}{\partial x \partial y} = \frac{\partial^2}{\partial y \partial y} \frac{\partial u}{\partial x} + \frac{\partial^2 v}{\partial x \partial x} \frac{\partial v}{\partial y} \qquad (2.31b)$$

From which it follows that

$$\frac{\partial^2 \gamma_{xy}}{\partial x \partial y} = \frac{\partial^2 \epsilon_x}{\partial y^2} + \frac{\partial^2 \epsilon_y}{\partial x^2} \qquad (2.32)$$

This equation, called the compatibility equation, provides the necessary restriction on the state of strain to ensure that the body will remain continuous under the action of any arbitrary system of applied forces. The physical interpretation of Eq. (2.32) can be visualized by imagining the undeformed body to be composed of mosaic tiles, as illustrated in Figure 2.9. The compatibility equation ensures that the tiles will remain conforming, as shown on the right and not leave open gaps in the deformation field, as depicted by the nonconforming tiles on the left.

With the addition of the constraint condition imposed by Eq. (2.32), there is now a complete set of eight equations, consisting of the two equilibrium equations

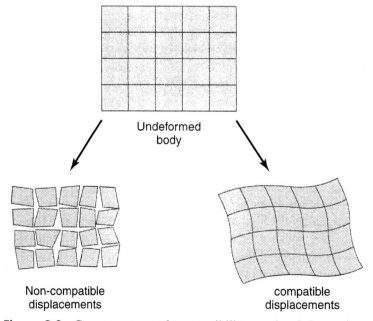

Undeformed
body

Non-compatible
displacements

compatible
displacements

Figure 2.9 Consequences of compatibility on the deformation of an arbitrary body.

[Eqs. (2.29)], the three strain-displacements equations [Eqs. (2.30)], and the three stress-strain relations for either plane strain [Eqs. (2.23)], or plane stress [Eqs. (2.26)]. This set of equations is sufficient to solve the in-plane problem of elasticity for any given system of boundary conditions. Once the in-plane solution is obtained, the antiplane quantities can be determined from the appropriate form of Hooke's law.

In principle these eight partial differential equations in eight unknowns could be solved for each specific boundary-value problem we might wish to consider. However, the mechanics of retaining eight unknowns in the mathematical formulation is awkward at best. It is computationally more efficient, and more enlightening, to reduce the number of unknowns by combining some of the governing equations and solving the reduced set of equations. A procedure for accomplishing this task that has gained universal acceptance is the method attributed to G. B. Airy, a British astronomer [Airy, 1863]. Airy defined a function, $F(x, y)$, such that

$$\sigma_x = \frac{\partial^2 F}{\partial y^2} \qquad \sigma_y = \frac{\partial^2 F}{\partial x^2} \quad \text{and} \quad \tau_{xy} = -\frac{\partial^2 F}{\partial x \partial y} \tag{2.33}$$

Substitution of these definitions into Eqs. (2.29) shows that, if the stresses are continuous, the Airy stress function, F, satisfies the equilibrium equations identically.

Recalling that the strain-displacement relations in Eqs. (2.30) were used to derive the compatibility equation in Eq. (2.32), we can incorporate the only remaining equations—i.e., the Hooke's law relations—into our formulation by substituting them into the compatibility equation. Consequently, for a state of plane strain, the compatability equation in terms of the in-plane stresses is

$$(1 - \nu) \cdot \left[\frac{\partial^2 \sigma_y}{\partial x^2} + \frac{\partial^2 \sigma_x}{\partial y^2} \right] - \nu \cdot \left[\frac{\partial^2 \sigma_x}{\partial x^2} + \frac{\partial^2 \sigma_y}{\partial y^2} \right] = 2 \cdot \frac{\partial^2 \tau_{xy}}{\partial x \partial y} \tag{2.34}$$

Using Airy's definitions from Eqs. (2.33), we find that the compatibility equation can be expressed in terms of the stress function, F, as

$$\frac{\partial^4 F}{\partial x^4} + 2 \cdot \frac{\partial^4 F}{\partial x^2 \partial y^2} + \frac{\partial^4 F}{\partial y^4} = 0 \tag{2.35a}$$

or

$$\left(\frac{\partial^2}{\partial x^2} + \frac{\partial^2}{\partial y^2} \right) \cdot \left(\frac{\partial^2}{\partial x^2} + \frac{\partial^2}{\partial y^2} \right) F = 0 \tag{2.35b}$$

where the term in the parentheses is the harmonic operator ∇^2. Therefore, we can write the single governing equation that combines all eight of the previous equations for the state of plane strain in terms of the Airy stress-function in compact form as

$$\nabla^2\nabla^2 F = 0 \qquad \text{or} \qquad \nabla^4 F = 0 \qquad\qquad (2.35c)$$

Using the Airy stress function approach, the plane problem of elasticity for plane strain has been reduced to the task of finding a suitable function $F(x, y)$ that satisfies the boundary conditions for each problem of interest. Once F is obtained, the stresses can be obtained from Eqs. (2.33) by differentiation. The in-plane strains are determined from the in-plane stresses by using the plane strain form of Hooke's law [Eqs. (2.23)] and the displacements u and v computed by integration of the strains. Finally, the antiplane stress is determined from the in-plane stresses by using Eq. (2.24). By this process, all of the stresses, strains, and displacements can be obtained once a suitable Airy stress function has been found.

For the case of plane stress, the procedure to be followed for determining the single governing equation parallels that for plane strain. Namely, the plane stress form of Hooke's law [Eqs. (2.26)] is substituted into the compatibility equation, Eq. (2.32), and the Airy stress function definitions of the stresses in Eqs. (2.33) are used to replace the stresses. The procedure is straightforward and leads directly to the condition that $\nabla^4 F = 0$—i.e.,the governing equation for the Airy stress function in plane stress is the same as that for plane strain. This result for plane stress follows directly from the assumption that the in-plane stresses are independent of z. However, there are compatibility equations in addition to Eq. (2.32) that need to be satisfied and, in the strictest sense, cannot be satisfied unless the in-plane stresses are allowed to vary through the thickness. If the plate is thin, this variation must necessarily be small and can be neglected [Wang, 1953]. Alternatively, we can eliminate the z-direction dependence in a more formal manner by considering the in-plane stresses to be averaged through the thickness. Using the averaged values for the stresses eliminates the incompatibility problem, and the governing equation for the Airy stress function is again $\nabla^4 F = 0$. The use of averaged values in the plane stress formulation is referred to as "generalized plane stress" [Timoshenko, 1951].

We can now generalize our earlier statement and state that, for *either* plane strain *or* plane stress (including generalized plane stress), the two-dimensional problem of elasticity has been reduced to the task of obtaining an Airy stress function, F, that satisfies Eqs. (2.35), subject to the boundary conditions in the xy-plane. The significance of this result is that the in-plane stresses will be the same for either plane strain or plane stress conditions. However, the antiplane stress, the strains, and the displacements will, as a consequence of the different forms of Hooke's law, be different, but can be directly computed from the in-plane stress results. It is also important to note that, in the absence of body forces (such as gravity or rapid rotation), the governing differential equation, Eq. (2.35), is independent of the material properties. As a result, the distribution of the in-plane stresses does not depend on the choice of material for a given set of boundary conditions (at least for simply connected bodies).

2.3 EQUATIONS OF ELASTICITY IN POLAR COORDINATES

In the previous section we developed the Cartesian form of the general equations
of two-dimensional elasticity. Unfortunately, many problems of interest in fracture
mechanics are more conveniently expressed in polar coordinates. As a consequence,
we will need to develop the polar forms of the governing equations in two dimensions.

The procedure to be used is precisely the same as we have previously followed for
the Cartesian coordinate system. Namely, we need to develop eight partial differential or algebraic equations and a compatibility constraint in the eight polar unknowns,
σ_r, σ_θ, $\tau_{r\theta}$, ϵ_r, ϵ_θ, $\gamma_{r\theta}$, u_r, and u_θ, and then seek a method to reduce the number of
governing equations to a smaller set. Recall that the strain-displacement equations
have previously been described in polar form in Eqs. (2.10). We also recognize that
the polar coordinate system is orthogonal and that the Hooke's law relations in plane
strain and plane stress have precisely the same form as their Cartesian counterparts.
All that is required is that we replace the coordinate variables x and y by r and θ,
respectively, in Eqs. (2.23) and (2.26) to obtain the polar forms. The same is not true
for the equilibrium equations in the polar coordinate system. Whereas the Hooke's
law relations describe the behavior of the material at a point in any orthogonal coordinate system, the equilibrium equations are the result of a rectilinear force balance
over a finite volume, and in the polar coordinate system the force vectors are no
longer uncoupled, and each must be resolved into its components along the two
perpendicular directions.

Using the notation of Figure 2.3, consider a cylindrical differential volume in a
slowly varying stress field (assuming unit thickness in the z-direction). The resultant
free body in the $r\theta$-plane is shown in Figure 2.10. Note that, for clarity, only those
force vectors that have components in the radial direction have been shown. Also,
body forces have been neglected. Applying Newton's first law in the radial direction
yields

$$\left(\sigma_r + \frac{\partial \sigma_r}{\partial r}dr\right)(r + dr)d\theta \cdot 1 - \sigma_r r d\theta \cdot 1$$

$$- \left(\sigma_\theta + \frac{\partial \sigma_\theta}{\partial \theta}d\theta\right)dr \cdot 1 \cdot sin\frac{d\theta}{2} - \sigma_\theta dr \cdot 1 \cdot sin\frac{d\theta}{2} \qquad (2.36)$$

$$+ \left(\tau_{r\theta} + \frac{\partial \tau_{r\theta}}{\partial \theta}d\theta\right)dr \cdot 1 \cdot cos\frac{d\theta}{2} - \tau_{r\theta}dr \cdot 1 \cdot cos\frac{d\theta}{2} = 0$$

Since we are free to let $d\theta$ and dr be as small as we like, provided the volume does
not shrink to zero, the small-angle approximations, $sin\frac{d\theta}{2} \cong \frac{d\theta}{2}$, and $cos\frac{d\theta}{2} \cong 1$, can
be employed and $r + dr \cong$ r. With these approximations and dividing by the nonzero

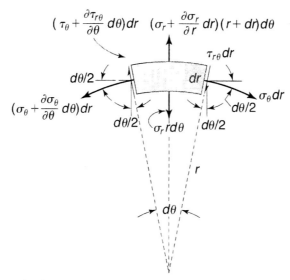

Figure 2.10 Resultant forces acting on a differential volume in polar coordinates. (Only those forces having components in the radial direction are shown.)

volume, $r\,dr\,d\theta \cdot 1$, we find that Eq. (2.36) has the form

$$\frac{\partial \sigma_r}{\partial r} + \frac{1}{r}\frac{\partial \tau_{r\theta}}{\partial \theta} + \frac{\sigma_r - \sigma_\theta}{r} = 0 \qquad (2.37a)$$

Similarly, summation of the forces in the direction perpendicular to the radial direction yields

$$\frac{\partial \tau_{r\theta}}{\partial r} + \frac{1}{r}\frac{\partial \sigma_\theta}{\partial \theta} + \frac{2\tau_{r\theta}}{r} = 0 \qquad (2.37b)$$

Following our earlier procedure, we can obtain the compatibility equation in polar form from the definitions of the in-plane strains in terms of the polar components of displacement, Eqs. (2.10) and (2.11), by suitable differentiation of the strains and interchange of the order of differentiation. The result is

$$\frac{\partial^2 \epsilon_\theta}{\partial r^2} + \frac{1}{r^2}\frac{\partial^2 \epsilon_r}{\partial \theta^2} + \frac{2}{r}\frac{\partial \epsilon_\theta}{\partial r} - \frac{1}{r}\frac{\partial \epsilon_r}{\partial r} = \frac{1}{r}\frac{\partial^2 \gamma_{r\theta}}{\partial r\,\partial \theta} + \frac{1}{r^2}\frac{\partial \gamma_{r\theta}}{\partial \theta} \qquad (2.38)$$

With the development of the compatibility equation in polar form, we now have a complete set of equations for the determining the eight polar unknowns. Using Airy's procedure, we define a function, $\mathcal{F}(r,\theta)$, with the property that

$$\sigma_r = \frac{1}{r}\frac{\partial \mathcal{F}}{\partial r} + \frac{1}{r^2}\frac{\partial^2 \mathcal{F}}{\partial \theta^2}$$

$$\sigma_\theta \;=\; \frac{\partial^2 \mathcal{F}}{\partial r^2} \tag{2.39}$$

$$\tau_{r\theta} \;=\; -\frac{\partial}{\partial r}\!\left(\frac{1}{r}\frac{\partial \mathcal{F}}{\partial \theta}\right)$$

With this choice of stress function, the equilibrium equations in polar form, Eqs. (2.37), are satisfied identically. If we now substitute the polar equivalents of either Eqs. (2.23) for plane strain or Eqs. (2.26) for plane stress (subject to the same understanding about the antiplane behavior as previously discussed) into the compatibility equation, Eq. (2.38), and use the Airy stress function definitions in Eqs. (2.39), we find that the two-dimensional problem of elasticity in polar form reduces to the solution of a single partial differential equation of the form

$$\left(\frac{\partial^2}{\partial r^2} + \frac{1}{r}\frac{\partial}{\partial r} + \frac{1}{r^2}\frac{\partial^2}{\partial \theta^2}\right)\cdot\left(\frac{\partial^2}{\partial r^2} + \frac{1}{r}\frac{\partial}{\partial r} + \frac{1}{r^2}\frac{\partial^2}{\partial \theta^2}\right)\mathcal{F} \;=\; 0 \tag{2.40}$$

However, the term in parentheses is the harmonic operator in polar form; therefore,

$$\nabla^2\nabla^2\mathcal{F} \;=\; 0 \quad \text{or} \quad \nabla^4\mathcal{F} \;=\; 0 \tag{2.41}$$

Comparing this result with the corresponding result for Cartesian coordinates, Eq. (2.35c), we see that $F(x, y) = \mathcal{F}(r, \theta)$, as we could have expected from the invariance of the harmonic operator.

2.4 SOLUTION OF THE BIHARMONIC EQUATION

In the previous section we showed that the plane problem of elasticity in either plane stress or plane strain reduced to finding an Airy stress function F (or $\mathcal{F}$) that satisfied the biharmonic equation, Eqs. (2.35), and satisfied all of the boundary conditions for the posed problem. Direct solutions of the governing equation are, for the most part, not available. Consequently, an indirect approach called the *semi-inverse method* is often employed to solve a specific elasticity problem. With this method the functional form of the solution is assumed. Usually, these solutions are the result of insight or are a variation on the known solution of a related problem. This assumed solution, which must satisfy the biharmonic equation, is checked against the boundary conditions. If all of the boundary conditions are satisfied, the assumed solution is the true solution. Moreover, as a consequence of the uniqueness theorem, the assumed solution is the only solution to the problem. On the other hand, if the boundary conditions are not satisfied, the assumed solution is modified and the problem re-solved. This process is repeated until an acceptable solution is found. While the semi-inverse method does not always yield a valid solution, it is a practical approach to an otherwise intractable

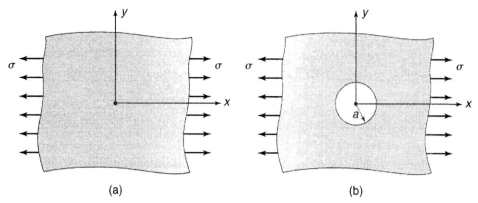

Figure 2.11 An infinite plate subjected to uniform remote stress, σ, in the x-direction: (a) uniform stress field; (b) with a central hole.

problem and has been used extensively in plane elasticity. The methodology of the semi-inverse method will be demonstrated by solving several simple problems.

Consider the elementary problem of an infinite plate subjected to uniform remote stress, σ, in the x-direction, as illustrated in Figure 2.11a. The stress solution can be deduced by inspection:

$$\sigma_x = \sigma \qquad \sigma_y = 0 \qquad \tau_{xy} = 0 \qquad \text{for all } x \text{ and } y \qquad (2.42)$$

Also, from the first of Eqs. (2.33), it is obvious that the stress function for this problem is

$$F = \frac{1}{2}\sigma y^2 \qquad (2.43)$$

This solution can be cast in a form more useful for the semi-inverse method by noting that $y = r \sin \theta$. Accordingly, Eq. (2.43) can be expressed as

$$\mathcal{F} = \frac{1}{2}\sigma r^2 \sin^2 \theta = \frac{1}{4}\sigma r^2 (1 - \cos 2\theta) \qquad (2.44)$$

which is of the form

$$\mathcal{F} = f(r) - f(r) \cdot \cos 2\theta \qquad (2.45)$$

In this latter form the twofold symmetry of the problem is brought out by the $\cos 2\theta$ term.

Let us now introduce a hole of radius, a, centered at the origin into the uniformly loaded plate, as illustrated in Figure 2.11b, and ask the following question: What might be a suitable stress function for this problem? Using the previous result for guidance, we will generalize the stress function given by Eq. (2.45) while retaining the twofold symmetry structure and assume a solution of the form

$$\mathcal{F} = f_1(r) + f_2(r) \cdot \cos 2\theta \qquad (2.46)$$

Substituting this trial stress function into $\nabla^4 \mathcal{F} = 0$ and requiring that the solution be valid for all θ, results in a pair of ordinary differential equations for the functions $f_1(r)$ and $f_2(r)$:

$$\left(\frac{d^2}{dr^2} + \frac{1}{r}\frac{d}{dr} \right) \cdot \left(\frac{d^2}{dr^2} + \frac{1}{r}\frac{d}{dr} \right) f_1 = 0 \qquad (2.47a)$$

$$\left(\frac{d^2}{dr^2} + \frac{1}{r}\frac{d}{dr} - \frac{4}{r^2} \right) \cdot \left(\frac{d^2}{dr^2} + \frac{1}{r}\frac{d}{dr} - \frac{4}{r^2} \right) f_2 = 0 \qquad (2.47b)$$

These two ordinary differential equations with variable coefficients can be converted into ordinary differential equations with constant coefficients by introducing the change of variable, $\zeta = \ln r$, and can then be readily solved. The results in terms of the original variable, r, can be written as

$$f_1(r) = C_1 r^2 \ln r + C_2 r^2 + C_3 \ln r + C_4 \qquad (2.48a)$$

$$f_2(r) = C_5 r^2 + C_6 r^4 + \frac{C_7}{r^2} + C_8 \qquad (2.48b)$$

The general form for the stresses in polar coordinates becomes

$$\sigma_r = C_1(1 + 2\ln r) + 2C_2 + \frac{C_3}{r^2} - \left(2C_5 + \frac{6C_7}{r^4} + \frac{4C_8}{r^2} \right) \cos 2\theta \qquad (2.49a)$$

$$\sigma_\theta = C_1(3 + 2\ln r) + 2C_2 - \frac{C_3}{r^2} + \left(2C_5 + 12C_6 r^2 + \frac{6C_7}{r^4} \right) \cos 2\theta \qquad (2.49b)$$

$$\tau_{r\theta} = \left(2C_5 + 6C_6 r^2 - \frac{6C_7}{r^4} - 2\frac{C_8}{r^2} \right) \sin 2\theta \qquad (2.49c)$$

Equations (2.49) follow directly from the assumed form of the solution given by Eq. (2.46), and, at this point, we have no assurance that they represent the general form of the stresses for the problem we posed. Following the methodology of the semi-inverse method, we will need to determine whether or not all of the boundary conditions can be satisfied by suitable choice of the constants C_1 through C_8.

Since we require that the stresses remain finite as $r \to \infty$, the constants C_1 and C_6 must be set equal to zero. Also, the constant C_4 does not appear in the expressions for the stresses and can be arbitrary in the stress solution. In order to determine the five remaining unknown coefficients, we will need to establish five boundary conditions. From the traction-free conditions around the boundary of the unloaded hole, we require that

$$\sigma_r = 0 \quad \text{and} \quad \tau_{r\theta} = 0 \quad \text{on} \quad r = a \tag{2.50}$$

for all θ. Also, at large distances from the hole, St. Venant's principle requires that the stresses approach those of a plate uniformly loaded in the x-direction, i.e., $\sigma_x = \sigma$ and $\sigma_y = \tau_{xy} = 0$. This condition, expressed in polar form, becomes

$$\left. \begin{aligned} \sigma_r &= \tfrac{1}{2}\sigma(1 + \cos 2\theta) \\ \sigma_\theta &= \tfrac{1}{2}\sigma(1 - \cos 2\theta) \\ \tau_{r\theta} &= -\tfrac{1}{2}\sigma \sin 2\theta \end{aligned} \right\} \quad \text{as } r \to \infty \tag{2.51}$$

The boundary conditions expressed by Eqs. (2.50) and (2.51) provide five independent equations from which the five remaining coefficients can be uniquely determined. After some algebra, the final results for the polar stresses are

$$\sigma_r = \frac{\sigma}{2}\left(1 - \frac{a^2}{r^2}\right) + \frac{\sigma}{2}\left(1 + \frac{3a^4}{r^4} - \frac{4a^2}{r^2}\right)\cos 2\theta \tag{2.52a}$$

$$\sigma_\theta = \frac{\sigma}{2}\left(1 + \frac{a^2}{r^2}\right) - \frac{\sigma}{2}\left(1 + \frac{3a^4}{r^4}\right)\cos 2\theta \tag{2.52b}$$

$$\tau_{r\theta} = -\frac{\sigma}{2}\left(1 - \frac{3a^4}{r^4} + \frac{2a^2}{r^2}\right)\sin 2\theta \tag{2.52c}$$

Since we were able to satisfy all of the boundary conditions by using the assumed form for the Airy stress function in Eq. (2.46) as well as the biharmonic equation, we are now in a position to state that the resulting stresses in Eqs. (2.52) are the solution to the posed problem and are unique. This solution was first obtained by Kirsch in 1898.

Although the problem of the plate with a hole in uniaxial tension is a relatively simple problem as far as elasticity problems go, the basic features of its solution are characteristic of all elastic behavior around isolated stress raisers, and a closer examination of the solution is warranted. First, around the boundary of the hole the only nonzero component of stress, σ_θ, becomes

$$\sigma_\theta = \sigma(1 - 2\cos 2\theta) \tag{2.53}$$

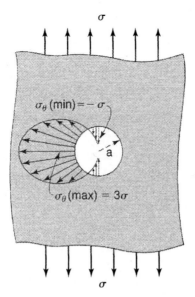

Figure 2.12 Distribution of the tangential stress σ_θ around the boundary of the hole.

Accordingly, the magnitude of the tangential stress, as plotted in Figure 2.12, varies in a systematic manner around the boundary of the hole and reaches a maximum value of three times the applied remote stress at the transverse diameter ($\theta = \pm\frac{\pi}{2}$). At the longitudinal diameter ($\theta = 0, \pi$) the tangential stress is equal in magnitude to the applied stress, but opposite in sign; that is, even though the applied stress is tensile, the tangential stress over a region of the boundary is compressive. From this observation it can easily be inferred that the hole will deform into an ellipse with its major axis in line with the applied load.

Across the net section ($\theta = \pm\frac{\pi}{2}$), the stresses become

$$\sigma_r = \sigma_y = \frac{\sigma}{2}\left(\frac{3a^2}{r^2} - \frac{3a^4}{r^4}\right) \qquad (2.54a)$$

$$\sigma_\theta = \sigma_x = \frac{\sigma}{2}\left(2 + \frac{a^2}{r^2} + \frac{3a^4}{r^4}\right) \qquad (2.54b)$$

$$\tau_{r\theta} = \tau_{xy} = 0 \qquad \text{(from symmetry)} \qquad (2.54c)$$

The distribution of these stresses is shown in Figure 2.13. The stress in the direction of the applied load, σ_x, is elevated at the hole boundary but decays away quickly as the distance from the hole increases and eventually approaches the magnitude of

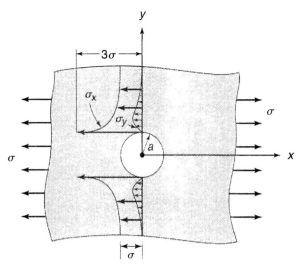

Figure 2.13 Distribution of the stresses across the net section for a uniaxially loaded plate with a circular hole.

the remote stress. In other words the effect of the stress raiser on the distribution of stress is localized to the region around the stress raiser. In this case the stress field returns to a uniform field at a distance of approximately 10 radii. From Eqs. (2.54) we note that the distribution of stress depends on the normalized distance, r/a, and, consequently, the distribution of stress is independent of the size of the hole.

It is customary for machine design applications to define a quantity called the *stress concentration factor* (SCF) as the ratio of the maximum stress to the applied stress. For this problem the SCF is

$$\text{SCF} = \frac{\sigma_\theta \mid_{\text{max}}}{\sigma_{\text{applied}}} = \frac{3\sigma}{\sigma} = 3 \tag{2.55}$$

Note that the stress concentration factor is independent of the hole diameter. Furthermore, the SCF is a measure of the stress elevation at a single point in the body and provides no information on the distribution of the stresses in the interior of the body.

Although the exact magnitudes and functional forms of the results will vary with the nature of the geometry being studied, these observations for the circular hole are typical of the behavior of the stress field around stress raisers with smooth contours. In later chapters we will compare these observations with those for bodies containing sharp discontinuities. Finally, we can see that the failure theories for brittle materials that are based on a maximum-tensile-stress argument, such as the Rankine or Coulomb–Mohr theories described in Chapter 1, would predict the same failure stress regardless of the size of the stress raiser. These theories would predict that, for geometrically similar stress raisers, the stress at failure would depend solely on the stress concentration factor, which is independent of the size of that stress raiser. This result is inconsistent with experimental observations of failures in brittle materials.

2.5 THE PROBLEM OF THE ELLIPTICAL HOLE

The rectangular and polar coordinate systems are special cases of the more general class of orthogonal coordinate systems. As a general rule, the solution to an elasticity problem is simplified if one or more of its boundaries coincide with the contour of the coordinate system. This was the approach adopted by Inglis [1913] to solve the problem of the elliptical hole subjected to uniform remote stresses.

In an arbitrary curvilinear coordinate system the in-plane strains can be written as

$$\epsilon_{\alpha\alpha} = h_1 \frac{\partial u_\alpha}{\partial \alpha} + h_1 h_2 u_\beta \frac{\partial}{\partial \beta}\left(\frac{1}{h_1}\right)$$

$$\epsilon_{\beta\beta} = h_2 \frac{\partial u_\beta}{\partial \beta} + h_1 h_2 u_\alpha \frac{\partial}{\partial \alpha}\left(\frac{1}{h_2}\right) \tag{2.56}$$

$$\frac{\gamma_{\alpha\beta}}{2} = \frac{h_1}{h_2}\frac{\partial}{\partial \alpha}(h_2 u_\beta) + \frac{h_2}{h_1}\frac{\partial}{\partial \beta}(h_1 u_\alpha)$$

where u_α and u_β are the displacements along coordinate directions α and β, respectively [Love, 1994]. The quantities, h_1 and h_2, which are formally defined as

$$h_1^2 = \left(\frac{\partial \alpha}{\partial x}\right)^2 + \left(\frac{\partial \beta}{\partial y}\right)^2 \quad \text{and} \quad h_2^2 = \left(\frac{\partial \beta}{\partial x}\right)^2 + \left(\frac{\partial \alpha}{\partial y}\right)^2 \tag{2.57}$$

provide the relationship between coordinate measures of distance and physical lengths, i.e.,

$$ds^2 = \left(\frac{\partial \alpha}{h_1}\right)^2 + \left(\frac{\partial \beta}{h_2}\right)^2 \tag{2.58}$$

For the elliptical–hyperbolic coordinate system shown in Figure 2.14,

$$x = c \cosh \alpha \cos \beta$$

$$y = c \sinh \alpha \sin \beta \tag{2.59}$$

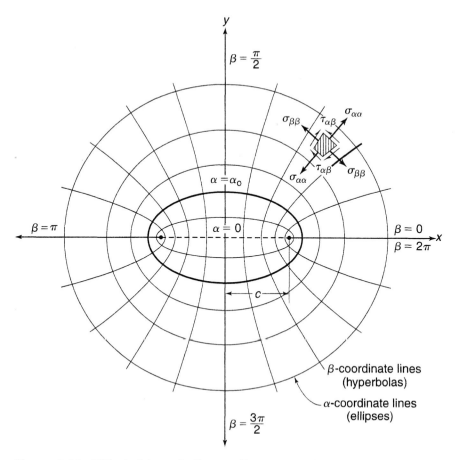

Figure 2.14 Elliptical hyperbolic coordinate system.

where c is the position of the common foci for both the ellipses and the hyperbolae. For this coordinate system it follows from Eqs. (2.57) that

$$h_1^2 = h_2^2 = h^2 = \frac{2}{c^2(\cosh 2\alpha - \cos 2\beta)} \qquad (2.60)$$

Let α_o denote the boundary of an elliptical hole with a and b as the lengths of the semimajor and semiminor axes, respectively. Then, from Eqs. (2.59), we find that

$$a = c \cosh \alpha_o \qquad (\beta = 0) \qquad (2.61a)$$

$$b = c \sinh \alpha_o \qquad (\beta = \pi/2) \qquad (2.61b)$$

Inglis did not use the Airy stress function approach to solve the elliptical hole problem, but rather combined the equilibrium equations (in this coordinate system) with Hooke's law and the strain-displacement equations in Eqs. (2.56) to obtain a pair

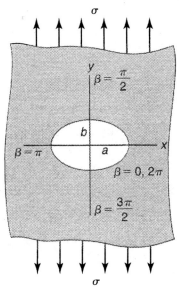

Figure 2.15 An infinite plate with an elliptical hole subjected to remote uniform normal stress.

of partial differential equations for the displacements along the coordinate directions. He was able to solve these equations in general infinite-series form involving arbitrary constants that have to be determined from the boundary conditions at the inner and outer boundaries. From the general solution Inglis was able to obtain exact solutions in finite form for an infinitely large plate with an elliptical hole for several cases of remotely applied stress.

One of the problems that Inglis solved was the case of a plate with an elliptical hole, as illustrated in Figure 2.15, subjected to uniform normal stresses at infinity. For the problem of uniform biaxial stress, the boundary conditions can be satisfied identically by retaining only three terms in the series solution and the exact solution for the stresses have relatively simple forms given by

$$\sigma_{\alpha\alpha} = \sigma \frac{\sinh 2\alpha[\cosh 2\alpha - \cosh 2\alpha_o]}{[\cosh 2\alpha - \cos 2\beta]^2} \tag{2.62a}$$

$$\sigma_{\beta\beta} = \sigma \frac{\sinh 2\alpha[\cosh 2\alpha + \cosh 2\alpha_o - 2\cos 2\beta]}{[\cosh 2\alpha - \cos 2\beta]^2} \tag{2.62b}$$

$$\tau_{\alpha\beta} = \sigma \frac{\sin 2\beta[\cosh 2\alpha - \cosh 2\alpha_o]}{[\cosh 2\alpha - \cos 2\beta]^2} \tag{2.62c}$$

For the case of uniaxial remote stress perpendicular to the major axis of the ellipse, Inglis found that an exact solution could be obtained with five nonzero constants from the general series solution.[1] The resulting expressions for the individual stresses are considerably more complex than those of Eqs. (2.62) and will not be shown here; however, the sum of the normal stresses at the boundary of the hole (i.e., when $\alpha = \alpha_o$) can be written compactly as

$$\sigma_{\alpha\alpha} + \sigma_{\beta\beta} = \sigma e^{2\alpha} \left[\frac{(1 + e^{-2\alpha}) \sinh 2\alpha}{\cosh 2\alpha - \cos 2\beta} - 1 \right] \qquad (2.63)$$

The stress normal to the boundary, $\sigma_{\alpha\alpha} = 0$, and Eq. (2.63) provides the distribution of the tangential stress at the hole:

$$\sigma_{\beta\beta}|_{\alpha=\alpha_o} = \sigma e^{2\alpha_o} \left[\frac{(1 + e^{-2\alpha_o}) \sinh 2\alpha_o}{\cosh 2\alpha_o - \cos 2\beta} - 1 \right] \qquad (2.64)$$

For comparison with our earlier results for a circular hole, it is instructive to examine the variation of the tangential stress, $\sigma_{\beta\beta}$, around the boundary of the hole. This result, shown in Figure 2.16 for the case $a/b = 3$, has features that are similar to those of the circular hole, except that the maximum stress is more elevated. As was the case for the circular hole, the maximum stress occurs at the edge of the elliptical hole along the net section, i.e., for $\beta = 0$. Therefore, from Eq. (2.64), the stress concentration factor for an elliptical hole in uniaxial tension is

$$\text{SCF} = 1 + 2 \coth \alpha_o = 1 + 2\left(\frac{a}{b}\right) \qquad (2.65)$$

which reduces to Eq. (2.55) for the circular hole, i.e. when $a = b$.

Inglis also noted that, from the geometric property of elliptical sections, the radius of curvature at the end of the major axis was related to the semiaxes of the ellipse by

$$\rho = \frac{b^2}{a} \qquad (2.66)$$

and the stress concentration factor could also be expressed as

$$\text{SCF} = 1 + 2\sqrt{\frac{a}{\rho}} \qquad (2.67)$$

Inglis undertook the problem of obtaining the exact solution for the elliptical hole with the expressed intent of considering the limiting case of an ellipse as representative of an internal crack. Since the solution is exact, he was assured that

---

[1] The solution is presented in detail in Coker and Filon [1957].

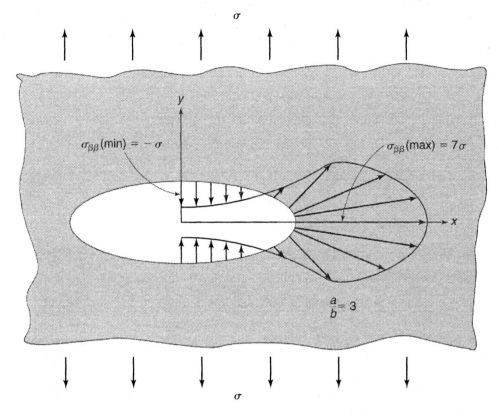

Figure 2.16   Distribution of the tangential stress around an elliptical hole for $\frac{a}{b} = 3$.

the results would be valid even if the semiminor axis, b, was taken to zero in the limit—provided that the shape of the cavity remained elliptical.

As important as the general solution of the elliptical hole problem was to the advancement of the theory of elasticity in the early 20th century, an even more significant accomplishment of the Inglis paper was the insight into material behavior in the presence of a crack that he presented in the discussion of his results. All of his observations and interpretations are represented in modern fracture mechanics theory, which was formulated some 40 years after the appearance of this paper.

Inglis acknowledged that the low strength of brittle materials was due to the presence of cracks and described the significance of a fine scratch in reducing the strength of glass in terms of the high tensile stresses along the line of the scratch compared with "insignificant stresses" in the rest of the plate. This interpretation of the weakness of glass in tension, as well as Inglis's exact mathematical results, were later used by Griffith [1921] to formulate the latter's energy theory of failure of glass and other brittle materials. (The Griffith theory is presented in detail in Chapter 7.)

For ductile materials Inglis reasoned that the maximum stress at the tip of a crack would be limited by localized yielding, but, in the case of alternating loads (i.e., fatigue

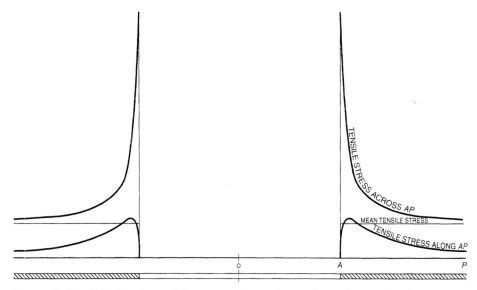

Figure 2.17 Distribution of the stress along the major axis of the ellipse for $\frac{a}{b} = 3$ [Inglis, 1913].

loading), reversing the load so as to close the crack would produce high compressive stresses. He argued that even small alternating loads would produce localized alternating plastic strains and would result in crack growth. In primitive form this explanation can be compared with contemporary models of damage accumulation in low-cycle fatigue.

Inglis was the first to demonstrate that, although the stresses at the tip of a highly oblate ellipse are very high, they decay rapidly with distance from the hole boundary. His results for $a = 3b$ are reproduced in Figure 2.17. He also pointed out that, at least for remote loading, any crack growth (a "tear" in his terminology) would increase the stress concentration factor even further, and the crack would continue to spread. In modern terminology, now some 40 years after Inglis, this process is called *unstable crack propagation* and is now explained in terms of a rise in a new parameter called the *stress intensity factor* (This concept will be fully developed in the next chapter.)

In addition to the foregoing observations, which could be deduced from the exact solution for an elliptical hole, Inglis went further to speculate on the influence of variations in the profile of the crack geometry on the state of stress at the ends of the cavity. He argued that, since the stress concentration factor could be expressed equally as well in terms of the total length of the cavity, $2a$, and the localized shape of the ends as represented by its radius of curvature, ρ, as given in Eq. (2.67), his results would apply acceptably well to a cavity of *any* shape, provided that the ends of the cavity transitioned smoothly into an ellipse. He presented several extreme examples, reproduced in Figure 2.18, to illustrate his point.

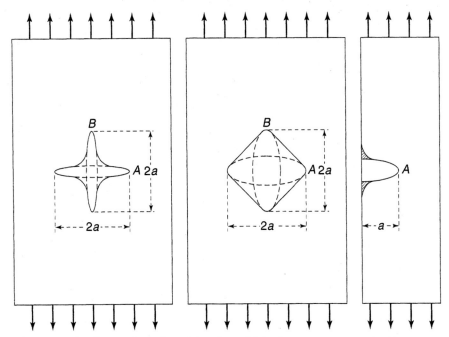

Figure 2.18 Examples of cavities for which the stress concentration factor can be expressed as $1 + 2\sqrt{\frac{a}{\rho}}$ [Inglis, 1913].

We can interpret this latter result in even more general terms. If we are willing to relax the requirement of exact similitude and require instead that the end profile of the cavity (i.e., ρ) remain fixed as we vary the overall length, Inglis's result predicts that the SCF is proportional to the square root of the cavity (crack) length and suggests a size dependence on fracture not predicted by classical elasticity theory.

Finally, Inglis examined the case depicted in Figure 2.19 of a small crack or notch emanating from the end of a larger notch or hole. He stated that, in this case, the maximum stress could be obtained from the product of the individual stress concentration factors for each cavity. He used this result to explain the observed weakness of punched holes in excess of the Kirsch prediction. The effect of punching a hole, he explained, introduces small cracks around the perimeter, and the influence of these cracks is aggravated by localized hardening of the material. These cracks would propagate through the hardened material when a load was applied and would continue into the base material, even if it was ductile. To prevent brittle failure, he recommended removing the hardened material (and the cracks) after punching the hole.

As significant as Inglis's results are, they provide only limited information about the behavior of cracks in bodies with arbitrary geometries. In order to advance the theory of fracture mechanics further, it was necessary to examine the nature of the stress field in the neighborhood of a crack tip in detail and to understand the

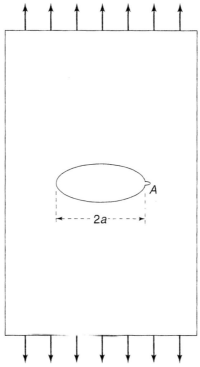

Figure 2.19 Example of a compounded stress concentration [Inglis, 1913].

role of geometry on these stresses. For this purpose the Inglis approach of solving problems involving continuous cavities and, hopefully, degenerating the solution to a crack-like defect has had only limited success. (The work of Neuber [1958] should be noted.) A more fruitful approach has been to solve the elasticity problem of a body containing a line defect directly. Since a line discontinuity has no well-defined end shape, these solutions produce singular results at the crack tip, and the concept of a stress concentration factor loses meaning. In the next chapter we will develop the theory of elasticity for singular stress fields by using several different approaches and then go on to define a new parameter to represent the influence of geometry on the state of stress to replace the meaningless stress concentration factor when the stress field contains singularities.

REFERENCES

Airy, G. B., 1863, "On the Strains in the Interior of Beams," *Phil. Trans. Royal Society*, 153, London, pp. 49–79.

Boresi, A. P., and Chong, K. P., 1987, *Elasticity in Engineering Mechanics*, Elsevier Science Pub. Co., Inc., New York.

Coker, E. G., and Filon, L. N. G., 1957, *A Treatise on Photoelasticity*, 2nd ed. (revised by H. T. Jessop), University Press, Cambridge, United Kingdom.

Griffith, A. A., 1921, "The Phenomena of Rupture and Flow in Solids," *Philosophical Trans. Royal Society*, London, Vol. A221, pp. 163–198.

Inglis, C. E., 1913, "Stresses in a Plate Due to the Presence of Cracks and Sharp Corners," *Trans. Inst. Naval Architects*, Vol. 55, pp. 219–230.

Kirsch, G, 1898, "Die Theorie der Elastizität und die Bedürfnisse der Festigkeits-lehre," *Zeit. Vereines deutscher Ingenieure*, 42, pp. 797–807.

Love, A. E. H., 1944, *A Treatise on the Mathematical Theory of Elasticity*, 4th ed., Dover Pub., New York.

Neuber, H., 1958, *Theory of Notch Stresses*, 2nd ed., Springer Publishers, Berlin.

Timoshenko, S., and Goodier, J. N., 1951, *Theory of Elasticity*, 2nd ed., McGraw-Hill Book Co., Inc., New York.

Wang, C. T., 1953, *Applied Elasticity*, McGraw Hill Book Co., Inc., New York, p. 45.

EXERCISES

2.1 At a point in a thin plate the Cartesian stresses are $\sigma_x = 60$ ksi, $\sigma_y = 6$ ksi, and $\tau_{xy} = 13$ ksi. If the plate is made of aluminum ($E = 10 \times 10^6$ psi, $\nu = 1/3$), determine the components of the strain tensor, ϵ_{ij}. In the stress-free state, the thickness of the plate is exactly 0.25000 inch. What is the thickness at the point for the given stress state?

2.2 Given the Airy stress function, $F = -\frac{P}{d^3}(xy)^2(3d - 2y)$, (a) determine the corresponding Cartesian components of stress; and (b) state the problem that is solved by this stress function over the region bounded by the lines $y = 0$, $y = d$, and $x = 0$ on the x-positive side.

2.3 Using the polar form of the strain definitions in Eqs. (2.10) and (2.11), verify Eq. (2.38).

2.4 Using the polar definitions of the Airy stress function in Eqs. (2.39), derive the polar form of the biharmonic equation from the compatibility equation Eq. (2.38).

2.5 The general form of the Airy stress function for radially symmetric problems is of the form, $\mathcal{F} = A \log r + Br^2 \log r + Cr^2 + D$. Derive expressions for the polar stress components, σ_r, σ_θ, and $\tau_{r\theta}$, for a large plate with a central hole of radius, a, that is subjected to a uniform radial stress, σ_o, at a large distance from the hole.

2.6 Plot the distribution of $\sigma_{\beta\beta}$ along the major axis of an ellipse in biaxial tension for $\frac{a}{b} = 2, 5$, and 10. What conclusions can you draw from these results?

2.7 Plot the distribution of $\sigma_{\beta\beta}$ around the boundary of an elliptical hole for $\frac{a}{b} = 2$, 5, and 10 for equal biaxial remote tension.

2.8 Plot the distribution of $\sigma_{\beta\beta}$ around the boundary of an elliptical hole for $\frac{a}{b} = 2$, 5, and 10 for uniaxial remote tension perpendicular to the major axis of the ellipse.

2.9 Show that Eq. (2.65) follows from Eq. (2.64).

2.10 A large plate containing a circular hole of radius 1.0 inch has a straight slot *on one side only*, terminating in a smooth root radius of 0.05 inch. (a) If the overall length of the slot is 0.2 inch, estimate the stress concentration factor at the tip of the slot. (b) If the slot is long relative to the diameter of the hole, estimate the stress concentration factor at the tip of the slot?

CHAPTER 3

Elasticity of Singular Stress Fields

3.1 OVERVIEW

The solutions to elasticity problems containing localized stress raisers, such as the Inglis solution discussed in the previous chapter or the notch problems solved by Neuber [1958], can provide useful information about the behavior of crack-like discontinuities when the limiting cases of the smooth geometry are considered. A more direct approach would be to formulate the problem with a line discontinuity that represents the crack as part of the boundary geometry. The resulting zero-thickness cut ends abruptly in a crack tip with an undefined geometry (often assumed to be semi-elliptical with a root radius of zero). As a consequence of the nature of this type of problem, called a *singular problem*, the stresses at the tip of the crack tend to infinity.

The modern view of analytical fracture mechanics uses singular problems to represent the elastic behavior of bodies containing one or more cracks. The dilemma of infinite stresses resulting from finite applied loads is clearly inconsistent with our perception of real-world behavior, and the first-order theory accepts this deficiency by excluding a small region surrounding the crack tip. It is argued that highly localized processes or damage mechanisms control the exact behavior in this region and are beyond the range of applicability of elasticity theory. If ductile behavior can be assumed, a nonlinear (i.e., plastic) zone is introduced to keep in the stresses in the very near region finite.

In any case, the behavior of the body everywhere except in an isolated region very near the crack tip is dominated by classical elasticity theory of singular fields, and this approach greatly broadens the range of problems that can be treated. Moreover, it brings out a striking similarity in the stress field behavior of all of the problems within this class of problems. In Chapter 6 we will use the characterizing feature of these problems as a basis for a failure theory whose predictions agree remarkably well with observed behavior for a wide range of materials. In the present chapter we will focus on the analytical approaches used to solve singular problems and will defer the fracture theory implications until later in the book.

3.2 THE WILLIAMS PROBLEMS

In a sequence of two papers, M. L. Williams [1952, 1957] developed the general solution to a particular singular problem. This solution has played an important role in the early development of LEFM theory. Not only are the results significant in their own right, but the timing of the paper, concurrent with the landmark paper by Irwin [1957], heightened the interest of researchers both in the United States and the United Kingdom to explore this new approach to explaining fracture behavior.

In the first paper, Williams [1952] solved the problem of a wedge of arbitrary apex angle, 2α, shown in Figure 3.1, that has stress-free faces, but otherwise allows any system of forces on the remaining boundary. Using the semi-inverse method, Williams assumed an Airy stress function in polar coordinates of the form

$$\mathcal{F}(r,\theta) = r^{\lambda+1} \cdot f(\theta) \tag{3.1}$$

where the value(s) of λ are to be determined as part of the solution. With this choice of stress function, the stresses can be expressed in polar coordinates from Eqs. (2.39) as

$$\sigma_r = r^{\lambda-1}[(\lambda + 1)f(\theta) + f''(\theta)]$$

$$\sigma_\theta = \lambda(\lambda + 1)r^{\lambda-1}f(\theta) \tag{3.2}$$

$$\tau_{r\theta} = -\lambda r^{\lambda-1}f'(\theta)$$

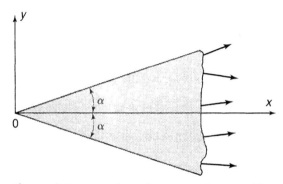

Figure 3.1 A wedge of apex angle 2α with traction-free flanks.

where the superscript primes refer to ordinary differentiation. The stress-free faces of the wedge require that

$$\sigma_\theta = 0 \quad \text{and} \quad \tau_{r\theta} = 0 \quad \text{on} \quad \theta = \pm\alpha \quad \text{for all } r \tag{3.3}$$

From Eqs. (3.2), these boundary conditions are equivalent to restrictions on $f(\theta)$ given by

$$f(\alpha) = f(-\alpha) = 0 \tag{3.4a}$$

and

$$f'(\alpha) = f'(-\alpha) = 0 \tag{3.4b}$$

provided $\lambda \neq 0$. Substituting the assumed Airy stress function, Eq. (3.1), into the biharmonic equation in polar form, Eq. (2.40) and requiring that $f(\theta)$ be defined for arbitrary values of the coordinate, r, results in an ordinary differential equation for $f(\theta)$ that can be written in compact form as

$$\left[\frac{d^2}{d\theta^2} + (\lambda - 1)^2\right] \cdot \left[\frac{d^2}{d\theta^2} + (\lambda + 1)^2\right] f(\theta) = 0 \tag{3.5}$$

From this equation it follows that

$$f(\theta) = C_1 \cos(\lambda - 1)\theta + C_2 \sin(\lambda - 1)\theta + C_3 \cos(\lambda + 1)\theta + C_4 \sin(\lambda + 1)\theta \tag{3.6}$$

where the constants C_1, C_2, C_3, and C_4 are to be determined from the boundary conditions on the faces of the wedge. Applying the boundary condition in Eqs. (3.4) to the general form of $f(\theta)$ from Eq. (3.6) produces a system of four simultaneous equations for the unknown constants:

$$C_1 \cos(\lambda - 1)\alpha + C_2 \sin(\lambda - 1)\alpha + C_3 \cos(\lambda + 1)\alpha + C_4 \sin(\lambda + 1)\alpha = 0 \tag{3.7a}$$

$$C_1 \cos(\lambda - 1)\alpha - C_2 \sin(\lambda - 1)\alpha + C_3 \cos(\lambda + 1)\alpha - C_4 \sin(\lambda + 1)\alpha = 0 \tag{3.7b}$$

$$-C_1(\lambda - 1)\sin(\lambda - 1)\alpha + C_2(\lambda - 1)\cos(\lambda - 1)\alpha$$
$$-C_3(\lambda + 1)\sin(\lambda + 1)\alpha + C_4(\lambda + 1)\cos(\lambda + 1)\alpha = 0 \tag{3.7c}$$

$$C_1(\lambda - 1)\sin(\lambda - 1)\alpha + C_2(\lambda - 1)\cos(\lambda - 1)\alpha$$
$$+C_3(\lambda + 1)\sin(\lambda + 1)\alpha + C_4(\lambda + 1)\cos(\lambda + 1)\alpha = 0 \tag{3.7d}$$

These equations can be separated into two independent sets of equations for the constants C_1, C_3, and C_2, C_4, respectively, by elementary algebraic operations. The resulting equations, in matrix form, are

$$
\begin{pmatrix}
\cos(\lambda - 1)\alpha & \cos(\lambda + 1)\alpha \\
(\lambda - 1)\sin(\lambda - 1)\alpha & (\lambda + 1)\sin(\lambda + 1)\alpha
\end{pmatrix}
\cdot
\begin{pmatrix} C_1 \\ C_3 \end{pmatrix}
=
\begin{pmatrix} 0 \\ 0 \end{pmatrix}
\tag{3.8a}
$$

$$
\begin{pmatrix}
\sin(\lambda - 1)\alpha & \sin(\lambda + 1)\alpha \\
(\lambda - 1)\cos(\lambda - 1)\alpha & (\lambda + 1)\cos(\lambda + 1)\alpha
\end{pmatrix}
\cdot
\begin{pmatrix} C_2 \\ C_4 \end{pmatrix}
=
\begin{pmatrix} 0 \\ 0 \end{pmatrix}
\tag{3.8b}
$$

Since Eqs. (3.8) are homogeneous, it is clear that a meaningful solution can be obtained only if the determinants of the coefficient matrices in Eqs. (3.8a) and (3.8b) are each equal to zero. From this requirement, it follows that

$$
\sin 2\lambda\alpha \;=\; \pm\lambda \sin 2\alpha
\tag{3.9}
$$

Since α is a fixed parameter for a specific wedge, Eq. (3.9) provides the value λ, called an *eigenvalue*, necessary to ensure a nontrivial solution.

In his 1957 paper, Williams considered the limiting case of $\alpha = \pi$—i.e., a wedge that is fully closed upon itself, as shown in Figure 3.2. This "wedge" has all of the attributes of a single-ended crack with stress-free crack faces subjected to an arbitrary system of forces over an as-yet-undefined boundary. In this sense, Williams has approached the crack problem in a manner similar to that of Inglis: Both started with the solution to some other problem and arrived at the crack problem as the limiting case. The difference, however, is that for the Williams approach the limiting case of a wedge is exactly a stress-free crack. Stated another way, had Williams chosen to predefine the wedge angle to be π in his initial formulation of the problem, he would have obtained the solution to the single-ended crack problem directly. In either event, the resulting eigenequation for the crack problem is

$$
\sin 2\pi\lambda \;=\; 0
\tag{3.10}
$$

the roots of which are

$$
\lambda \;=\; \frac{n}{2} \quad \text{for all integer values of } n \ (0, \pm 1, \pm 2, \cdots)
\tag{3.11}
$$

Some of these roots yield physically unacceptable results for the crack problem and must be rejected. In particular, for any $n < 0$, the displacement calculated from the

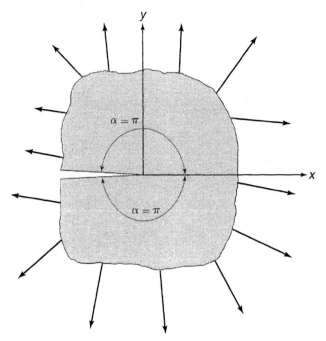

Figure 3.2 A single-ended crack with traction-free crack faces as the limiting case of a wedge ($\alpha = \pi$).

corresponding term in the Airy stress-function would be of the form

$$u_r = \frac{1}{r^{|\frac{n}{2}|}} g(\theta) \tag{3.12}$$

where $g(\theta)$ is a known function of θ, which predicts unbounded displacements at the origin ($r = 0$). Similarly, for $n = 0$, the stress and strain are of the forms

$$\sigma_{ij} = \frac{1}{r} m(\theta) \quad \text{and} \quad \epsilon_{ij} = \frac{1}{r} n(\theta) \tag{3.13}$$

respectively, where $m(\theta)$ and $n(\theta)$ are known functions of θ only.

As a consequence, the strain energy density of Eq. (2.17) is of the form

$$W = \frac{1}{r^2} m(\theta) \cdot n(\theta) \tag{3.14}$$

Integrating Eq. (3.14) over a closed region surrounding the crack tip results in the observation that, if the stresses and strains have the form given by Eqs. (3.13), it would be possible to store an infinite amount of strain energy in a finite volume. Since this observation is clearly contrary to natural law, we must also reject the root $n = 0$. None of the remaining eigenvalues present any physically unrealizable conditions, and consequently, the general solution consists of the sum of all of the terms containing acceptable eigenvalues, i.e., all $\lambda = n/2$ for $n > 0$.

Before proceeding to write out the general solution, we need to note that not all of the coefficients are independent. Letting $\lambda = n/2$ and $\alpha = \pi$ in Eqs. (3.8) yields

$$C_{1n} \cos\left(\frac{n}{2} - 1\right)\pi = -C_{3n} \cos\left(\frac{n}{2} + 1\right)\pi \tag{3.15a}$$

$$\left(\frac{n}{2} - 1\right)C_{1n} \sin\left(\frac{n}{2} - 1\right)\pi = -\left(\frac{n}{2} + 1\right)C_{3n} \sin\left(\frac{n}{2} + 1\right)\pi \tag{3.15b}$$

$$C_{2n} \sin\left(\frac{n}{2} - 1\right)\pi = -C_{4n} \sin\left(\frac{n}{2} + 1\right)\pi \tag{3.15c}$$

$$\left(\frac{n}{2} - 1\right)C_{2n} \cos\left(\frac{n}{2} - 1\right)\pi = -\left(\frac{n}{2} + 1\right)C_{4n} \cos\left(\frac{n}{2} + 1\right)\pi \tag{3.15d}$$

These equations can be satisfied for all values of n if

$$C_{3n} = -\frac{n-2}{n+2}C_{1n} \quad \text{and} \quad C_{4n} = -C_{2n} \text{ for } n = 1, 3, 5, \cdots \tag{3.16a}$$

$$C_{3n} = -C_{1n} \quad \text{and} \quad C_{4n} = -\frac{n-2}{n+2}C_{2n} \text{ for } n = 2, 4, 6, \cdots \tag{3.16b}$$

Substituting Eqs. (3.16) into Eq. (3.6), setting $\lambda = n/2$ in Williams' assumed Airy stress function, and collecting terms, we arrive at the general solution for the single-ended crack with traction-free faces

$$
\begin{aligned}
\mathcal{F}(r, \theta) = &\sum_{n=1,3,\cdots} r^{1+\frac{n}{2}}\left[C_{1n}\left(\cos\frac{n-2}{2}\theta - \frac{n-2}{n+2}\cos\frac{n+2}{2}\theta\right)\right.\\
&\left. +C_{2n}\left(\sin\frac{n-2}{2}\theta - \sin\frac{n+2}{2}\theta\right)\right]\\
&+ \sum_{n=2,4,\cdots} r^{1+\frac{n}{2}}\left[C_{1n}\left(\cos\frac{n-2}{2}\theta - \cos\frac{n+2}{2}\theta\right)\right.\\
&\left. +C_{2n}\left(\sin\frac{n-2}{2}\theta - \frac{n-2}{n+2}\sin\frac{n+2}{2}\theta\right)\right]
\end{aligned} \tag{3.17}
$$

Even before we compute the stresses from the Airy stress function for this problem, we can make an important observation. By comparing the form of the stresses from Eqs. (3.2) with the general solution in Eq. (3.17), it becomes obvious that, because of the symmetry of the cosine function, the terms involving C_{1n} will correspond

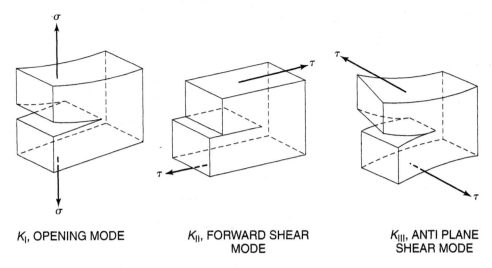

K_I, OPENING MODE K_{II}, FORWARD SHEAR K_{III}, ANTI PLANE
                                   MODE   SHEAR MODE

Figure 3.3   Independent modes of crack deformation.

to stress fields that are symmetric with respect to the crack plane. Similarly, the C_{2n} terms produce antisymmetric stress fields. These two types of stress fields are mutually exclusive and, by using the principle of superposition, can be uncoupled and treated independently. The deformation fields produced by the symmetric and antisymmetric stress fields result in characteristically different motions of the crack faces. The symmetric part results in an opening displacement of the crack faces, whereas the antisymmetric part imparts a shearing motion. These displacements are illustrated in Figure 3.3 and have been given the names, *opening mode* and *forward shear mode*, respectively. To complete the picture there is a third deformation mode that produces a scissoring motion of the crack faces. This mode of deformation, called the *antiplane shear mode*, does not occur in the plane problem of elasticity and will not be treated here. The opening mode of deformation is by far the most important mechanism controlling failure of homogeneous, isotropic materials. It is well established that, unless constrained otherwise, a growing crack will turn itself so as to minimize or totally eliminate the forward shear mode of deformation at the crack tip. Throughout the remainder of this text, the primary emphasis will be on the development of linear elastic fracture mechanics for opening mode deformation, with only the key results for the forward shear mode given. Since the mathematical developments for opening mode deformation and forward shear mode are parallel, the reader can easily fill in the details for himself or herself as the need arises.

Before proceeding to derive the stress field around the crack tip for the Williams problem in general terms, it is instructive to look at the stress field produced by one specific term, C_{12}, in the series expansion. This term has been the source of some misunderstanding, particularly in the formative years of LEFM theory, and its significance is only now being fully appreciated. For $n = 2$, the symmetric part of

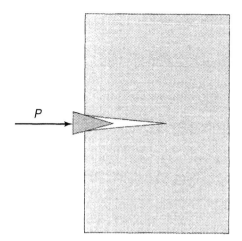

Figure 3.4 An example of a
crack problem with a stress-free
vertical edge at a finite distance
from the crack tip.

the general solution for the Airy stress function becomes

$$\mathcal{F}(r, \theta) = r^2 C_{12}(1 - \cos 2\theta) \qquad (3.18)$$

which we recognize as the Airy stress function for a uniform stress field in the x-direction. [See Eq. (2.44).] Accordingly, the general solution for the opening mode stress field in any body containing a single-ended crack (a temporary restriction) will include a term corresponding to a uniform stress, $\sigma_{ox}(= 4C_{12})$, parallel to the crack plane. Williams notes the presence of this term in his solution, but goes on to make the unfortunate observation that, for bodies containing an unloaded vertical edge, such as the one depicted in Figure 3.4, the boundary conditions along that edge require that $C_{12} = 0$. This statement is, of course, in error. Since the general solution contains an infinite number of terms, there exists some combination of values for all C_{1n} such that the boundary conditions along all of the external boundaries can be satisfied and C_{12} is only one term in this infinite set. In other words the value of C_{12} is determined by the geometry and loading of the body taken as a whole and cannot be *a priori* assigned.

Substituting Eq. (3.17) into Eqs. (2.39) and retaining explicitly only those terms that do not vanish as we approach the crack tip, we obtain

$$\sigma_r = \frac{C_{11}}{4\sqrt{r}} \left(5 \cos \frac{\theta}{2} - \cos \frac{3\theta}{2} \right) + \frac{1}{2}\sigma_{ox}(1 + \cos 2\theta) + \text{H.O.T.}$$

$$\sigma_\theta = \frac{C_{11}}{4\sqrt{r}} \left(3 \cos \frac{\theta}{2} + \cos \frac{3\theta}{2} \right) + \frac{1}{2}\sigma_{ox}(1 - \cos 2\theta) + \text{H.O.T.} \qquad (3.19)$$

$$\tau_{r\theta} = \frac{C_{11}}{4\sqrt{r}} \left(\sin \frac{\theta}{2} + \sin \frac{3\theta}{2} \right) - \frac{1}{2}\sigma_{ox} \sin 2\theta + \text{H.O.T.}$$

for the opening mode and

$$\sigma_r = \frac{C_{21}}{4\sqrt{r}} \left(-5\sin\frac{\theta}{2} + 3\sin\frac{3\theta}{2} \right) + \text{H.O.T.}$$

$$\sigma_\theta = \frac{C_{21}}{4\sqrt{r}} \left(-3\sin\frac{\theta}{2} - 3\sin\frac{3\theta}{2} \right) + \text{H.O.T.} \qquad (3.20)$$

$$\tau_{r\theta} = \frac{C_{21}}{4\sqrt{r}} \left(\cos\frac{\theta}{2} + 3\cos\frac{3\theta}{2} \right) + \text{H.O.T.}$$

for the forward shear mode, where H.O.T. denotes "higher order terms" of order $r^{1/2}$ and beyond. Williams examined several features of the stress field around the tip of the crack that follow directly from the above equations. Probably the most important observation was his interpretation of the implication of these equations on brittle failure of materials that might otherwise tend to fail by yielding. He noted that as the crack tip is approached, the stress state approaches a nearly hydrostatic condition (particularly when plane strain conditions prevail). Since hydrostatic stress states do not favor plastic flow (Bridgeman, 1952), the material has the opportunity to seek an alternative mechanism of failure, namely cleavage fracture. Equations (3.19) and (3.20) are called the *near-field equations*, or more properly, the *modified near-field equations* when the σ_{ox} term is included. These equations describe the distribution of the stress field over some, albeit small, region around the crack tip. It is significant that all of the stresses have the same characteristic form. In particular, they all contain the same inverse-square-root singularity multiplied by *known* functions of θ, regardless of the geometry or details of the loading. As a consequence, the *only* difference in the state of stress between differing geometric problems is the amplitude of the stresses; the *distribution* of the stress in the immediate neighborhood of the crack is always the same.

This observation is in marked contrast to our previous experience with bodies containing smooth discontinuities, such as holes or other cavities. In these cases the distribution of stress depends on the geometric parameters of the problem, and each problem has a unique stress distribution that can be determined only by formally solving the boundary-value problem. On the other hand the singular crack problem in elasticity, as typified by the Williams solution for single-ended cracks, has the dominant feature that the distributions of stress, strain, and displacement are always the same and depend only on the magnitude of the characterizing parameters C_{11} and σ_{ox}. As a consequence, provided that we restrict our attention to the near-field region (alternatively called the *singularity-dominated-zone*) around the crack tip, the crack problem in elasticity has been reduced to finding the magnitude of these characterizing parameters by whatever technique we choose, and unless we desire to know the distribution of stress over some more global region, the formalities of solving the elasticity problem need not be performed. This form invariance of the near-field equations was first observed by Irwin [1957] and has provided a cornerstone of the theory of linear elastic fracture mechanics.

Despite the significance of the Williams solution to early fracture mechanics theory, it is not the only solution to all fracture problems. It represents the complete solution to only one class of problems, namely that of a single-ended crack with stress-free crack faces subjected to smoothly varying boundary conditions over finite boundaries. The solutions to other classes of crack problems, including, for example, multiple crack tips and/or crack-face loading, have not been forthcoming in real-variable elasticity theory. Fortunately, the introduction of complex variables provides an infinite number of potential Airy stress functions that might be brought to bear on the fracture problem. This approach will be developed in detail in the next section. For the benefit of the reader a brief introduction to the complex-variable method in elasticity theory is presented in Appendix B.

3.3 THE GENERALIZED WESTERGAARD APPROACH

The introduction of complex variables into the formulation of the two-dimensional elasticity problem offers significant advantages. In addition to algebraic compactness, the use of complex analytic functions ensures that the biharmonic equation (Eqs. 2.35), are satisfied. (See Appendix B for details.) In the adaptation of this approach to the crack problem, two formulations have been adopted. The classical Goursat–Kolosov approach, popularized by Muskhelishvili [1953], offers powerful tools, including conformal mapping, to solve a wide variety of singular problems. However, this approach is mathematically very demanding.

Fortunately, there is another complex-variable method, limited to straight crack problems only, that offers the advantages of algebraic compactness and simplicity without the disadvantages of the more arduous classical approach. This method, introduced by Westergaard in 1939, applies the semi-inverse method to the Airy stress function expressed in the complex domain. As originally formulated by Westergaard, this method could be applied only to infinite body problems with uniform (including zero) remote boundary conditions. These restrictions were known at the outset. Nonetheless, the method was widely used in the late 1950s and throughout the 1960s to solve a number of significant problems during the early development of fracture mechanics theory. The method was modified by Sih [1966] and Eftis and Liebowitz [1972] to accommodate unequal remote biaxial loads, but otherwise the method still had restricted application. More recently, the Goursat–Kolosov representation of the Airy stress-function for planar crack problems was recast into Westergaard notation [Sanford, 1979], and a second analytic function was introduced into the analysis. This generalized Westergaard formulation retains both the completeness of the Goursat–Kolosov approach and the algebraic simplicity of Westergaard's formulation. As a result, the method can now be applied to both infinite- and finite-body problems with arbitrary boundary conditions subject only to the condition that the crack(s) is (are) constrained to lie along the $y = 0$ plane. In the upcoming sections of this chapter, the generalized Westergaard method will be developed in full detail, and the earlier formulations will be shown to be special cases.

In contrast to real-function theory, where the selection of functions that satisfy the biharmonic equation is difficult to obtain, the principal advantage of the formulation of the theory of elasticity in complex variables is that all analytic functions are potential Airy stress functions. Although the number of analytic functions is infinite, not all complex functions are analytic. A function is analytic if and only if it satisfies the Cauchy–Riemann conditions,

$$\frac{\partial Re\, Z}{\partial y} = -\frac{\partial Im\, Z}{\partial x} = -Im\, Z' \tag{3.21a}$$

$$\frac{\partial Im\, Z}{\partial y} = \frac{\partial Re\, Z}{\partial x} = Re\, Z' \tag{3.21b}$$

where the prime superscript denotes differentiation with respect to the complex variable, z. These relations have two important properties. First, they demonstrate that the real and imaginary parts of potential Airy stress-functions are not independent. Second, they provide the rules that relate the derivatives of complex functions with respect to real variables to their complex counterparts. We will use this latter property of the Cauchy–Riemann conditions repeatedly.

With these comments as background let us consider an Airy stress function of the form

$$F(z) = Re\, \tilde{\tilde{Z}}(z) + y(Im\, \tilde{Z}(z) + Im\, \tilde{Y}(z)) \tag{3.22}$$

where

$$\frac{d\tilde{\tilde{Z}}}{dz} = \tilde{Z}, \quad \frac{d\tilde{Z}}{dz} = Z, \quad \text{and} \quad \frac{d\tilde{Y}}{dz} = Y$$

and $Z(z)$ and $Y(z)$ are as-yet-undefined analytic functions [Sanford, 1979]. Recall, from Eqs. (2.33), that the Cartesian stress components are obtained from an Airy stress function through the second derivatives with respect to the real variables, x and y. These relations are still valid even when the stress function is expressed in terms of the complex coordinate, z. Using the Cauchy–Riemann relations and noting that the second term of Eq. (3.22) is a product of functions when differentiating with respect to y, we obtain for the first partial derivatives of $F(z)$

$$\frac{\partial F}{\partial x} = Re\, \tilde{Z} + y(Im\, Z + Im\, Y) \tag{3.23a}$$

$$\frac{\partial F}{\partial y} = Im\, \tilde{Y} + y(Re\, Z + Re\, Y) \tag{3.23b}$$

Repeating this process for the second partial derivatives, we obtain the Cartesian stress components as

$$\sigma_x = Re\,Z - y(Im\,Z' + Im\,Y') + 2Re\,Y \qquad (3.24a)$$

$$\sigma_y = Re\,Z + y(Im\,Z' + Im\,Y') \qquad (3.24b)$$

$$\tau_{xy} = -Im\,Y - y(Re\,Z' + Re\,Y') \qquad (3.24c)$$

These equations are a generalization of the results presented by Westergaard in 1939.

From the in-plane stress components, we can obtain the Cartesian strain components by direct substitution into the appropriate form of Hooke's law. For the case of plane stress [Eqs. (2.26)], the resulting strain components are

$$E\epsilon_x = (1 - v)Re\,Z - (1 + v)[yIm\,Z' + yIm\,Y'] + 2Re\,Y \qquad (3.25a)$$

$$E\epsilon_y = (1 - v)Re\,Z + (1 + v)[yIm\,Z' + yIm\,Y'] - 2vRe\,Y \qquad (3.25b)$$

$$E\gamma_{xy} = -2(1 + v)Im\,Y - 2(1 + v)[yRe\,Z' + yRe\,Y'] \qquad (3.25c)$$

Finally, we can complete the general form of the solution for the Airy stress function of Eq. (3.22) by integrating the strains to obtain the displacements.[1]Using the strain-displacement relations from Eqs. (2.30) and the rightmost equality in the Cauchy–Riemann conditions of Eqs. (3.21), we obtain for the in-plane displacements (neglecting terms which represent rigid body motions)

$$Eu = (1 - v)Re\,\tilde{Z} - (1 + v)[yIm\,Z + yIm\,Y] + 2Re\,\tilde{Y} \qquad (3.26a)$$

$$Ev = 2Im\,\tilde{Z} - (1 + v)[yRe\,Z + yRe\,Y] + (1 - v)Im\,\tilde{Y} \qquad (3.26b)$$

The strain and displacement results for the case of plane strain are very similar. The particular details are left as an exercise for the reader. Recognizing that our goal is to solve crack problems, we can impose restrictions on acceptable choices for the functions Z and Y to ensure that at least some of the boundary conditions for this class of problem are satisfied. Since we are free to place the coordinate axes wherever we choose, there is no loss in generality by considering the crack to lie along the x-axis, but otherwise unconstrained, as shown in Figure 3.5. Then, along the crack plane, the stresses become

---

[1]Note that the terms in square brackets in Eq. (3.25b) need to be integrated by parts to obtain the v-component of displacement. Specifics of the procedure are described in Appendix B.

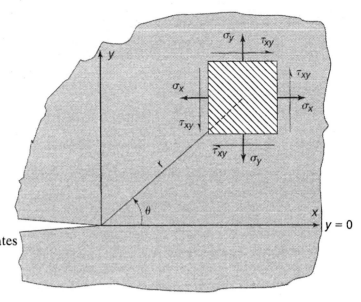

Figure 3.5 Coordinates
for an arbitrary
crack in the complex
domain.

$$
\left.
\begin{array}{rcl}
\sigma_x & = & Re\,Z + 2Re\,Y \\[2mm]
\sigma_y & = & Re\,Z \\[2mm]
\tau_{xy} & = & -Im\,Y
\end{array}
\right\} \quad \text{for } y = 0 \qquad (3.27)
$$

If we now restrict our attention to opening mode crack problems only, the stresses must be symmetric with respect to the crack plane, and accordingly, $\tau_{xy} = 0$ on $y = 0$ for all x. From Eqs. (3.27), this symmetry condition can be satisfied by imposing a constraint on our possible choices for $Y(z)$ such that

$$
Im\,Y(z) = 0 \text{ on } y = 0 \qquad (3.28)
$$

This constraint condition can be satisfied in several ways. Westergaard did not include the function Y in his formulation of the problem. This exclusion is equivalent to imposing the condition that $Y \equiv 0$. Sih [1966] proposed a "modified" Westergaard theory in which $Y = A$, a real constant. This modification removed an objection to the original Westergaard formulation, which predicts equal biaxial stresses at infinity. The modified formulation accommodates unequal (including uniaxial) remote stresses, but otherwise is still overly restrictive. Finally, the generalized Westergaard formulation presented here requires only that $Y = a\ real\ function$ on $y = 0$ (i.e., $Im\,Y = 0$ on $y = 0$) and may be fully complex for $y \neq 0$. Later we will demonstrate that it is always possible to define such a function with these minimal properties.

The necessity of retaining the generalized formulation can be demonstrated by a simple experiment. The photoelastic method of experimental stress analysis (see Chapter 5) produces optical fringe patterns that are contour lines of constant maximum shearing stress, τ_{max}. From elementary concepts, τ_{max} can be expressed in terms of the Cartesian stresses as

$$\tau_{max} = \sqrt{\left(\frac{\sigma_x - \sigma_y}{2}\right)^2 + \tau_{xy}^2} \qquad (3.29)$$

Along the crack line, Westergaard's formulation predicts that $\tau_{max} = 0$ for all x, which would appear in the photoelastic pattern as a uniform zeroth-order black fringe all along the crack line for all acceptable choices of the function $Z(z)$. The modified Westergaard formulation results in a uniform (nonzero) value for the maximum shearing stress along the crack line, and again, a constant (but nonzero) fringe order would be predicted. In contrast, the generalized Westergaard formulation imposes no such restrictions on the distribution of maximum shearing stress along the crack line. Instead, the photoelastic fringe-order distribution along $y = 0$ is determined by a real function whose form depends on the particular problem being solved. The photoelastic fringe pattern shown in Figure 3.6 for the compact tension geometry widely used in fracture testing contains a series of fringe loops of different order that cross the crack plane ahead of the crack tip. Clearly, of the three choices, only the generalized Westergaard formulation is capable of analyzing this problem. As

Figure 3.6 Typical photoelastic pattern for a compact-tension specimen geometry with a deep crack.

a general rule, the generalized formulation is required whenever geometries with finite boundaries are to be studied.

Having investigated fully the constraints on possible choices for the analytic function $Y(z)$, we now turn our attention to the corresponding constraints on the function $Z(z)$. Whereas the constraint condition of Eq. (3.28) was necessary to ensure symmetry, the constraint on acceptable choices for the $Z(z)$ function will be used to satisfy the boundary conditions on the traction-free portions of the crack faces. This boundary condition is equivalent to the two conditions $\tau_{xy} = 0$ and $\sigma_y = 0$ on those portions of the crack faces on which there are no applied forces. The first condition ($\tau_{xy} = 0$) is automatically satisfied by the symmetry requirement of Eq. (3.28). From the second of Eqs. (3.27), the second boundary condition ($\sigma_y = 0$) can be satisfied by requiring that

$$Re \; Z(z) \; = \; 0 \tag{3.30}$$

on traction-free crack faces.

From the foregoing discussions, we are now in a position to formulate the means of solution of an opening mode crack problem in generalized Westergaard form using the semi-inverse approach. Namely, if we select two functions, $Z(z)$ and $Y(z)$, that are analytic over some domain (except, possibly, at isolated singularities) subject to the constraints imposed by Eqs. (3.28) and (3.30), we *may* have the solution to a problem of practical interest. In accordance with the methodology of the semi-inverse method, we would substitute our chosen functions into the stress-field equations in Eqs. (3.24) and examine the behavior of these stresses along the selected boundaries of the problem. Since we are already assured that the biharmonic equation and the boundary conditions along the crack plane are satisfied, we need only examine the behavior of our chosen functions at the remaining boundaries (or at infinity for infinite-body problems). If the boundary conditions are satisfied exactly, then the chosen functions are the appropriate stress functions for that particular problem, and the stresses, strains and displacements predicted by these functions are the exact solution to the problem.

For the forward-shear (mode II) fracture problem the Airy stress function in complex form is

$$F_{II}(z) \; = \; Im \, \tilde{\tilde{Y}} \; - \; y \left[Re \, \tilde{Y}(z) \; + \; Re \, \tilde{Z}(z) \right] \tag{3.31}$$

Following the same procedure as performed previously, we find that the Cartesian stresses are of the form

$$\sigma_x \; = \; Im \, Y_{II} \; + \; y \left(Re \, Y'_{II} \; + \; Re \, Z'_{II} \right) \; + \; 2Im \, Z_{II} \tag{3.32a}$$

$$\sigma_y \; = \; Im \, Y_{II} \; - \; y \left(Re \, Y'_{II} \; + \; Re \, Z'_{II} \right) \tag{3.32b}$$

$$\tau_{xy} = -y \left(Im\, Y'_{II} + Im\, Z'_{II} \right) + Re\, Z_{II} \tag{3.32c}$$

where the subscript, II, denotes stress functions applicable to the forward shear mode of deformation.[2] In order to ensure that the deformation conforms to the in-plane shear mode of deformation, the functions $Z_{II}(z)$ and $Y_{II}(z)$ (which often have forms similar to their opening mode counterparts) are subject to the constraints that

$$Re\, Z_{II} = 0 \text{ on the crack faces} \tag{3.33a}$$

to ensure that $\tau_{xy} = 0$ and

$$Im\, Y_{II} = 0 \text{ on } y = 0 \tag{3.33b}$$

to ensure that $\sigma_y = 0$ for all x.

Unlike the solution to crack problems in real coordinates—for example, the Williams solution derived earlier in this chapter, wherein the mixed mode problem is treated as a single problem—the Westergaard approach separates the problem into its symmetric and antisymmetric parts and solves them independently. Then, by using the principle of superposition, these results are combined in the appropriate proportions to determine the overall stress, strain and displacement fields. (To be strictly correct a subscript, I, should have been added to the opening-mode stress functions previously discussed; however, unless there is a possibility of confusion, this subscript is customarily omitted when only the opening mode of deformation is being considered.) Because of the similarity between the mathematical developments of the opening and forward shear modes, only the opening-mode results will be presented in the upcoming sections. The corresponding results for the forward-shear case can be easily developed as the need arises.

3.4 THE CENTRAL CRACK PROBLEM

Recall from Chapter 2 that Inglis was able to obtain an approximate solution to the problem of an internal crack of length, $2a$, under remote biaxial tension by considering the degenerate case of an ellipse of similar length. Westergaard examined the same problem in his 1939 paper and obtained an exact solution by suitable choice of the complex function Z. (The function Y was not included in his formulation of the method.) Following the methodology of the semi-inverse method, let us determine the problem (if any) solved by the functions

$$Z(z) = \frac{\sigma z}{\sqrt{z^2 - a^2}} \quad \text{and} \quad Y(z) = 0 \tag{3.34}$$

---

[2] ASTM Standard E 1823–96 recommends the use of the Arabic subscripts, 1 and 2, to denote the opening and forward-shear modes, respectively. However, the preponderance of literature uses Roman numerals, I and II, for this purpose. That historical practice will be followed here as well.

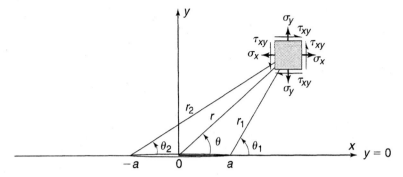

Figure 3.7 Multiple coordinate systems for the function $Z = \dfrac{\sigma z}{\sqrt{(z^2 - a^2)}}$.

From these functions, it follows that

$$Z'(z) = -\frac{\sigma a^2}{(z^2 - a^2)^{3/2}} \qquad (3.35)$$

We note that Eqs. (3.34) has isolated singularities at $z = \pm a$, and we will find it convenient to introduce additional coordinates centered at each singularity. These auxiliary polar coordinates, illustrated in Figure 3.7, are defined by the relations

$$z - a = (x - a) + iy = r_1 e^{i\theta_1} \qquad (3.36a)$$

$$z + a = (x + a) + iy = r_2 e^{i\theta_2} \qquad (3.36b)$$

Factoring the denominator of $Z(z)$ in Eqs. (3.34) and introducing the auxiliary coordinates yields

$$\sigma_x = \frac{\sigma r}{\sqrt{r_1 r_2}} \cos\left(\theta - \frac{\theta_1 + \theta_2}{2}\right) - \frac{\sigma a^2}{(r_1 r_2)^{\frac{3}{2}}} r_1 \sin\theta_1 \sin\frac{3}{2}(\theta_1 + \theta_2) \qquad (3.37a)$$

$$\sigma_y = \frac{\sigma r}{\sqrt{r_1 r_2}} \cos\left(\theta - \frac{\theta_1 + \theta_2}{2}\right) + \frac{\sigma a^2}{(r_1 r_2)^{\frac{3}{2}}} r_1 \sin\theta_1 \sin\frac{3}{2}(\theta_1 + \theta_2) \qquad (3.37b)$$

$$\tau_{xy} = \frac{\sigma a^2}{(r_1 r_2)^{\frac{3}{2}}} r_1 \sin\theta_1 \cos\frac{3}{2}(\theta_1 + \theta_2) \qquad (3.37c)$$

The next step in the semi-inverse method is to determine the boundary conditions predicted by the derived stress components. To this end we note that $Z(z)$ can be expressed as

$$Z(x) = \frac{-i\sigma x}{\sqrt{a^2 - x^2}} \quad \text{along the crack line } (y = 0) \tag{3.38}$$

which, in the interval $-a < x < a$, is a pure imaginary number. Since $Re\, Z$ is zero in this interval, the boundary condition of Eq. (3.30) reveals that there is a traction-free crack over this interval (and nowhere else).

Now let us examine the behavior of the stresses at great distances from the origin. For a crack of finite length, $2a$, it is always possible to determine a distance such that

$$z - a \approx z + a \approx z \tag{3.39a}$$

or, in real variables,

$$r_1 \approx r_2 \approx r \quad \text{and} \quad \theta_1 \approx \theta_2 \approx \theta \tag{3.39b}$$

Under these conditions, the stresses reduce to

$$\sigma_x = \sigma \qquad \sigma_y = \sigma \qquad \tau_{xy} = 0 \quad \text{for large } z \tag{3.40}$$

which correspond to a uniform biaxial stress state at an infinite distance.

Combining our observations in the near and far regions, we are now able to state that the functions Z and Y defined in Eqs. (3.34) solve *exactly* the problem of an internal crack of length, $2a$, in an infinite plate that is subjected to equal biaxial tension, σ, as illustrated in Figure 3.8a. The stresses computed from these functions, Eqs. (3.37), are exact and, accordingly, are valid everywhere in the body.

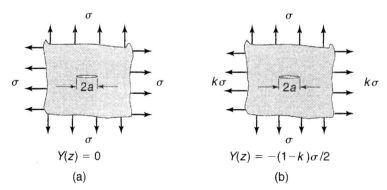

Figure 3.8 Remote loading on an infinite plate with a central crack of length $2a$ for two choices of stress function, $Y(z)$.

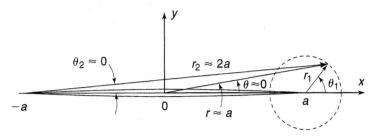

Figure 3.9   Localized crack-tip coordinate system.

Now consider the problem solved by

$$Z(z) = 0 \quad \text{and} \quad Y(z) = -\frac{(1-k)}{2}\sigma \tag{3.41}$$

It can readily be determined that this pair of functions corresponds to a uniaxial stress in the direction parallel to the crack plane. Using superposition, we can add this result to that of the previous case to arrive at the exact solution to the internal crack problem under arbitrary biaxial stress. (See Figure 3.8b.) The significance of the biaxial load factor, k, on the stress field in center-cracked panels has been discussed by Liebowitz, Lee, and Eftis [1978]. For $k = 0$, the uniform-stress solution exactly balances the remote applied stress in the x-direction in the biaxial solution, and we obtain the exact solution for the central-cracked panel under uniaxial remote stress.

It is important to observe that the uniform σ_x solution does not introduce any singular stresses at the crack tip. This is not to say that the stress applied parallel to the crack line does not affect the stress state in the body. Clearly, the magnitude and direction of the principal stresses are influenced by the additional stress, but the crack opening stress, σ_y, in the neighborhood of the crack tip that dominates the behavior of a cracked body is not affected, nor is the singular character of the stress state. These observations will prove to be important in later discussions of a stress-based theory of fracture in Chapter 6.

Let us now confine our attention to a small region (relative to the crack length) around one of the crack tips, as shown in Figure 3.9. If the distance to the other crack tip is great compared with the region of interest, then it is reasonable to approximate these distances by

$$r_2 \approx 2a \quad \text{and} \quad \theta_2 \approx \theta \approx 0 \tag{3.42}$$

Substituting these approximations in Eqs. (3.37), we obtain

$$\sigma_x = \frac{\sigma\sqrt{\pi a}}{\sqrt{2\pi r_1}} \cos\frac{\theta_1}{2} \left[1 - \sin\frac{\theta_1}{2}\sin\frac{3\theta_1}{2}\right] \tag{3.43a}$$

$$\sigma_y = \frac{\sigma\sqrt{\pi a}}{\sqrt{2\pi r_1}} \cos\frac{\theta_1}{2}\left[1 + \sin\frac{\theta_1}{2}\sin\frac{3\theta_1}{2}\right] \tag{3.43b}$$

$$\tau_{xy} = \frac{\sigma\sqrt{\pi a}}{\sqrt{2\pi r_1}} \cos\frac{\theta_1}{2}\sin\frac{\theta_1}{2}\cos\frac{3\theta_1}{2} \tag{3.43c}$$

where the factor $\sqrt{\pi}$ has been introduced to make the result conform to current notation. These results can be written in a more general form by making several observations. First, since there is only one germane coordinate system in the small region around a crack tip, we can delete the subscripts in Eqs. (3.43) and reinterpret the coordinates r and θ as crack tip coordinates. Second, for a specific geometry and amplitude of applied load, the term $\sigma\sqrt{\pi a}$ has a fixed value and can be replaced by a constant, K. Finally, Eqs. (3.43) were obtained from the exact results for the equal biaxial loading condition. But, from our earlier discussion, we know that we can accommodate arbitrary degrees of biaxiality by superposing a constant stress, σ_{ox}, in the x-direction. Incorporating these generalities, we can now write the Cartesian stress components in the neighborhood of a crack tip as

$$\sigma_x = \frac{K}{\sqrt{2\pi r}} \cos\frac{\theta}{2}\left[1 - \sin\frac{\theta}{2}\sin\frac{3\theta}{2}\right] + \sigma_{ox} \tag{3.44a}$$

$$\sigma_y = \frac{K}{\sqrt{2\pi r}} \cos\frac{\theta}{2}\left[1 + \sin\frac{\theta}{2}\sin\frac{3\theta}{2}\right] \tag{3.44b}$$

$$\tau_{xy} = \frac{K}{\sqrt{2\pi r}} \cos\frac{\theta}{2}\sin\frac{\theta}{2}\cos\frac{3\theta}{2} \tag{3.44c}$$

If we compare these near-field equations in Cartesian form with the opening mode near-field equations in polar form [Eqs. (3.19)] that we obtained from the solution to the Williams problem, we find that they are exactly the same (where $K = \sqrt{2\pi}C_{11}$). As a consequence, we have now demonstrated that the form invariance of the near-field equations discussed earlier extends beyond single-ended crack problems and, at least in this one case, includes internal cracks as well. In fact, we can now state that the distribution of the stresses, strains and displacements in some small region around a crack tip are always the same for *any* cracked body in which the stress state around the crack tip is dominated by the inverse-square-root singularity (this restriction excludes, for example, cracks at interfaces between dissimilar materials) the form of the stresses is given in polar form by Eqs. (3.19) and in Cartesian form by Eqs. (3.44).

For the forward-shear mode of deformation this observation also holds, and the near-field stresses in polar form are given by Eqs. (3.20) and by the following equations in Cartesian form,

$$\sigma_x = \frac{K_{II}}{\sqrt{2\pi r}} \sin\frac{\theta}{2}\left[2 + \cos\frac{\theta}{2}\cos\frac{3\theta}{2}\right] \tag{3.45a}$$

$$\sigma_y = \frac{K_{II}}{\sqrt{2\pi r}} \sin\frac{\theta}{2}\cos\frac{\theta}{2}\cos\frac{3\theta}{2} \tag{3.45b}$$

$$\tau_{xy} = \frac{K_{II}}{\sqrt{2\pi r}} \cos\frac{\theta}{2}\left[1 - \sin\frac{\theta}{2}\sin\frac{3\theta}{2}\right] \tag{3.45c}$$

where K_{II} $(= \sqrt{2\pi}C_{21})$ is the forward-shear counterpart to the parameter K in Eqs. (3.44).

3.5   SINGLE-ENDED CRACK PROBLEMS

In the previous section the near-field equations for the opening mode of deformation were obtained by considering a small region around one of the crack tips of an internal crack. These same equations can be obtained more directly by examining the behavior of the stress functions

$$Z(z) = \frac{K}{\sqrt{2\pi z}} = \frac{K}{\sqrt{2\pi}}z^{-1/2} \quad \text{and} \quad Y(z) = 0 \tag{3.46}$$

From these functions, it follows that

$$Z'(z) = -\frac{1}{2}\frac{K}{\sqrt{2\pi}}z^{-3/2} \tag{3.47}$$

In terms of the real variables, r and θ, these equations become

$$Z(z) = \frac{K}{\sqrt{2\pi r}}\cos\frac{\theta}{2} - i\frac{K}{\sqrt{2\pi r}}\sin\frac{\theta}{2} \tag{3.48}$$

$$Z'(z) = -\frac{K}{2\sqrt{2\pi}}r^{-3/2}\cos\frac{3\theta}{2} + i\frac{K}{2\sqrt{2\pi}}r^{-3/2}\sin\frac{3\theta}{2} \tag{3.49}$$

When $\theta = \pm\pi$ in Eq. (3.48), $Re\, Z = 0$, and the entire negative half-plane ($x < 0$) corresponds to a traction-free crack face.

Substituting the real and imaginary parts of Eqs. (3.48) and (3.49) into Eqs. (3.24) yields expressions for the stresses that are identical to the near-field equations developed previously (with $\sigma_{ox} = 0$). Therefore, whereas the near-field equations were an approximate result for the central-crack problem, they are the exact representation of the stresses for the stress function given by Eqs. (3.46). We have just shown that this simple stress function satisfies the boundary conditions along the symmetry plane for a single-ended crack, but we still must determine the remote boundary conditions matched by this stress function. For this purpose, let $r \rightarrow \infty$ in the near-field equations, and observe that the stress state approaches zero. Consequently, since this stress function predicts no applied tractions on the crack faces or at the remote boundaries, it is not an exact solution for any crack problem of practical interest.

Despite the failure of Eqs. (3.46) to solve (in the exact sense) any crack problem, it does produce the near-field equations in a simple manner. It is intriguing to ask whether or not we can attach any additional significance to this stress function. To address this question, consider the stress functions

$$Z(z) = \sum_{j=0}^{\infty} A_j z^{j-1/2} \tag{3.50a}$$

$$Y(z) = \sum_{m=0}^{\infty} B_m z^{m} \tag{3.50b}$$

in which the leading terms ($j = 0$ and $m = 0$) are of the same form as in Eqs. (3.46) and therefore will reproduce the modified near-field equations.

Following the procedure of the semi-inverse method, we ask if there is some geometry and associated surface tractions for which these series stress functions are an exact solution. To answer this question, we rewrite Eq. (3.50a) in terms of real variables as

$$Z(r, \theta) = \sum_{j=0}^{\infty} A_j r^{j-1/2} \left[\cos\left(j - \frac{1}{2}\right)\theta + i \sin\left(j - \frac{1}{2}\right)\theta \right] \tag{3.51}$$

When $\theta = \pm\pi$, the term $\cos(j - \frac{1}{2})\theta$ is zero for all j, and therefore, $Re\,Z = 0$ for all $x < 0$ and all j. Similarly, on $y = 0$,

$$Y(z) = Y(x) = \sum_{m=0}^{\infty} B_m x^{m} \tag{3.52}$$

which is a *real* function and satisfies the requirement of Eq. (3.28).

Consequently, these two series stress functions satisfy the boundary conditions for a single-ended, traction-free crack and contain as their leading terms functions that result in the modified near-field equations. That is, we can rewrite Eqs. (3.50) as

$$Z(z) = \frac{K}{\sqrt{2\pi z}} + \sum_{j=1}^{\infty} A_j z^{j-1/2} \tag{3.53a}$$

$$Y(z) = \frac{\sigma_{ox}}{2} + \sum_{m=1}^{\infty} B_m z^m \tag{3.53b}$$

More importantly, whereas the near-field equations fail to represent the stresses in the far field for any meaningful problem, the series-type stress functions contain an infinite number of coefficients $(A_1, A_2,\ldots, B_1, B_2,\ldots)$ which can be determined to satisfy the boundary conditions on any remote boundary. As a consequence, the stress function of Eqs. (3.50) or, alternatively, Eqs. (3.53) is a general solution and solves exactly the same class of problems as does the symmetric part of the Williams stress function analyzed in Chapter 2 in real variables—namely, a single-ended crack with traction-free crack faces subjected to arbitrary surface tractions in a finite body. From the uniqueness theorem of elasticity we know that there can be only one solution to every problem, and accordingly, the Airy stress function of Eq. (3.22) with the complex functions, Z and Y, defined by Eqs. (3.50) or (3.53) is functionally equivalent to the symmetric part of the Williams stress function of Eq. (3.17).

The results of this section further reinforce the concept that a characteristic feature of crack-tip stress fields is that the distribution of the stress in the neighborhood of a crack tip (as represented by the near-field equations) is universal and that only the magnitude of the singular term (i.e., K) varies with the geometery and type of loading. Because of the importance of this concept to the formulation of a stress-based theory of fracture, it is instructive to estimate the size of the region around the crack tip for which the singular term describes the distribution of the stresses to a reasonable degree of accuracy.

Chona, Irwin and Sanford [1983] studied the stress state around the crack tip in the modified compact tension (MCT) and the rectangular double-cantilever beam (RDCB) geometries, as illustrated in Figure 3.10. These geometries are of interest for dynamic fracture testing. Using the photoelastic method to visualize the stresses in conjunction with a local collocation procedure,[3] they were able to accurately represent the state of stress over an extended region around the crack tip with a truncated form of Eqs. (3.53). They compared the relative contributions of the singular and nonsingular terms in the series solution and constructed diagrams that defined the zone

---

[3] A detailed discussion of this approach is presented in Chapter 5 on experimental measurement of the stress intensity factor.

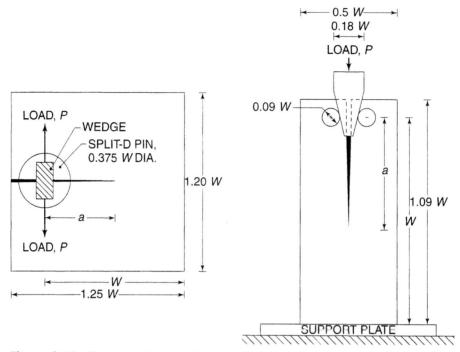

Figure 3.10 Configurations of the modified-compact-tension (MCT) and the rectangular-double-cantilever-beam (RDCB) specimens often used for dynamic fracture studies.

of dominance of the singular stresses. Typical results for the σ_y stress in the modified compact-tension geometry for various crack lengths are shown in Figure 3.11. The size of the singularity-dominated-zone (SDZ) for the other stress components and the other geometry is similar. We note in passing that the stress behind the crack tip is low, and the enlarged size of the SDZ in this region is of little practical significance.

If it is argued that the fracture-process zone is entirely ahead of the crack, then a meaningful measure of the singularity-dominated-zone is the radius, r_{min}, of the largest circle lying within the shaded region of Figure 3.11. Using this measure, Chona, et al., plotted the SDZ as a function of crack length for both geometries, as shown in Figure 3.12. They discussed the differences in zone size for the two geometries and observed that the governing parameter was not the length of the crack, but rather the distance to the nearest boundary. For the MCT geometry, the net ligament length is the germane parameter. For the RDCB configuration the sides of the specimen are the controlling parameter over most of the crack length, changing over to the net ligament length for long cracks. When normalized by this new parameter (see Figure 3.13) the SDZ is shown to be nearly constant for all crack lengths and reasonably independent of specimen type. This latter observation has been confirmed in additional unpublished studies by some of the same researchers. Based on this result, we can state, as a rule of thumb, that the size of the singularity-

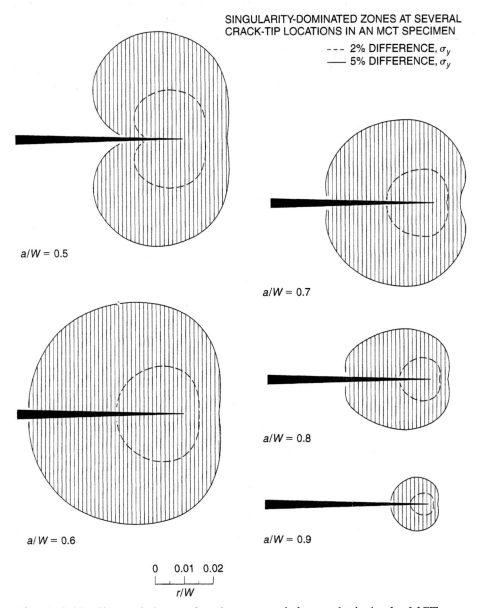

Figure 3.11 Size and shape of regions around the crack tip in the MCT geometry for which the singular term dominates the σ_y stress.

dominated-zone is of the order of two percent of the distance to the nearest boundary in typical specimen geometries.

These studies of the size of the SDZ demonstrate that there is a finite, albeit small, region around a crack tip in which the state of stress (except possibly for an additional uniform stress parallel to the crack line) is adequately represented by

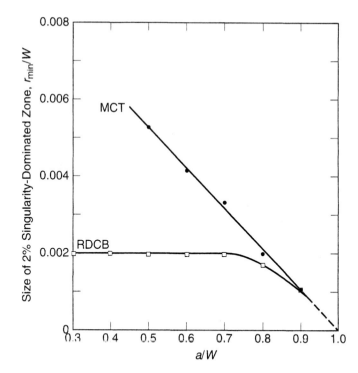

Figure 3.12 Singularity-dominated zone r_{min}/W versus nondimensional crack length for the MCT and RDCB geometries.

the single parameter, K. The significance of this observation cannot be overstated. Unlike the concept of a stress concentration factor that determines the magnitude of the stress at a single point, as discussed in Chapter 2, the parameter K, called the *stress intensity factor* (SIF), provides a *complete* description of the state of stress,

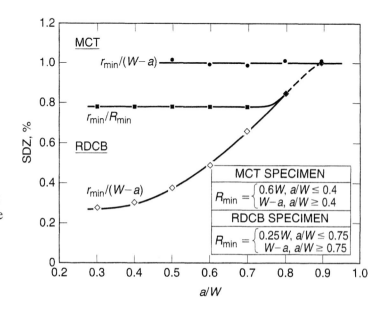

Figure 3.13 Two-percent singularity-dominated zone r_{min} as a fraction of the distance to the nearest boundary, R_{min}, versus nondimensional crack length.

strain and displacement over some *region* of the body. Since the stress intensity factor represents the strength of the singularity—i.e., the rate at which the stresses approach infinity—we can formally define K as

$$K = \lim_{\delta^+ \to 0} \sigma_y|_{\theta=0} \cdot \sqrt{2\pi\delta} \qquad (3.54)$$

where δ is the distance measured from the crack tip, and the limit is taken from the material (+) side.

Applying this definition to the σ_y stress for the central-crack problem [Eq. (3.37b)], we find that, for this geometry

$$K = \sigma\sqrt{\pi a} \qquad (3.55)$$

From this result we can make several observations about the general form of the stress intensity factor. First, the SIF is proportional to the applied stress. This relationship is a direct consequence of the linear nature of the theory of elasticity. Second, K contains the crack length as a parameter. Therefore, unlike the stress concentration factor, the stress intensity factor is size dependent. In addition, the stress intensity factor must be a function of the geometry of the body. Therefore, we can infer that, in general, the stress intensity factor must be of the form

$$K = \sigma\sqrt{\pi a} \cdot Y\left(\frac{a}{W}\right) \qquad (3.56)$$

where $Y(a/W)$ is a dimensionless shape factor that embodies the effects of all of the geometric parameters and W is any characteristic in-plane dimension (often the width of the body). For the central-crack problem, $Y(a/W) = 1$.

Finally, we note here that later in the book, we will introduce a material property that, unfortunately, is also called the "stress intensity factor." The quantity defined by Eq. (3.54) is not a material property, but rather a geometric property definable within the context of the linear theory of elasticity. To minimize the confusion over terminology, we should more precisely refer to the stress intensity factor described here as the *geometric stress intensity factor*, but, when the meaning is clear, we will continue to use the historical term.

3.6 THE EFFECT OF FINITE BOUNDARIES

Because of their inherent difficulties, the number of closed-form analytical solutions to crack problems with finite dimensions is limited. There are, however, several problems of historical and practical importance to the theory of linear elastic fracture mechanics that deserve special treatment.

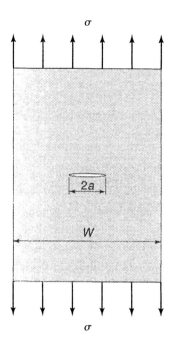

Figure 3.14 Geometry of the center-cracked tension (CCT) panel.

The earliest attempt to obtain a solution to a finite-body problem is attributed to Irwin [1957]. He was interested in the stress intensity factor for the problem shown in Figure 3.14—namely, a plate of width, W, with a central crack of length, $2a$, subjected to uniform uniaxial tension. The solution to this problem was of more than academic interest. Since fracture testing during the early years of linear elastic fracture mechanics used specimens of this type, it became vital to the development of LEFM to have an accurate representation of the geometric stress intensity factor for this specimen.

In order to solve the center-cracked tension (CCT) panel problem, Irwin first considered the problem of an infinite array of colinear cracks subjected to uniform remote stress, σ, illustrated in Figure 3.15. For this latter geometry, Westergaard [1939] proposed a stress function of the form

$$Z(z) = \frac{\sigma}{\sqrt{1 - \frac{\sin^2\left(\frac{\pi a}{W}\right)}{\sin^2\left(\frac{\pi z}{W}\right)}}} \tag{3.57}$$

The stress intensity factor can be determined directly from the stress function Z by noting that Eq. (3.54) can be written as

$$K = \lim_{\delta^+ \to 0} Re\, Z\big|_{y=0} \cdot \sqrt{2\pi\delta} \tag{3.58}$$

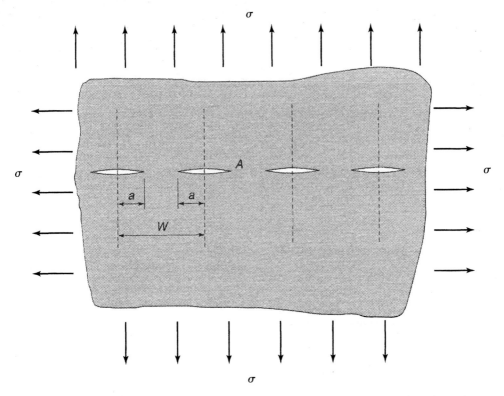

Figure 3.15   Infinite array of colinear cracks subjected to remote biaxial tension.

from which it follows that at crack tip A in Figure 3.15,

$$K_A = \frac{\sigma \sin\left(\frac{\pi x}{W}\right)}{\sqrt{\sin^2\left(\frac{\pi x}{W}\right) - \sin^2\left(\frac{\pi a}{W}\right)}} \cdot \sqrt{2\pi(x - a)} \qquad (3.59\text{a})$$

After some trigonometric manipulation, we obtain

$$K_A = \frac{\sigma\left[\sin\frac{\pi(x-a)}{W}\cos\left(\frac{\pi a}{W}\right) + \cos\frac{\pi(x-a)}{W}\sin\left(\frac{\pi a}{W}\right)\right]}{\sqrt{\sin\frac{\pi}{W}(x + a)}\sqrt{\sin\frac{\pi}{W}(x - a)}} \cdot \sqrt{2\pi(x - a)} \quad (3.59\text{b})$$

In the limit as $x \to a$, we note that

$$\sqrt{\sin\frac{\pi}{W}(x - a)} \approx \sqrt{\frac{\pi}{W}(x - a)} = \frac{\sqrt{2\pi(x - a)}}{\sqrt{2W}} \qquad (3.59\text{c})$$

It follows directly that

$$K_A = \sigma\sqrt{2W}\,\frac{\sin\frac{\pi a}{W}}{\sqrt{\sin\frac{2\pi a}{W}}} \tag{3.59d}$$

Because of the periodic nature of the geometry, the stress intensity factors are the same for all crack tips. Equation (3.59d) can be rewritten in the form of Eq. (3.56) as

$$K = \sigma\sqrt{\pi a}\,\sqrt{\frac{W}{\pi a}\tan\frac{\pi a}{W}} \tag{3.60}$$

Irwin observed that there were similarities between the problem of the periodic array of cracks and the CCT panel. In particular, along the vertical lines midway between the cracks in the periodic array, considerations of symmetry require that $\tau_{xy} = 0$. This condition agrees with the boundary condition for the CCT panel. However, the other boundary condition on the vertical edge of the panel ($\sigma_x = 0$) is not satisfied by the periodic array solution. Irwin reasoned that, for small a/W, the σ_x stress varied smoothly along the symmetry lines—i.e., it is nearly uniform, as shown in Figure 3.16. Since it had already been established that uniform stress parallel to the crack plane has no influence on the stress intensity factor, the effect on K due to the failure to satisfy this boundary condition on the free surfaces could only enter through the *gradients* of σ_x, which is a second-order effect. As a result, the same stress intensity factor applies to both a periodic array of cracks and a center-cracked panel, provided that $2a << W$.

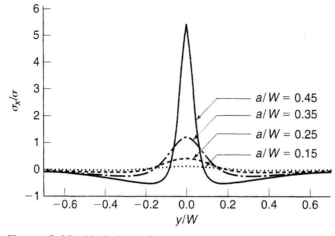

Figure 3.16 Variation of the σ_x stress component along the symmetry line of the colinear array of cracks for varying ratios of $\frac{a}{W}$.

Based on these arguments, the "tangent formula" became the standard for calculating the stress intensity factor from fracture tests of CCT panels. Later, Isida [1966] performed a detailed numerical stress analysis of the same problem using classical complex-variable formulation. His tabular results were fit by Brown and Srawley [1966] to a polynomial of the form

$$Y\left(\frac{a}{W}\right) = 1 + 0.256\left(\frac{a}{W}\right) - 1.152\left(\frac{a}{W}\right)^2 + 12.200\left(\frac{a}{W}\right)^3 \tag{3.61}$$

Independently, Fedderson [1966] observed that Isida's results could be represented over a wide range of a/W by the function

$$K = \sigma\sqrt{\pi a}\sqrt{\sec\frac{\pi a}{W}} \tag{3.62}$$

Fedderson's secant formula was based on heuristic arguments and attempts to justify it on theoretical grounds have been unsuccessful. However, Eftis and Liebowitz [1972] did propose an approximate Westergaard-type stress function that yields Eq. (3.62), but suffers from the same problem as the tangent formula in regard to the boundary conditions on the lateral edges. These various geometric shape factors for the center-cracked panel are compared in Figure 3.17. The tangent formula approximates Isida's results only up to an a/W ratio of about 0.4, whereas the secant approximation matches the more exact result extremely well up to a/W of 0.85 and is generally preferred over the tangent formula.

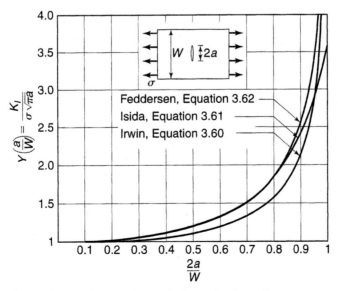

Figure 3.17   Comparison of dimensionless shape factors $Y\left(\frac{a}{W}\right)$ for the center-cracked tension panel.

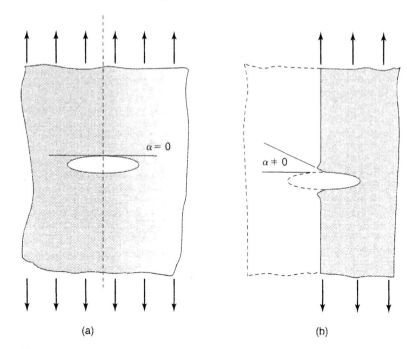

(a)   **(b)**

Figure 3.18   Effect of edge constraint on the shape of a crack.

In passing we note that Irwin's arguments on the similarities between the periodic array of cracks and the center-cracked panel apply equally well to the double-edge-cracked panel in tension. This comparison tends to break down as the edge cracks becomes small, due to a "free-edge effect." For the array of colinear cracks, the shape of the crack becomes elliptical when a load is applied, and at the midpoint of the ellipse, the tangent to the crack boundary must be zero, as illustrated in Figure 3.18a. However, for the free-edge crack no such constraint exists, and the crack faces have a finite slope, as shown in Figure 3.18b. The free-edge effect has been extensively studied by a variety of techniques, and the consensus of opinion is that for a small crack emanating from a free surface, the stress intensity factor has the form

$$K = 1.1215\sigma\sqrt{\pi a} \tag{3.63}$$

The factor 1.12, which appears in many derived stress intensity factor expressions can be traced to this free-surface effect.

3.7 DETERMINING THE GEOMETRIC STRESS INTENSITY FACTOR

In this section we will explore various methods to determine, either exactly or approximately, the geometric stress intensity factor for problems of practical interest. For this purpose the generalized Westergaard method provides powerful tools that are not available in real-variable theory. In particular, we will find that, in many

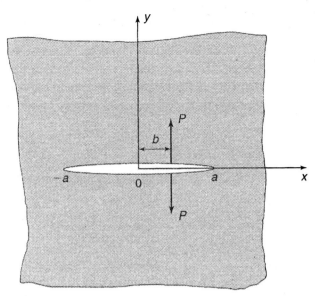

Figure 3.19 A central crack opened by a pair of splitting forces P.

cases, it is not necessary to analytically formulate global expressions for the stresses in order to extract their behavior in the neighborhood of a crack tip. Instead, we need only to develop a suitable stress function, $Z(z)$, that models the problem and satisfies equilibrium. The following example illustrates the approach.

Consider a central crack of length, $2a$, opened by a pair of splitting forces, P, applied at an arbitrary distance, b, from the origin, as illustrated in Figure 3.19. A suitable stress function for this geometry is

$$Z(z) = \frac{P}{\pi} \frac{\sqrt{a^2 - b^2}}{(z - b)\sqrt{z^2 - a^2}} \tag{3.64}$$

This function has all of the features needed to model the problem posed. In particular, the $\sqrt{z^2 - a^2}$ term, which we previously saw in the remotely loaded central-crack problem, leads to singularities of order $1/\sqrt{r}$ at each crack tip. From the elastic solution for a point load on a half-space [Timoshenko and Goodier, 1951], we know that a point force produces a singularity of order $1/r$ at the point of application of the load, and this feature is represented by the $(z - b)$ term in our proposed stress-function. Because this is a linear theory, the load, P, must appear to the first power only. Finally, the remaining constants are a consequence of the requirement that the system must be in static equilibrium, i.e.,

$$\int_{-\infty}^{\infty} \sigma_y\big|_{y=0} = P \tag{3.65}$$

The other stress function, $Y(z)$, need not be formulated if we are concerned only with determining the stress intensity factor, since it never contributes to the strength of the singularity. This is not to say that the $Y(z)$ function is unimportant. Whenever the stresses, strains, or displacements are required within the body, this function contributes and must be included. However, if our attention is confined to the crack plane, the $Y(z)$ function does not affect the σ_y stress component, i.e. ,

$$\sigma_y\big|_{y=0} = Re\, Z(x) \tag{3.66}$$

From Eq. (3.58), the stress intensity factor at the right crack tip, K_A, is

$$K_A = \lim_{x^+ \to a} \frac{P}{\pi} \frac{\sqrt{a^2 - b^2}}{(x - b)\sqrt{x^2 - a^2}} \cdot \sqrt{2\pi(x - a)} \tag{3.67a}$$

$$K_A = \frac{P}{\sqrt{\pi a}}\sqrt{\frac{a + b}{a - b}} \tag{3.67b}$$

Similarly, at the left crack tip,

$$K_B = \lim_{x^- \to -a} \frac{P}{\pi} \frac{\sqrt{a^2 - b^2}}{(x - b)\sqrt{x^2 - a^2}} \cdot \sqrt{2\pi(x + a)} \tag{3.67c}$$

$$K_B = \frac{P}{\sqrt{\pi a}}\sqrt{\frac{a - b}{a + b}} \tag{3.67d}$$

This solution for crack-face loading has a characteristic feature that differs from the problem of remote loading that we previously examined—namely, the stress intensity factor decreases with increasing crack length for a constant force. Consequently, should unstable crack propagation occur, the increase in crack length will result in a decrease in the stress state at the crack tip, and, eventually, the crack will arrest.

The point-load solution of Eqs. (3.67) can be used as the basis for the solution of a wide variety of crack-face-loaded problems, such as those illustrated in Figure 3.20, using the principle of superposition. In the limiting case of a continuous distribution of forces—i.e., a pressure distribution—integration of the point-load solution gives the desired result. For example, consider the case of a central crack symmetrically loaded by a uniform pressure, p, over a portion of the crack face from $-c$ to c, as shown in Figure 3.21. Then, at either crack tip,

$$K = \int_0^c \frac{p}{\sqrt{\pi a}}\left[\sqrt{\frac{a + b}{a - b}} + \sqrt{\frac{a - b}{a + b}}\right] db \tag{3.68a}$$

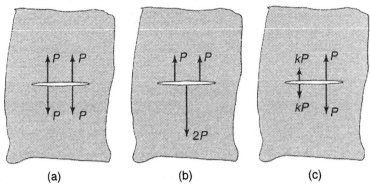

Figure 3.20 Examples of crack-face-loaded problems that can be solved by superposition.

$$K = 2p \sqrt{\frac{a}{\pi}} \cdot \left[\sin^{-1} \left(\frac{c}{a} \right) \right] \tag{3.68b}$$

In the limiting case of uniform pressure over the entire crack length $c = a$, and

$$K = p \sqrt{\pi a} \tag{3.69}$$

Equation (3.69) has the same form as Eq. (3.55) for the remotely-loaded central crack.

Let us now examine a new stress function that is similar in form to Eq. (3.64), namely,

$$Z(z) = C \frac{\sqrt{b^2 - a^2}}{(b - z)\sqrt{a^2 - z^2}} \tag{3.70}$$

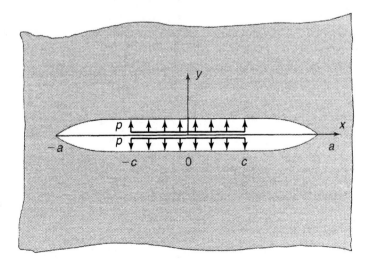

Figure 3.21 A central crack loaded by a uniform pressure distribution over a portion of the crack face.

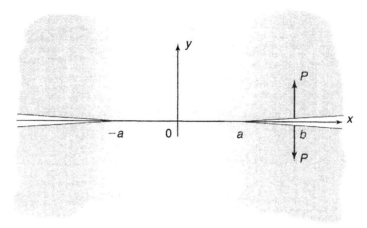

Figure 3.22 Boundary conditions along the crack plane
for the stress function $Z(z) = \dfrac{C\sqrt{b^2-a^2}}{(b-z)\sqrt{(a^2-z^2)}}$.

where C is a constant to be determined and $b > a$. Following the methodology of the
semi-inverse approach, we first examine the behavior of this function along $y = 0$ to
locate the crack face(s)—that is;

$$Z(z) = C\frac{\sqrt{b^2 - a^2}}{(b - x)\sqrt{a^2 - x^2}} \tag{3.71}$$

From the boundary condition requirement of Eq. (3.30), we observe that the domain,
$|x| > a$ along the crack plane, corresponds to a traction-free crack surface, except at
$x = b$, where a point load acts, as illustrated in Figure 3.22.

From these observations it would appear that the stress function of Eq. (3.70)
solves the "complementary" problem to the stress function of Eq. (3.64)—namely,
a central ligament (as opposed to a central crack) of length, $2a$, subjected to a pair
of splitting forces at $x = b$. However, before we can make this claim, we must first
ensure that all of the conditions of global equilibrium are satisfied.

From force equilibrium,

$$\int_{-a}^{a} \sigma_y\big|_{y=0}dx = P \tag{3.72}$$

we find that the constant C in Eq. (3.70) must equal P/π. However, from moment
equilibrium along the crack plane, we find that

$$M = \int_{-a}^{a} \sigma_y\big|_{y=0}x \, dx \neq Pb \tag{3.73}$$

as required if the loading on the body is due solely to the splitting force, P. Instead, Eq. (3.73) predicts that a remote moment of magnitude

$$M = -P \sqrt{b^2 - a^2} \tag{3.74}$$

must be present in addition to the splitting force if moment equilibrium is to be maintained.

Therefore, from a consideration of all relevant factors, the boundary-value problem solved by the proposed stress function, in Eq. (3.70) is that of a central ligament of length, $2a$, subjected to splitting forces on the crack line at $x = b$ *and a proportional remote bending moment*, as illustrated in Figure 3.23. The existence of the additional remote loading for stress functions of this type was first observed by Tada in 1985 [Tada, Paris, and Irwin 1985].[4] A discussion of the origin of the problem and additional results are presented by Jiang, et al. [1991].

The solution to the problem that is complimentary to that solved by the stress function of Eq. (3.64), i.e., a central ligament loaded by splitting forces *only*—is obtained by subtracting from Eq. (3.70) the stress function for a ligament with a remotely applied bending moment of suitable magnitude to balance out the unwanted loading

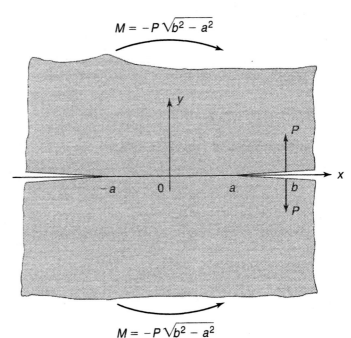

Figure 3.23 Opening mode crack problem solved by the stress function

$$Z(z) = \frac{C\sqrt{(b^2-a^2)}}{(b-z)\sqrt{(a^2-z^2)}}.$$

---

[4] The 1985 edition of *The Stress Analysis of Cracks Handbook* corrects the errors in the 1973 edition for this and similar problems, and the results of the earlier edition should not be cited as authoritative. The *Handbook* is now in its third edition [Tada, Paris, and Irwin, 2000].

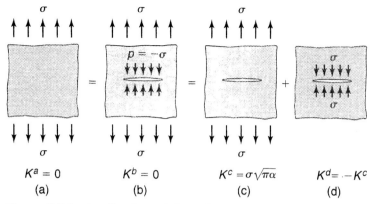

Figure 3.24 Application of the principle of superposition to the central crack with uniform internal pressure.

component. The resulting combined stress function provides the exact solution to the desired problem and can be used to determine the state of stress, strain and displacement everywhere throughout the body through Eqs. (3.24), (3.25), and (3.26). However, if only the stress intensity factor(s) (and, consequently, the near-field stress state) is desired, then the geometric stress intensity factors for the two problems can be algebraically combined without first forming the exact stress function.

The additive property of the geometric stress intensity factor that follows directly from the invariant form of the stress field in the near neighborhood of a crack provides a rapid method for constructing the stress intensity factor for a new problem through the application of the principle of superposition. As an illustration of the principles involved, let us reexamine the problem of the pressurized internal crack.

In Figure 3.24, the trivial problem of a uniform infinite plate under remote stress [body (a)] is recast as an infinite plate with a central crack of length, $2a$, subjected to both remote stress *and* crack-face forces just sufficient to close the crack [body (b)]. From a theory of elasticity viewpoint these two bodies are equivalent in that the state of stress everywhere throughout them is the same. Furthermore, since body (a) has no crack, the stress intensity factor must be zero, and likewise, $K_b = 0$ for the equivalent body. Now let us separate the loads in body (b) into two independent cases, bodies (c) and (d), as shown in Figure 3.24.

It follows that, from superposition,

$$K_b = K_c + K_d = 0 \tag{3.75a}$$

$$K_c = -K_d = \sigma \sqrt{\pi a} \tag{3.75b}$$

$$K_d = -\sigma \sqrt{\pi a} \tag{3.75c}$$

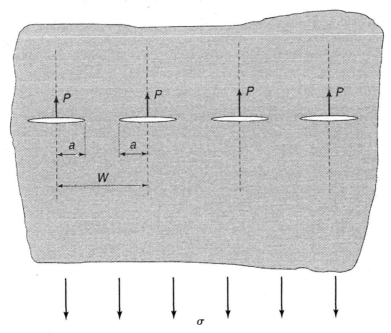

Figure 3.25 Fracture-mechanics idealization of the riveted construction of an aircraft fuselage.

Finally, if we define the crack face pressure, p, as being equal and opposite to the crack face closing stress, σ, then

$$K_d = p\sqrt{\pi a} \qquad (3.75\text{d})$$

which is the same result as previously obtained by the rigorous approach [Eq. (3.69)]. We note, however, that whereas both approaches yield the same stress intensity factor, only the rigorous approach will yield the stress state outside the singularity-dominated-zone.

An important role of the superposition approach is in estimating the stress intensity factor for problems of practical interest when rigorous solutions are not available or the effort to obtain such solutions cannot be justified. For example, the fracture mechanics behavior of an array of rivet-loaded holes in an aircraft fuselage, as depicted in Figure 3.25, can be obtained by combining the results for two classical infinite-crack-array problems for which the geometric stress intensity factors are readily available.

One of these problems is that of the remotely loaded plate with an infinite array of cracks previously shown in Figure 3.15, for which the stress intensity factor has been determined and is given by Eq. (3.60). The other problem is that of a geometrically similar body that is loaded by centrally located point loads on the crack line, as shown in Figure 3.26. From Appendix C, the geometric stress intensity factor for this

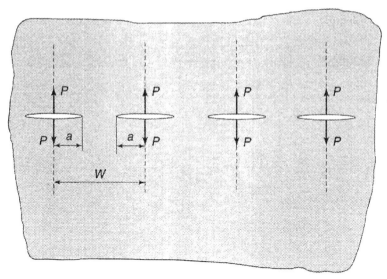

Figure 3.26 Infinite array of colinear cracks subjected to centrally located splitting forces on the crack line.

geometry/loading combination is

$$K = \frac{2P}{W} \sqrt{\pi a} \sqrt{\frac{W}{\pi a} \tan \frac{\pi a}{W}} \cdot \frac{1}{\sin(\frac{\pi a}{W})} \qquad (3.76)$$

An idealized representation of the rivet-hole problem in terms of these two known solutions is illustrated in Figure 3.27 and is described mathematically as

$$K_a = K_b + K_c - K_d \qquad (3.77)$$

However, since inverting the body has no effect on its K value, $K_a = K_d$, and, from equilibrium, $2P/W = \sigma$, the remote stress. Finally, we can express an appropriate expression for the stress intensity factor for an array of rivet holes as

$$K = \sigma \sqrt{\pi a} \, Y\left(\frac{a}{W}\right) \qquad (3.78a)$$

where

$$Y\left(\frac{a}{W}\right) = \frac{1}{2} \sqrt{\frac{W}{\pi a} \tan \frac{\pi a}{W}} \cdot \left[1 + \frac{1}{\sin\left(\frac{\pi a}{W}\right)}\right] \qquad (3.78b)$$

The geometric shape function, $Y(a/W)$, for this problem is plotted in Figure 3.28.

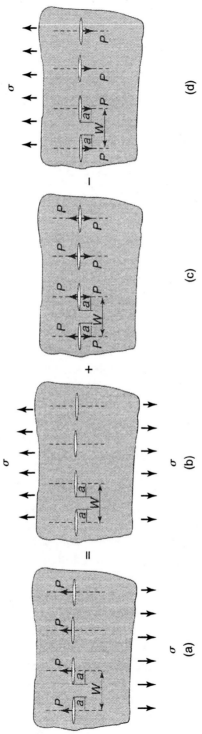

Figure 3.27 Superposition sequence for an infinite array of rivet-hole-loaded cracks.

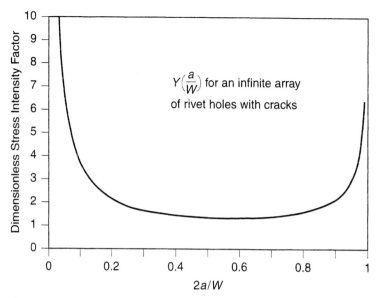

Figure 3.28 Dimensionless shape function $Y(a/W)$ for an infinite array of rivet holes with cracks.

Despite the idealizations involved in considering this solution as equivalent to an array of rivet holes (with assumed defects) in an aircraft fuselage, this solution provides an explanation of the origin of a problem of significant interest to the commercial aircraft industry. This industry has recently adopted the position that, with proper maintenance and inspection, there is no reason that an air frame cannot last indefinitely (or, at least, until the cost of repair and inspection becomes excessive). As a consequence, older aircraft that were originally designed for a finite life are being allowed to remain in service beyond their design values, and, as a direct result, the average age of the commercial aircraft fleet is increasing. These "aging aircraft" have begun to exhibit a mechanism of failure not considered during the initial design stage. This mechanism, called *multiple-site-damage*, occurs when small, nonvisual cracks grow from each hole in a long row of fastener holes simultaneously. At some point these cracks may join to form a long crack whose length far exceeds the critical crack length for a pressurized shell (a.k.a., the fuselage). Both of these features of crack growth from fastener holes can be predicted from the form of the shape function in Figure 3.28. From Figure 3.28, we can see that for small cracks, the geometric stress intensity factor is very high, and as a result, crack growth initiating from the holes is inevitable after some number of cycles. Following an initial period of rapid crack growth after initiation, there follows a long period of very slow crack growth, during which the cracks remain too small to detect. As a result, there is a long period of apparent safe service life. However, as Figure 3.28 shows, the crack growth will ultimately accelerate, and, in a relatively small number of cycles, the structure will become unstable. The failure of the Aloha 737 aircraft in 1988,

Figure 3.29   Failed fuselage of the Aloha 737 aircraft, attributed to multiple-site damage [courtesy National Aeronautics and Space Administration].

shown in Figure 3.29, has been attributed to multiple-site-damage. This aircraft, which was used for inter-island transportation, saw an unusually large number of takeoff and landing pressure cycles relative to its age (19 years) may be a precursor of a much larger problem throughout the aircraft industry. Fortunately, the same fracture-mechanics analysis that identified the problem is being used to quantify the growth rate and, ultimately, solve the problem.

There are situations in which a reasonable estimate of the stress intensity factor or its behavior with increasing crack length is sufficient, and the rigorous procedures of the preceding paragraphs are not justified or a suitable stress function cannot be found. For example, an engineer in the field may need to estimate the load carrying of a part in the presence of a defect just uncovered by visual inspection. Alternatively, in the failure analysis of an existing component, the size of a prior defect can often be determined by examining the fracture surface. In this case the load at the instant of failure could be estimated if a suitable expression for the geometric stress intensity factor were available. For these cases, an educated guess for the behavior of the stress intensity factor can quite often be obtained from basic solutions by invoking the engineer's best asset—ingenuity.

As a demonstration of the thought processes involved, consider the problem of estimating the behavior of K with increasing crack length for a remotely loaded tension panel containing a circular hole with cracks, as shown in Figure 3.30. Detailed analyses of this problem have been conducted [Newman, 1971; Bowie, 1956], but our focus here is to obtain an approximate solution by simple means. First, when the crack is very small compared to the hole diameter as illustrated in Figure 3.31, we can envision the behavior to be similar to that of the single-edge notched geometry

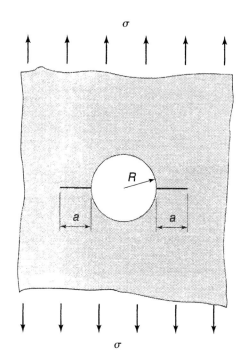

Figure 3.30 Remotely loaded plate with cracks emanating from a circular hole.

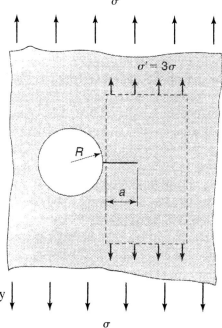

Figure 3.31 Similarity between a very small crack at the boundary of a hole and an equivalent single-edge notched plate.

(shown dotted in the figure), wherein the "remote stress" on the hypothetical SEN specimen is the local tangential stress near the hole boundary. Since the stress concentration factor (SCF) for a plate with a circular hole in tension is 3, the "effective" stress on the SEN geometry is three times the actual remote stress. Therefore, in the limit as $a \to 0$, Eq. (3.63) applies and

$$K \approx 1.12(3\sigma)\sqrt{\pi a} \qquad\qquad (3.79a)$$

or

$$\left(\frac{K}{\sigma}\right)^2 \approx 11.3\,\pi a \qquad\qquad (3.79b)$$

As the crack length increases, the effect of the SCF at the hole boundary on the stress at the crack tip decreases, and, from Inglis's argument, eventually the effects of the circular cutout are insignificant compared with the crack length, and the problem is indistinguishable from that of a wide plate with a central crack [Eq. (3.55)], in which the half-length of the crack is $R + a$. So,

$$K \approx \sigma \sqrt{\pi(R + a)} \qquad\qquad (3.80a)$$

or

$$\left(\frac{K}{\sigma}\right)^2 \approx \pi(R + a) \qquad\qquad (3.80b)$$

These two extremes provide the asymptotic behavior of the function we seek, and when plotted as $(K/\sigma)^2$ vs. a, as in Figure 3.32, represent straight lines with slopes of 11.3π and π for short and long cracks, respectively. It is then a simple task to construct a smooth curve with these asymptotes. Finally, when the crack length starts to approach the plate width, the finite-width correction of Fedderson in Eq. (3.62), predicts a vertical asymptote at $a = W$, and again a smooth transition curve can be constructed.

As we have seen in the preceding discussion, the use of auxiliary techniques such as superposition or approximating schemes greatly increases the range of problems that can be solved. Just as was the case for the Airy stress function approach, wherein the success of the method depended on the ability to select a suitable stress function, these additional techniques rely for their success on a library of fundamental solutions from which new solutions can be constructed. An abbreviated table of Westergaard stress functions and their associated geometric stress intensity factors to affect these superpositions for a variety of classical cases is presented in Appendix C. More extensive tables, including numerical and graphical results for special cases, are available in handbooks by Tada, Paris, and Irwin [1985, 2000], Sih [1973], and Rooke and Cartwright [1976].

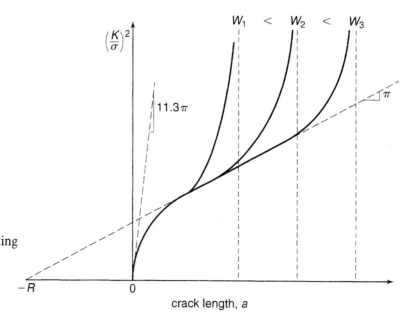

Figure 3.32 Estimating the stress intensity factor for a plate of finite width with a cracked hole.

3.8 THE THREE-DIMENSIONAL CRACK PROBLEM

Up until now, we have focused our attention on two-dimensional crack problems. In Chapter 2 we justified this focus on the basis of our assessment that the rate of change of stress tangent to a curved crack front varied slowly relative to that in the plane perpendicular to the crack. On the other hand, if the radius of curvature of the crack front is of the order of the other dimensions in the geometry, this assumption may not be justified. An important class of problems in which the curvature of the crack front does play a significant role is that of the semi-elliptical surface flaw (often called the *thumbnail crack*), such as those shown in Figure 3.33. This type of crack often leads to failure in structures that lack local stress raisers (such as threads, holes, or notches). Examples of this class of problem include welded pressure vessels, pipe and ship hulls. Since most fracture failures tend to start at a free surface, a mathematical description of the stress intensity factor and its variation along the crack front is important if we wish to predict the crack growth behavior of a surface defect. Unfortunately, a rigorous mathematical development of the theory of three-dimensional cracks requires the use of highly specialized mathematical tools and concepts that are beyond the scope of this book. Accordingly, the following discussion will present only the key historical results that led to the currently accepted expressions for the stress intensity factor.

The first significant progress toward the solution of the surface-crack problem was the exact solution by Sneddon [1946] for the embedded circular crack in an infinite medium, shown in Figure 3.34. This problem is often called the *penny-shaped crack problem*. Using transform methods, Sneddon obtained exact expressions for the state of stress around the crack tip for the case of remote-tension loading, but little insight into fracture mechanics was gained from the solution. Some years later, Sneddon's

Figure 3.33 Semi-elliptical fatigue crack in a propeller blade [courtesy Naval Surface Warfare Center, Carderock Division].

results were reexamined from the LEFM point of view, and the geometric stress intensity factor was extracted from Sneddon's solution. It was shown that, for the penny-shaped crack geometry

$$K = \frac{2}{\pi}\sigma\sqrt{\pi a} = 0.64\sigma\sqrt{\pi a} \tag{3.81}$$

Comparing this result with that of the corresponding through-crack problem—i.e., a planar crack of length, $2a$, in an infinite sheet ($K = \sigma\sqrt{\pi a}$)—we see that the

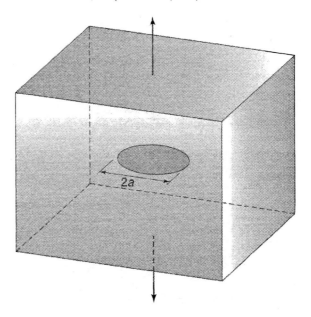

Figure 3.34 A penny-shaped crack in an infinite body.

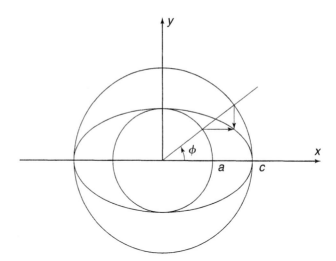

Figure 3.35 Parametric representation of a point on an ellipse.

effect of crack-front curvature is to decrease the stress intensity factor by 36%. In other words, the crack-front curvature tends to stiffen the structure as compared with a straight crack of the same dimension that provides an additional constraint on crack-opening displacement.

For the case of an embedded flat elliptical crack loaded in remote tension, Green and Sneddon [1950] showed that the crack-opening displacement could be expressed as

$$v = v_o \left(1 - \frac{x^2}{a^2} - \frac{y^2}{c^2} \right)^{\frac{1}{2}} \tag{3.82}$$

where $2a$ and $2c$ are the minor and major dimensions of the elliptical crack, respectively, and v_o is one half of the total separation of the two crack faces at the origin. From this result, Irwin [1962] used strain energy arguments (the details of which will be presented in Chapter 7) to derive an exact expression for the stress intensity factor at any point around the perimeter of the crack in terms of a parametric angle, ϕ, as illustrated in Figure 3.35.

From Figure 3.35, we can see that all points along the flaw border are obtained by the horizontal and vertical projections of points on the radial intercept of ϕ with circles whose radii represent the semiminor and semimajor axes of the ellipse, respectively. Note that only when $\phi = 0$, $\frac{\pi}{2}$, and π does the parametric angle agree with the polar-angle measure of the crack perimeter position. In terms of this parametric angle, Irwin's resulting expression for the geometric stress intensity factor at any point around the perimeter of the crack is:

$$K = \frac{\sigma \sqrt{\pi a}}{\Phi} \left[\left(\frac{a}{c} \right)^2 \cos^2 \phi + \sin^2 \phi \right]^{\frac{1}{4}} \tag{3.83}$$

where

$$\Phi = \int_0^{\frac{\pi}{2}} \left[1 - \left(\frac{c^2 - a^2}{c^2} \right) \sin^2 \phi \right]^{\frac{1}{2}} d\phi$$

is the elliptic integral of the second kind that is tabulated in handbooks. Alternatively, the elliptic integral can be computed to any degree of accuracy from the series expansion (for $a < c$) [Abramowitz and Stegun, 1964]

$$\Phi = \frac{\pi}{2} \left[1 - \frac{1}{1} \left(\frac{1}{2} \right)^2 \left(\frac{c^2 - a^2}{c^2} \right) - \frac{1}{3} \left(\frac{1 \cdot 3}{2 \cdot 4} \right)^2 \left(\frac{c^2 - a^2}{c^2} \right)^2 \right.$$

$$\left. - \frac{1}{5} \left(\frac{1 \cdot 3 \cdot 5}{2 \cdot 4 \cdot 6} \right)^2 \left(\frac{c^2 - a^2}{c^2} \right)^3 - \cdots \right]$$

(3.84)

From inspection of Eq. (3.83), we observe that the stress intensity factor is not constant along the crack border, but rather varies from a minimum along the major axis ($\phi = 0, \pi$) of

$$K = \frac{\sigma \sqrt{\pi a}}{\Phi} \sqrt{\frac{a}{c}}$$

(3.85a)

to a maximum along the minor axis ($\phi = \frac{\pi}{2}$) of

$$K = \frac{\sigma \sqrt{\pi a}}{\Phi}$$

(3.85b)

Therefore, in the absence of mitigating factors, an embedded elliptical crack will tend to grow into a circular crack (i.e., $a = c$). In the latter case, the value of the elliptic integral is $\Phi = \frac{\pi}{2}$, and Eq. (3.83) reduces to that of a penny-shaped crack. At the other extreme, $a << c$ (i.e., a very flat ellipse), $\Phi \to 1$, and K approaches that of a two-dimensional crack of length, $2a$, in an infinite sheet as in Eq. (3.55). From these observations, we see that Eq. (3.83) is the general equation for an embedded crack in tension, and our earlier results could be extracted as special cases.

Tada and Paris [1979] have extended the applicability of Eq. (3.83) still further by noting that the quantity, $a \left[(a/c)^2 \cos^2 \phi + \sin^2 \phi \right]^{\frac{1}{2}}$, is the length, ℓ, of the normal to the ellipse, as illustrated in Figure 3.36. In terms of this length Eq. (3.83) takes on the particularly simple form

$$K = \frac{\sigma \sqrt{\pi \ell}}{\Phi}$$

(3.86)

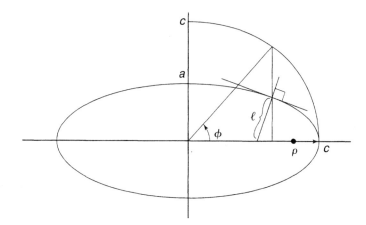

Figure 3.36 The normal length l at a point on an ellipse.

Tada and Paris argued that for internal flaws of nonelliptical shape, such as those depicted in Figure 3.37, the geometric stress intensity factor could be estimated with the aid of Eq. (3.86).

For all of its rigor and broad range of applicability, Irwin's equation for the embedded elliptical flaw in an infinite body is of only limited practical utility. Of greater importance is the semi-elliptical (or nearly so) surface-breaking crack in a finite thickness sheet, as shown in Figure 3.38. For this case Irwin suggested three empirical corrections to Eq. (3.83) to account for the following factors

(a) insertion of a free surface normal to the crack through the major axis,

(b) the presence of a free surface opposite the crack opening, and

(c) plastic strain at the crack tip.

Irwin reasoned that, for a long shallow crack (i.e., a/c is small), the free-edge correction factor, 1.12, applied. As a/c increases, this correction should decrease but an increase in a/c usually implies a deeper crack with a corresponding approach of the back surface. The lack of material constraint from the back surface also has the tendency to elevate the value of K. According to Irwin, the combination of these competing factors should result in a nearly constant correction for the two free-surface effects, at least for $a/t < 0.5$.

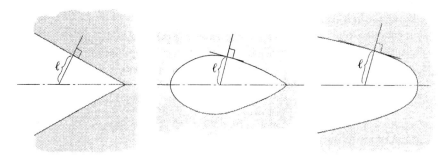

Figure 3.37 Nonelliptical embedded cracks.

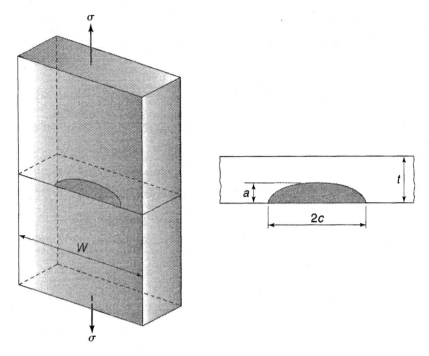

Figure 3.38 Semi-elliptical surface-flaw parameters in a finite plate.

To compensate for the plastic strain at the crack tip, Irwin used the arguments of a previous study [Irwin, 1960] to propose that the actual crack depth, a, needed to be augmented by a plastic-zone correction, r_Y, to account for the loss of load-carrying ability in a small plastically deformed region ahead of the crack tip. For plane strain conditions this correction has the following form:[5]

$$r_Y = \frac{1}{4\pi\sqrt{2}} \frac{K^2}{\sigma_{ys}^2} \tag{3.87}$$

Rewriting Eq. (3.85b) for the deepest penetration of the crack ($\phi = \frac{\pi}{2}$) and replacing a with $a + r_Y$, we obtain

$$K^2\Phi^2 = (1.12)^2\sigma^2\pi\left(a + \frac{1}{4\pi\sqrt{2}} \frac{K^2}{\sigma_{ys}^2}\right) \tag{3.88}$$

---

[5] The expression used here conforms to the definition advanced by Irwin in 1960 and differs from currently accepted definitions. Chapter 6 provides a thorough discussion of the origin of the plastic-zone correction and more recent estimates of its value.

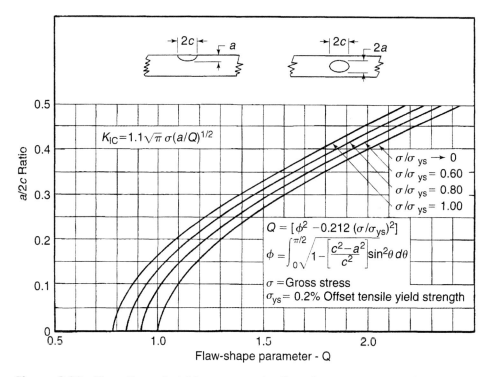

Figure 3.39 The effect of yield stress on the flaw-shape parameter Q.

Solving for K in Eq. (3.88), we obtain an approximate expression for the geometric stress intensity factor for a semi-elliptical surface flaw of the form

$$K = 1.12\sigma\sqrt{\frac{\pi a}{Q}} \qquad (3.89)$$

where

$$Q = \Phi^2 - 0.212\left(\frac{\sigma}{\sigma_{ys}}\right)^2 \qquad (3.90)$$

The flaw-shape parameter, Q, which is affected by both the crack aspect ratio, a/c, and the yield stress, is plotted in Figure 3.39.

While the Irwin equation generally produces good results for shallow surface flaws in semi-infinite bodies in tension, it does not fully consider the influence of the back surface in finite thickness plates nor does it include the effects of the stress gradient for plates in bending. Attempts to incorporate these variables by either analytical [Smith and Sorensen, 1974; Kobayashi, 1975; Browning and Smith, 1976] or numerical [Kathiresan, 1976; Raju and Newman, 1979] methods have produced widely differing results, as shown in Figure 3.40.

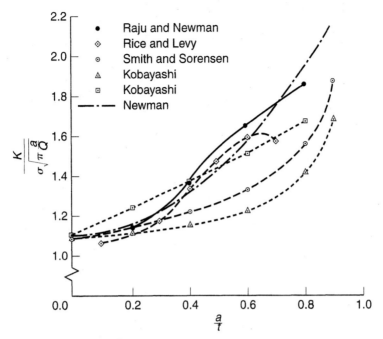

Figure 3.40 Comparison of the results from various studies for the maximum stress intensity factor of a semi-elliptical surface flaw with $\frac{a}{c} = 0.2$ [Raju and Newman, 1979].

In an effort to resolve the discrepancies among the prior results, Raju and Newman [1979] undertook a detailed, comprehensive three-dimensional finite element analysis of the semi-elliptical surface-flaw problem for both tension and antiplane bending. The finite-element models for various profiles of elliptical cracks were obtained by using elliptical transformations of circular crack profiles and incorporated both singular- and linear-strain isoparametric elements. A typical model is shown in Figure 3.41. The stress intensity factor was extracted from the numerical results by a nodal-force method developed by Raju and Newman [1979].[6] In order to ensure convergence, these models contained up to 6900 degrees of freedom. For the limiting cases for which analytical results are available, the Raju and Newman results are within 0.4 percent to 1 percent of the analytical results. Considering all factors, these results are generally considered by the engineering community to be definitive.

In contrast to an analytical solution, the finite-element approach provides only discrete numerical values for the geometries that are modeled. In an effort to generalize their results Newman and Raju [1981] fit the extensive data they had generated with double-series polynomials to produce an empirical equation for the geometric

---

[6] Chapter 4 presents details of this and other methods for determining the geometric stress intensity factor from finite-element results.

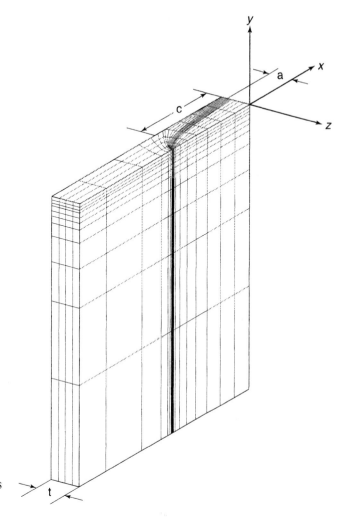

Figure 3.41 Finite-element idealization of a semi-elliptical surface flaw in three dimensions [Raju and Newman, 1979].

stress intensity factor for the surface-flaw problem. Their resulting equation is

$$K = (\sigma_t + H\sigma_b)\sqrt{\frac{\pi a}{Q}} \cdot F\left(\frac{a}{t}, \frac{a}{c}, \frac{c}{W}, \phi\right) \qquad (3.91)$$

where σ_t is the remotely applied tensile stress and $\sigma_b = (6M)/(Wt^3)$ is the extreme fiber stress due to the bending moment M.

Since the numerical analysis was for purely elastic conditions, the flaw shape parameter, Q, in Eq. (3.91) does not include the plastic zone correction component.

In addition, Newman and Raju use an empirical expression for Q (attributed to Rawe) with a stated maximum error of 0.13 percent for all values of a/c in lieu of the series expression of Eq. (3.84). The expressions are

$$Q = \Phi^2 = 1 + 1.464 \left(\frac{a}{c}\right)^{1.65} \qquad \text{for } \frac{a}{c} \leq 1 \qquad (3.92a)$$

and

$$Q = \Phi^2 = 1 + 1.464 \left(\frac{c}{a}\right)^{1.65} \qquad \text{for } \frac{a}{c} \geq 1 \qquad (3.92b)$$

The fitting function chosen by Newman and Raju has the form

$$F = \left[M_1 + M_2 \left(\frac{a}{t}\right)^2 + M_3 \left(\frac{a}{t}\right)^4\right] f_\phi \cdot g \cdot f_W \qquad (3.93a)$$

where:

$$M_1 = 1.13 - 0.09 \left(\frac{a}{c}\right) \qquad (3.93b)$$

$$M_2 = -0.54 + \frac{0.89}{0.2 + (a/c)} \qquad (3.93c)$$

$$M_3 = 0.5 - \frac{1.0}{0.65 + (a/c)} + 14 \left(1.0 - \frac{a}{c}\right)^{24} \qquad (3.93d)$$

$$f_\phi = \left[\left(\frac{a}{c}\right)^2 \cos^2 \phi + \sin^2 \phi\right]^{\frac{1}{4}} \qquad (3.93e)$$

$$g = 1 + \left[0.1 + 0.35 \left(\frac{a}{t}\right)^2\right] (1 - \sin \phi)^2 \qquad (3.93f)$$

$$f_W = \left[\sec \left(\frac{\pi c}{W} \sqrt{\frac{a}{t}}\right)\right]^{\frac{1}{2}} \qquad (3.93g)$$

The combined effects of all of these polynomial expressions on the fitting function, F, are shown graphically in Figure 3.42 for the case of a wide plate (i.e., $f_W \approx 1$).

If bending stresses are present, a function, H, that modifies the tension function, F, also needs to be computed. Again, using engineering judgment, Newman and Raju expressed this function as

$$H = H_1 + (H_2 - H_1) \sin^P \phi \tag{3.94a}$$

where

$$p = 0.2 + \left(\frac{a}{c}\right) + 0.6 \left(\frac{a}{t}\right) \tag{3.94b}$$

$$H_1 = 1 - 0.34 \left(\frac{a}{t}\right) - 0.11 \left(\frac{a}{c}\right) \left(\frac{a}{t}\right) \tag{3.94c}$$

$$H_2 = 1 + G_1 \left(\frac{a}{t}\right) + G_2 \left(\frac{a}{t}\right)^2 \tag{3.94d}$$

$$G_1 = -1.22 - 0.12 \left(\frac{a}{c}\right) \tag{3.94e}$$

$$G_2 = 0.55 - 1.05 \left(\frac{a}{c}\right)^{3/4} + 0.47 \left(\frac{a}{c}\right)^{3/2} \tag{3.94f}$$

The combined modifying function, HF, for bending stress is plotted in Figure 3.43. As might be expected, the effect of the stress gradient in antiplane bending has the most influence on the stress intensity factor at the maximum depth of penetration of the crack ($\phi = \frac{\pi}{2}$), due to the decrease in fiber stress at that location. This feature is illustrated in Figure 3.44 for a typical crack profile. In this figure the extreme fiber stress is the same for both tension and antiplane bending.

Newman and Raju report that their continuous function for the stress intensity factor at all points along the flaw border agrees with their finite-element results to within 5 percent for the full range of a/t and a/c (0 to 1) and for $2c/W < 0.5$. Despite their empirical origin, these equations have gained wide acceptance and are employed for various fracture-related calculations whenever surface flaws are encountered.

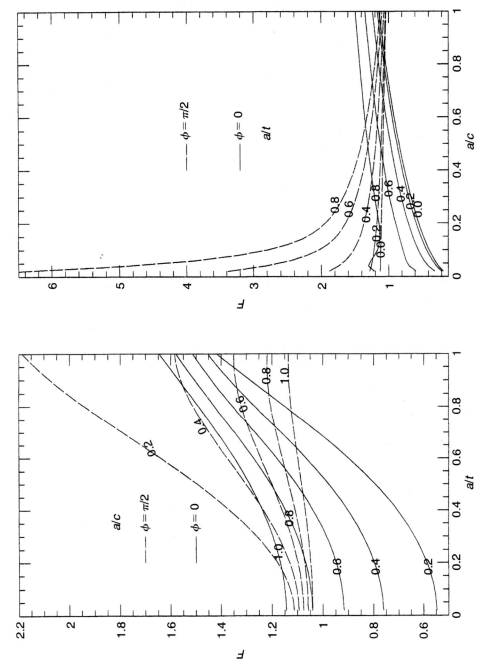

Figure 3.42 The Newman–Raju tension-fitting function F for a wide plate of finite thickness.

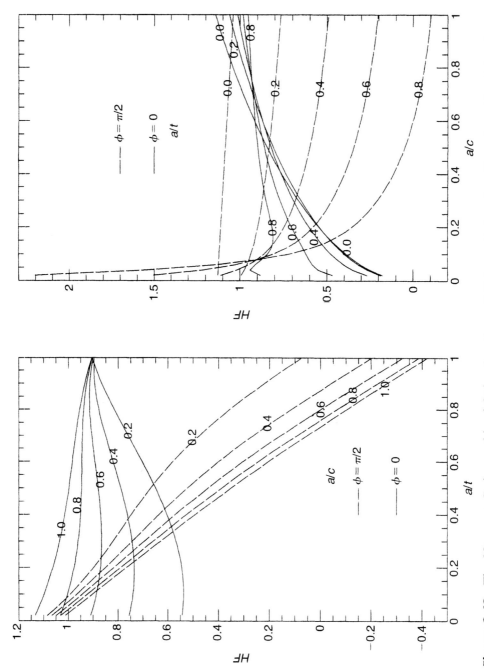

Figure 3.43 The Newman–Raju combined-fitting functions HF for antiplane bending of a wide plate of finite thickness.

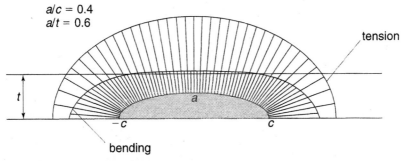

Figure 3.44 A comparison of the distribution of the stress intensity factor around the border of a flaw for equal tension and bending stress for $\frac{a}{c} = 0.4$ and $\frac{a}{t} = 0.6$ in a wide plate.

REFERENCES

Abramowitz, M., and Stegun, I. A., Eds, 1964, *Handbook of Mathematical Functions,* NBS Applied Mathematics Series #55, U.S. Government Printing Office, Washington, DC.

Bowie, O. L., 1956, "Analysis of an Infinite Plate Containing Radial Cracks Originating at the Boundary of an Internal Circular Hole," *J. Mathematics and Physics*, 35, pp. 60–71.

Bridgman, P. W., 1952, *Studies in Large Plastic Flow and Fracture with Special Emphasis on the Effects of Hydrostatic Pressure*, McGraw-Hill, New York.

Brown, W. F., and Srawley, J. E., 1966, *Plane Strain Crack Toughness Testing of High Strength Metallic Materials*, STP 410, ASTM, Philadelphia, pp. 1–65.

Browning, M. W., and Smith, F. W., 1976, "An Analysis for Complex Three-Dimensional Crack Problems," *Developments in Theoretical and Applied Mechanics*, Vol. 8, Proc. 8th SECTAM Conference.

Chona, R., Irwin, G. R., and Sanford, R. J., 1983, "The Influence of Specimen Size and Shape on the Singularity-Dominated Zone," *Proceedings, 14th National Symposium on Fracture Mechanics*, STP 791, Vol. 1, ASTM, Philadelphia, pp. I1–I23.

Eftis, J., and Liebowitz, H., 1972, "On the Modified Westergaard Equations for Certain Plane Crack Problems," *Int. J. Fracture Mechanics*, 8, pp. 383–392.

Fedderson, C. E., 1966, *Plane Strain Crack Toughness Testing of High Strength Metallic Materials*, STP 410, ASTM, Philadelphia, pp. 77–79.

Green, A. E., and Sneddon, I. N., 1950, "The Distribution of Stress in the Neighborhood of a Flat Elliptical Crack in an Elastic Solid," *Proceedings of the Cambridge Philosophical Society*, 46, Part 1, pp. 159–163.

Irwin, G. R., 1957, "Analysis of Stress and Strains Near the End of a Crack Traversing a Plate," *J. Applied Mechanics*, 79, pp. 361–364.

Irwin, G. R., 1960, "Plastic Zone Near a Crack and Fracture Toughness," *Mechanical and Metallurgical Behavior of Sheet Materials*, Proc. Seventh Sagamore Ordinance Materials Research Conference, pp. IV–63–IV–78.

Irwin, G. R., 1962, "Crack-Extension Force for a Part-Through Crack in a Plate," *J. Applied Mechanics*, 29, pp. 651–654.

Isida, N., 1966, "Stress-Intensity Factors for the Tension of an Eccentrically Cracked Strip," *J. Applied Mechanics*, 33, pp. 674–675.

Jiang, C. P., Zou, Z. Z., Wang, D., and Liu, Y. W., 1991, "A Discussion about a Class of Stress Intensity Factors and Its Verification," *Int'l J. Fracture*, 49, pp. 141–157.

Kathiresan, K., 1976, *Three-Dimensional Linear Elastic Fracture Mechanics Analysis by a Displacement Hybrid Finite Element Model*, Ph.D. Thesis, Georgia Inst. of Technology.

Kobayashi, A. S., 1975, "Surface Flaws in Plates in Bending," *Proc. 12th Annual Meeting*, Society of Engineering Science, Austin, Texas.

Liebowitz, H., Lee, J. D., and Eftis, J., 1978, "Biaxial Load Effects in Fracture Mechanics," *Engineering Fracture Mechanics*, 10, pp. 315–335.

Muskhelishvili, N. I., 1953, *Some Basic Problems of the Mathematical Theory of Elasticity*, 3rd ed., P. Noordhoff, Groningen, Holland.

Neuber, H., 1958, *Theory of Notch Stresses*, 2nd ed., Springer Publishers, Berlin.

Newman, J. C., Jr., 1971, "An Improved Method of Collocation for the Stress Analysis of Cracked Plates with Various Shaped Boundaries," NASA TN D–6376, National Aeronautics and Space Administration, Washington, DC.

Newman, J. C., Jr., and Raju, I. S., 1981, "An Empirical Stress-Intensity Factor Equation for the Surface Crack," *Engineering Fracture Mechanics*, 15, pp. 185–192.

Raju, I. S., and Newman, J. C., Jr., 1979, "Stress-Intensity Factors for a Wide Range of Semi-elliptical Surface Flaws in Finite-Thickness Plates," *Engineering Fracture Mechanics*, 11, pp. 817–829.

Rooke, D. P., and Cartwright, D. J., 1976, *Compendium of Stress Intensity Factors*, Her Majesty's Stationery Office, London.

Sanford, R. J., 1979, "A Critical Re-examination of the Westergaard Method for Solving Opening-Mode Crack Problems," *Mechanics Research Communications*, 6(5), pp. 289–294.

Sih, G. C., 1966, "On the Westergaard Method of Crack Analysis," *Int. J. Fracture Mechanics*, 2, pp. 628–631.

Sih, G. C., 1973, *Handbook of Stress-Intensity Factors for Researchers and Engineers*, Institute of Fracture and Solid Mechanics, Lehigh University, Bethlehem, PA.

Smith, F. W., and Sorensen, D. R., 1974, "Mixed-Mode Stress Intensity Factors for Semi-elliptical Surface Cracks," NASA CR–134684, National Aeronautics and Space Administration, Washington DC.

Sneddon, I. N., 1946, "The Distribution of Stress in the Neighborhood of a Crack in an Elastic Solid," *Proceedings of the Royal Society A*, 187, pp. 229–260.

Tada, H., and Paris, P., 1979, "Discussion on Stress-Intensity Factors for Cracks," *Part-Through Crack Fatigue Life Prediction*, STP 687, American Soc. for Testing and Materials, Philadelphia, pp. 43–46.

Tada, H., Paris, P. C., and Irwin, G. R., 1985, *The Stress Analysis of Cracks Handbook*, 2nd ed., Del Research Corp., St. Louis, MO.

Tada, H., Paris, P. C., and Irwin, G. R., 2000, *The Stress Analysis of Cracks Handbook*, 3rd ed., American Society of Mechanical Engineers, New York.

Timoshenko, S., and Goodier, J. N., 1951, *Theory of Elasticity*, 2nd ed., McGraw-Hill, New York.

Westergaard, H. M., 1939, "Bearing Pressures and Cracks," *J. Applied Mechanics*, 6, pp. A49–A53.

Williams, M. L., 1952, "Stress Singularities Resulting From Various Boundary Conditions in Angular Corners of Plates in Extension," *J. Applied Mechanics*, 19, pp. 526–528.

Williams, M. L., 1957, "On the Stress Distribution at the Base of a Stationary Crack," *J. Applied Mechanics*, 24, pp. 109–114.

EXERCISES

3.1 Show that Eqs. (3.8) follow from Eqs. (3.7).

3.2 Derive eigenequation Eq. (3.9) from the determinants in Eqs. (3.8).

3.3 Show that the eigenvalue $n = 0$ in Eq. (3.11) yields an unacceptable solution to the single-ended crack problem posed by M. L. Williams.

3.4 The plate shown in Figure E3.1 contains a sharp notch with an included angle, β. Assuming that the plate is subjected to in-plane forces that produce only symmetric displacements with respect to the bisector of the wedge angle, (a) use the Williams formulation to determine the order of the singularity at the tip of the notch, and (b) derive expressions for the stresses in polar coordinates, retaining only the first three nonzero terms. Select $\beta = 30°$, $60°$, $90°$, or $120°$.

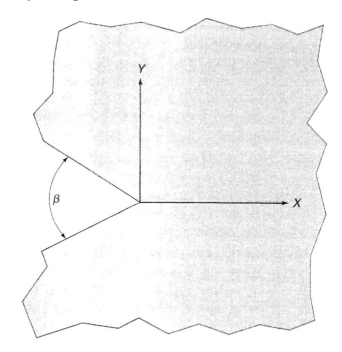

Figure E3.1

3.5 Verify that the Airy stress function of Eq. (3.22) satisfies the compatibility equations, Eqs.(2.35).

3.6 Derive the general form of the displacement equations in plane strain corresponding to the Airy stress function of Eq. (3.22).

3.7 Derive the general form of the displacements in plane stress for the forward-shear mode of deformation corresponding to the stresses given by Eqs. (3.32).

3.8 Derive the general form of the displacements in plane strain for the forward-shear mode of deformation corresponding to the stresses given by Eqs. (3.32).

3.9 Verify that the Airy stress function of Eq. (3.31) satisfies the compatibility equations, Eqs. (2.35).

3.10 Prove that the series solution in Westergaard form for the traction-free, single-ended crack, [Eqs. (3.50)] is equivalent to the Williams solution to the same problem. What is the relationship between the constants C_{1n} in the Williams formulation and A_j and B_m in the Westergaard formulation?

3.11 Repeat Exercise 3.10 for the forward-shear mode of deformation.

3.12 The Westergaard stress functions that solve the opening-mode problem of a crack of length, $2a$, in an infinite body subjected to four equal point loads, P, on the crack faces, as shown in Figure E3.2, are given by

$$Z(z) = \frac{2P}{\pi} \frac{z\sqrt{a^2 - b^2}}{(z^2 - b^2)\sqrt{z^2 - a^2}}$$

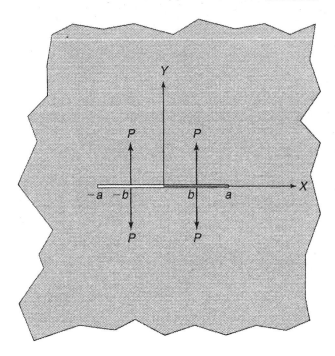

Figure E3.2

and

$$Y(z) = 0$$

(a) Derive an exact expression for the Cartesian stress σ_y valid everywhere in the body. **(b)** From the results of part (a), derive the geometric stress intensity factors for this combination of geometry and loading.

3.13 From the results of Exercise 3.12, derive an expression for K for the problem shown in Figure E3.3, using the principle of superposition.

3.14 The Westergaard stress functions that solve the opening-mode problem of a semi-infinite crack subjected to a crack-line force, P, as shown in Figure E3.4, are given by

$$Z(z) = \frac{P}{\pi} \frac{1}{z + b} \sqrt{\frac{b}{z}}$$

and

$$Y(z) = 0$$

(a) Derive an exact expression for the Cartesian stress σ_y valid everywhere throughout the body. **(b)** Using the formal definition of the geometric stress intensity factor, derive an expression for K for this combination of geometry

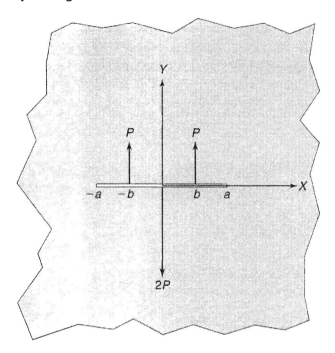

Figure E3.3

and loading. **(c)** Show that the exact expression for σ_y reduces to the near-field equation within some suitable region around the crack tip. **(d)** Determine the size of the singularity-dominated-zone for σ_y, based on a five percent error definition. Present your results in a suitable illustration.

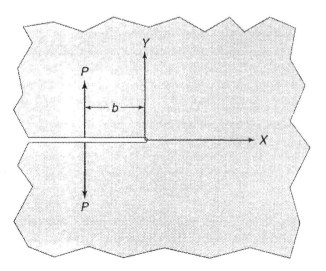

Figure E3.4

3.15 Using the Newman–Raju empirical equation for a semi-elliptical surface flaw [Eq.(3.91)], tabulate and plot the distribution of $Y(\frac{a}{W})$ around the flaw border (i.e., as a function of the flaw parameter, ϕ) for both tension and pure bending for the following flaw-geometry parameters:

(a) $a/c = 1.0$ and $a/t = 0.3$

(b) $a/c = 1.0$ and $a/t = 0.7$

(c) $a/c = 0.5$ and $a/t = 0.3$

(d) $a/c = 0.5$ and $a/t = 0.7$

CHAPTER 4

Numerical Methods for *K* Determination

4.1 INTRODUCTION

In the last chapter we observed that the invariant nature of the stress-field equations in the neighborhood of a crack tip reduced the general elasticity problem for geometries containing cracks to the lesser problem of determining the geometric stress intensity factor for the problem of interest. Ideally, we would like to have an analytical expression for the value of *K* that contains all of geometric variables as parameters, but, to do so requires that we be able to solve the corresponding elasticity problem in full detail. Previously, the Williams approach and the generalized Westergaard method were presented as possible means to the desired end, but they are not the only methods available.

One such approach is the *alternating method*, in which the exact solutions to two problems related to the problem of interest are available, each of which satisfies some of the boundary conditions for the new problem. The approach is illustrated schematically in Figure 4.1. In this example, the solution to a finite body problem (Figure 4.1d) is obtained from the general solutions to several semi-infinite body problems by alternately matching the boundary conditions along each of the free edges. In Figure 4.1, the geometry of the body in part (a) satisfies the boundary conditions on the remote boundary, the leading free edge, and the crack plane, but predicts nonzero stresses along the back free edge. By adding an appropriate solution for the body depicted in part (b), the boundary conditions on the back edge are now satisfied. However, there is now a residual error on the surface of the leading edge that must be removed by the addition of a second correction solution, shown in part (c), to restore the leading edge to a stress-free state. This addition produces a new (but hopefully smaller) residual error on the back edge. The process of alternating between solutions (b) and (c) to reduce the residual error is repeated until the overall

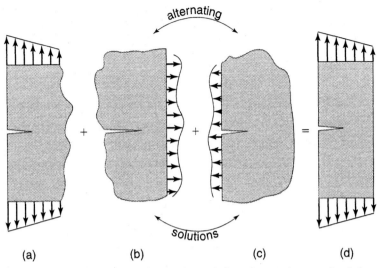

Figure 4.1 Schematic illustration of the alternating method for solving finite-body problems.

errors are of an acceptably low value. Although tedious, the alternating method does approach the exact solution in the limit. The method has proven to be particularly useful for three-dimensional problems [e.g., Grandt, Harter, and Heath, 1984].

A related method, called the *compounding method*, in which the stress intensity factor for a complex geometry is obtained as the sum of those for a sequence of ancillary problems, has been proposed by Cartwright and Rooke [1974]. In this method, illustrated in Figure 4.2, the K value due to each of the geometric features is

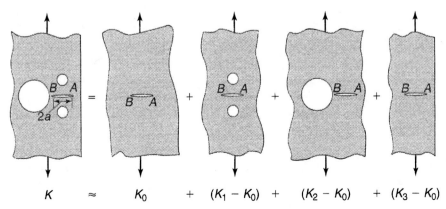

$$K \approx K_0 + (K_1 - K_0) + (K_2 - K_0) + (K_3 - K_0)$$

Figure 4.2 The compounding method for estimating the stress intensity factor for geometries with multiple stress raisers.

determined independently, and the results are "compounded" to obtain the desired stress intensity factor from the relation

$$K = K_o + \sum_{n=1}^{N}(K_n - K_o) + K_e \tag{4.1}$$

where K_o is the fundamental solution to the infinite-body problem (e.g., $\sigma\sqrt{\pi a}$ for a remotely loaded internal crack). K_n is the stress intensity factor for the ancillary problem, and K_e is the contribution to the stress intensity factor due to the interactions between boundaries (generally unknown). Conceptually, the compounding method is similar to the alternating method, except that only the K value is considered. However, the influences of the residual boundary stresses that were removed in the alternating solution scheme are now embodied in the value of K_e which is not computed. If the ancillary boundaries are far removed from each other in comparison with the crack dimension, K_e is neglected or may be estimated [Rooke, 1977, 1986]. Upon casual observation it would appear that the compounding method is equivalent to the method of superposition; however, this is not the case. In the method of superposition the geometry is held fixed, and the effects of the addition of several *independent* loading conditions are added to obtain the combined effect. This methodology is in strict accordance with the mathematical laws governing the linear partial differential equations of elasticity, and, as a consequence, the combined K values are exact in the mathematical sense. On the other hand, the compounding method utilizes *different* geometries that are subjected to the same applied loads. This approach is not amenable to the principle of linear superposition.

Despite all of the mathematical methods previously described, as well as more advanced mathematical procedures that are beyond the scope of this text (including, for example, the transform methods of Sneddon and Lowengrub [1969] and Laurent series expansions [Isida, 1973]), the number of exact solutions for bodies of finite dimensions is limited. The situation is even worse for geometries containing concentrated loads. Nonetheless the need for accurate values of the stress intensity factor remains. For example, an accurate calculation of the geometric stress intensity factor and its variation with crack length for specimens of certain standardized geometries[1] is necessary if the measurements of the material's fracture toughness are to be geometry independent. Also, it will be shown in Chapter 9 that predictions of fatigue life are sensitive to inaccuracies in the K values since the inaccuracies tend to be accentuated in the integration process used to determine lifetime from growth rate equations. With the wide availability of cheap, high-speed digital computation, including desktop computers and dedicated workstations, the focus in contemporary computational fracture mechanics is now directed toward numerical methods for ex-

---

[1] The particulars of standardized material tests to determine fracture-resistance properties are described in Chapter 8.

tracting the stress intensity factor. In particular, the boundary collocation method, the finite element method (FEM), and, most recently, the boundary element method (BEM), the latter of which combines some features of the previous two, have become the methods of choice for many analysts. In particular, when the immediate need for an accurate estimate of the geometric stress intensity factor outweighs the desire for an analytical expression in terms of the geometric variables (which could take months or years to obtain and verify), the use of these numerical methods is justified. The following sections of this chapter present the basic concepts of some of these numerical methods and the procedures by which the stress intensity factor is determined. However, to lay out the full details of the theory of each method would require complete books in their own right (and many such books have been written). Moreover, this level of understanding is not necessary if we are interested only in being intelligent consumers of the technology, rather than its developers.

4.2 BOUNDARY COLLOCATION

In its most general form the boundary collocation method is a numerical procedure for determining the coefficients (or, more precisely, a subset thereof) in the general, series-type solution for the combination of geometry and loading under study. For the solution of crack problems the mathematical developments in the previous chapter have demonstrated that the general solution was formulated such that the boundary conditions along the crack plane are satisfied identically. Since this solution satisfies the governing equations everywhere in the interior and along the crack plane, the only remaining requirement is to satisfy the boundary conditions on the remote boundaries. In an analytical solution the boundary conditions would also be satisfied at every point; however, in the boundary collocation method, we require only that they be satisfied at a discrete number of points.

Clearly, there is no one general solution for all crack problems; however, regardless of the problem, it is always possible to write a system of equations of the form

$$(\sigma)_1 = \sum_{j=0}^{J} A_j f_j(z_1) + \sum_{m=0}^{M} B_m g_m(z_1)$$

$$(\sigma)_2 = \sum_{j=0}^{J} A_j f_j(z_2) + \sum_{m=0}^{M} B_m g_m(z_2)$$

(4.2)

$$\vdots \qquad \vdots \qquad \vdots$$

$$(\sigma)_N = \sum_{j=0}^{J} A_j f_j(z_N) + \sum_{m=0}^{M} B_m g_m(z_N)$$

where $(\sigma)_i$ is the appropriate stress (or displacement) boundary condition[2] and $f_j(z_i)$ and $g_m(z_i)$ are known, real functions of the position coordinate, z_i. A_j and B_m are the unknown coefficients from a suitable series-type general solution. For example, they might be the coefficients from series forms of the generalized Westergaard functions, $Z(z)$ and $Y(z)$, respectively. Equations (4.2) can be written symbolically as

$$\left[f_j\,(r_i,\theta_i) \,\middle|\, g_m\,(r_i,\theta_i) \right] \left[\frac{A_j}{B_m} \right] = [\sigma_i] \tag{4.3a}$$

or

$$[f \mid g] \left[\frac{A}{B} \right] = [\sigma] \tag{4.3b}$$

If the total number of boundary equations N is equal to the combined number of unknowns (i.e., $N = J+M+2$), the matrix $[f \mid g]$ is square, and the set of coefficients that satisfy the boundary conditions at all of the chosen boundary stations is given by

$$\left[\frac{A}{B} \right] = [f \mid g]^{-1}[\sigma] \tag{4.4}$$

The earliest boundary collocation results used this solution scheme; however, Hulbert [1963] demonstrated that the difference between the stress (or displacement) computed from Eq. (4.4) and the required boundary value could be very large in the interval between collocated boundary-stations. To minimize these errors, Hulbert proposed using a larger number of boundary stations than unknown coefficients $(N > J + M + 2)$. In this case, $[f \mid g]$ is not square, and a solution of the form of Eq. (4.4) is not possible. Since the system of equations is overdetermined, it is necessary to statistically average the boundary values to make use of all of the available data. It can easily be demonstrated that multiplying Eq. (4.3b) from the left by $[f \mid g]^T$ (i.e., the transpose of $[f \mid g]$) is equivalent to constructing the least-squares solution to the overdetermined system of equations. Accordingly, the least-squares counterpart to Eqs. (4.3) for the best-fit constants A_j and B_m can be written as

$$[f \mid g]^T [f \mid g] \left[\frac{A}{B} \right] = [C] \left[\frac{A}{B} \right] = [f \mid g]^T [\sigma] \tag{4.5}$$

---

[2]Note that each chosen boundary station usually yields two equations of the type shown.

where $[C] = [f \mid g]^T [f \mid g]$. Since the matrix $[C]$ is square, the solution to Eq. (4.5) is

$$\begin{bmatrix} A \\ B \end{bmatrix} = [C]^{-1} [f \mid g]^T [\sigma] \qquad (4.6)$$

Experience with the application of the boundary collocation method for the solution of opening mode crack problems indicates that a redundancy factor of between 2 and 5 is sufficient to minimize the variation in boundary values between the collocated points. With typically 40 or more coefficients required to accurately determine the state of stress throughout the body, the number of equations to be solved can be of order 200 or greater, and a direct implementation of Eq. (4.6) may lead to numerical instabilities. In such cases Sanford and Berger [1990] have found that more refined solution schemes based on orthogonalization procedures, such as the QR decomposition scheme [Longley, 1984], are effective.

The key to successful implementation of the boundary collocation method depends on the selection of a suitable general solution for the geometry and loading of the problem. For simplicity we will discuss only general solutions for opening mode problems, but the procedure is identical for mode II problems. For the analysis of mixed mode problems, it is best to take advantage of the uncoupled nature of the governing equations and treat each mode separately to keep the size of the matrices as small as possible.

Historically, the stress function used for most of the early boundary-collocation analyses of crack problems was the Williams [1957] solution in Eq. (3.17) [Gross, Srawley, and Brown, 1964]. This stress function is applicable to single-edge notched (SEN) geometries with traction-free crack faces and no other stress raisers. We have previously shown that generalized Westergaard functions of the form

$$Z(z) = \sum_{j=0}^{J} A'_j \left(\frac{z}{W} \right)^{j-1/2} \qquad (4.7a)$$

$$Y(z) = \sum_{m=0}^{M} B'_m \left(\frac{z}{W} \right)^{m} \qquad (4.7b)$$

are functionally equivalent to the opening mode portion of the Williams solution and therefore are suitable for solving exactly the same class of problems. Equations (4.7) are a variation of Eqs. (3.50) in which the normalized variable, z/W, has been introduced as a convenience to limit numerical problems associated with the large exponents in the highest order terms in the truncated series expansions. Experience has shown that the normalizing dimension, W, (often the specimen width) should be chosen so as to make the coordinates of the boundary stations of $O(1)$.

Substituting Eqs. (4.7) into Eqs. (3.24), we find that the Cartesian stress components follow directly as jot=11pt

$$\sigma_x = \sum_{j=0}^{J} A_j' \left(\frac{r}{W}\right)^{j-1/2} \left\{ \cos\left[\left(j - \frac{1}{2}\right)\theta\right] - \left(j - \frac{1}{2}\right) \sin\theta \sin\left[\left(j - \frac{3}{2}\right)\theta\right] \right\}$$

$$+ \sum_{m=0}^{M} B_m' \left(\frac{r}{W}\right)^{m} \left\{ 2\cos(m\theta) - m\sin\theta \sin[(m - 1)\theta] \right\} \tag{4.8a}$$

$$\sigma_y = \sum_{j=0}^{J} A_j' \left(\frac{r}{W}\right)^{j-1/2} \left\{ \cos\left[\left(j - \frac{1}{2}\right)\theta\right] + \left(j - \frac{1}{2}\right) \sin\theta \sin\left[\left(j - \frac{3}{2}\right)\theta\right] \right\}$$

$$+ \sum_{m=0}^{M} B_m' \left(\frac{r}{W}\right)^{m} \left\{ m\sin\theta \sin[(m - 1)\theta] \right\} \tag{4.8b}$$

$$\tau_{xy} = \sum_{j=0}^{J} A_j' \left(\frac{r}{W}\right)^{j-1/2} \left\{ \left(j - \frac{1}{2}\right) \sin\theta \sin\left[\left(j - \frac{3}{2}\right)\theta\right] \right\}$$

$$+ \sum_{m=0}^{M} B_m' \left(\frac{r}{W}\right)^{m} \left\{ \sin(m\theta) + m\sin\theta \cos[(m - 1)\theta] \right\} \tag{4.8c}$$

where the geometric stress intensity factor is obtained from

$$K = A_o' \sqrt{2\pi W} \tag{4.8d}$$

From these equations it is a straightforward procedure to develop the system of boundary equations in Eqs. (4.2), from which the coefficients can be obtained from Eq. (4.6).

The usual procedure for conducting a boundary collocation analysis is to select a small number of terms from both series, compose the associated matrices, and solve for the corresponding set of coefficients, A_j and B_m. This process is repeated with increasingly larger number of terms while observing the convergence behavior of the stress intensity factor. As an example of the process, we will examine the results

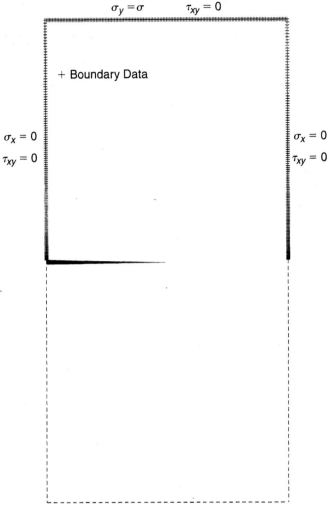

$$\sigma_y = \sigma \qquad \tau_{xy} = 0$$

+ Boundary Data

$\sigma_x = 0$ ⋮ ⋮ $\sigma_x = 0$

$\tau_{xy} = 0$ ⋮ ⋮ $\tau_{xy} = 0$

Figure 4.3 Boundary conditions and collocation points for a single-edge notched geometry (SEN).

of a boundary-collocation analysis of the single-edge notched geometry loaded in uniform remote tension shown in Figure 4.3. The stress intensity factor as a function of the number of terms retained in the series expansions is shown in Figure 4.4 for a normalized crack length a/W of 0.51. Clearly, convergence of the K value to the widely accepted empirical result due to Tada, et al. [1973] has occurred after approximately 20 terms from each series.

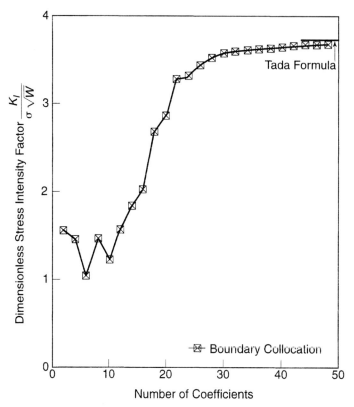

Figure 4.4 Effect of the number of retained terms in the series expansion on the convergence of the stress intensity factor for the SEN geometry.

Another measure of convergence is the RMS residual boundary error

$$\left[\frac{1}{N} \sum_{i=0}^{N} (\hat{\sigma}_i - \sigma_i)^2 \right]^{1/2} \tag{4.9}$$

where $\hat{\sigma}_i$ is the value of the boundary stress, σ_i, computed from the collocation coefficients. The RMS error for this sample problem is shown in Figure 4.5. The convergence behavior of the RMS error mirrors that of the stress intensity factor. If the behavior of either of these measures is other than that shown in this example, the problem is most likely associated with either the formulation of the global stress equations or numerical instabilities resulting from the finite arithmetic limitations of digital computation.

From the uniqueness theorem of elasticity, the coefficients in the series expansions for $Z(z)$ and $Y(z)$ should approach their exact values as the number of terms becomes infinitely large. Of course, computer limitations (not to mention time and

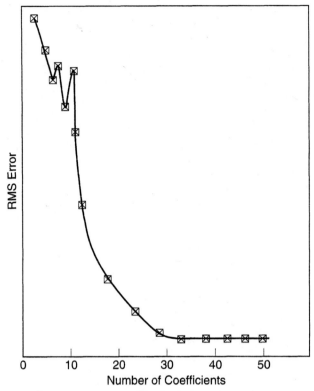

Figure 4.5 Corresponding decrease in the RMS error for the example shown in Figure 4.4.

money) prohibit us from taking the process to its ultimate limit, but experience has shown that the series truncated after the RMS error has stabilized to an acceptably low value provides a quite accurate representation of the state of stress throughout the entire body. For example, a comparison of the photoelastic isochromatic pattern (contours of constant τ_{max}) for the SEN specimen with $a/W = 0.6$, is shown in the lower half of Figure 4.6, along with a computer reconstruction of the experimental pattern based on a 40-term expansion of Eqs. (4.8) in the upper half of the figure. The agreement between these two patterns over the entire region graphically demonstrates the ability of the boundary collocation method to accurately model the state of stress in the body. The results can be used with confidence for a wide variety of purposes other than just determining the geometric stress intensity factor.

For the class of problems represented by a traction-free internal crack in a body of finite dimensions with remote loading, as illustrated in Figure 4.7, the general solution in Westergaard form can be expressed as

$$Z(z) = \sum_{j=0}^{J} A_j \frac{(z - z_o)^{j+1}}{\sqrt{(z - a)(z - b)}} \qquad (4.10a)$$

Figure 4.6 Comparison of the computer-generated isochromatic pattern (top) with the photoelastic pattern (bottom) for the SEN geometry, using a 40 term expansion.

and

$$Y(z) = \sum_{m=0}^{M} B_m z^m \quad \left[\text{or, alternatively,} \quad Y(z) = \sum_{m=0}^{M} \overline{B}_m (z - z_o)^m \right] \quad (4.10b)$$

where $z_o = (a + b)/2$ and $\overline{B}_m \neq B_m$. (Note that when $a = -b$, $z_o = 0$, and the general solution for the symmetric internal-crack problem is obtained.)

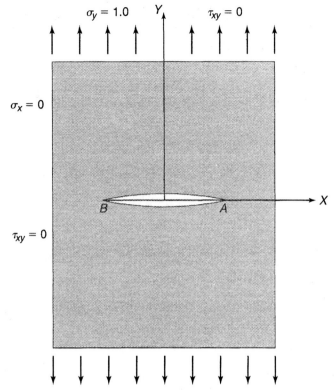

Figure 4.7 Boundary conditions for a rectangular plate with an eccentric internal crack.

When expressed in terms of multiple polar coordinates, similar to those used in deriving Eqs. (3.37) for the infinite-body problem, the Cartesian stress components are

$$\sigma_x = \sum_{j=0}^{J} A_j \frac{r_o^{j+1}}{(r_1 r_2)^{1/2}} \left\{ \cos\left[(j+1)\theta_o - \frac{\theta_1 + \theta_2}{2} \right] \right.$$

$$- (j+1)\sin\theta_o \sin\left[j\theta_o - \frac{\theta_1 + \theta_2}{2} \right]$$

$$+ \frac{1}{2}\left(\frac{r_o}{r_1}\right) \sin\theta_o \sin\left[(j+1)\theta_o - \frac{3\theta_1 + \theta_2}{2} \right]$$

$$\left. + \frac{1}{2}\left(\frac{r_o}{r_2}\right) \sin\theta_o \sin\left[(j+1)\theta_o - \frac{\theta_1 + 3\theta_2}{2} \right] \right\}$$

$$+ \sum_{m=0}^{M} B_m r_o^m [2\cos m\theta_o - m\sin\theta_o \sin(m-1)\theta_o] \qquad (4.11a)$$

$$\sigma_y = \sum_{j=0}^{J} A_j \frac{r_o^{j+1}}{(r_1 r_2)^{1/2}} \left\{ \cos\left[(j+1)\theta_o - \frac{\theta_1 + \theta_2}{2}\right] \right.$$

$$+ (j+1) \sin\theta_o \sin\left[j\theta_o - \frac{\theta_1 + \theta_2}{2}\right]$$

$$- \frac{1}{2}\left(\frac{r_o}{r_1}\right) \sin\theta_o \sin\left[(j+1)\theta_o - \frac{3\theta_1 + \theta_2}{2}\right]$$

$$\left. - \frac{1}{2}\left(\frac{r_o}{r_2}\right) \sin\theta_o \sin\left[(j+1)\theta_o - \frac{\theta_1 + 3\theta_2}{2}\right] \right\}$$

$$+ \sum_{m=0}^{M} B_m r_o^m [m \sin\theta_o \sin(m-1)\theta_o] \tag{4.11b}$$

$$\tau_{xy} = -\sum_{j=0}^{J} A_j \frac{r_o^{j+1}}{(r_1 r_2)^{1/2}} \left\{ (j+1) \sin\theta_o \cos\left[j\theta_o - \frac{\theta_1 + \theta_2}{2}\right] \right.$$

$$- \frac{1}{2}\left(\frac{r_o}{r_1}\right) \sin\theta_o \cos\left[(j+1)\theta_o - \frac{3\theta_1 + \theta_2}{2}\right]$$

$$\left. - \frac{1}{2}\left(\frac{r_o}{r_2}\right) \sin\theta_o \cos\left[(j+1)\theta_o - \frac{\theta_1 + 3\theta_2}{2}\right] \right\}$$

$$- \sum_{m=0}^{M} B_m r_o^m [\sin m\theta_o + m \sin\theta_o \cos(m-1)\theta_o] \tag{4.11c}$$

Applying the definition of K from Eq. (3.54) to each crack tip independently, we find that the geometric stress intensity factors at each end of the internal crack are obtained from the series coefficients as

$$K_A = \sum_{j=0}^{J} A_j \frac{\sqrt{2\pi}}{\sqrt{a-b}}\left(\frac{a-b}{2}\right)^{j+1} \quad \text{and} \quad K_B = \sum_{j=0}^{J}(-1)^j A_j \frac{\sqrt{2\pi}}{\sqrt{a-b}}\left(\frac{a-b}{2}\right)^{j+1}$$

$$\tag{4.12}$$

In contrast to the single-ended crack problem, in which the value of the stress intensity factor was determined from only the leading term of the $Z(z)$ series, the K values at each crack tip in the internal-crack problem depend on all of the A_j coefficients. This requirement places additional demands on a suitable convergence criterion.

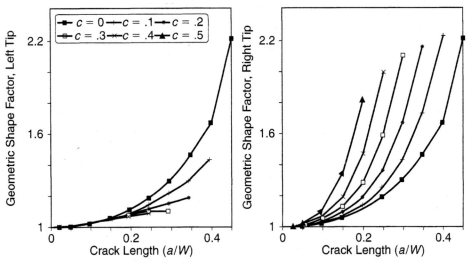

Figure 4.8   Results of a boundary collocation analysis of the internal-crack problem for a wide range of eccentricities [Drude, 1993].

The stresses in Eqs. (4.11) were used to perform a boundary collocation analysis [Drude, 1993] of the geometry in Figure 4.7 for a variety of eccentricity values[3], and the results were compared with those the Laurent series expansion solution of Isida [1973]. The results, shown in Figure 4.8, were in good agreement with those presented by Isida, provided that a large number of boundary stations were used for the collocation. Drude observed that the results approached stable values with as few as 20 terms in the expansions for the stresses, but the solution was sensitive to the number of boundary stations. Typical results for one case are shown in Figure 4.9. It appears that the form of Eqs. (4.12) places significantly greater demands on the accuracy of each coefficient, and, accordingly, a large number of boundary stations is required.

The general series-solution for the traction-free, single-edge crack problem [Eqs. (4.7)], and the general series-solution for the traction-free, internal-crack problem [Eqs. (4.10)] are special cases of a more general formulation of crack solutions for finite bodies of the form

$$Z(z) = Z_\infty(z) \cdot P(z) \qquad (4.13)$$

where $Z_\infty(z)$ is the Westergaard stress function for the infinite-body problem with the same crack geometry and applied loading (for example, any of those cases listed

---

[3] The symmetric problem was first studied by Kobayashi, Cherepy, and Kinsel [1964].

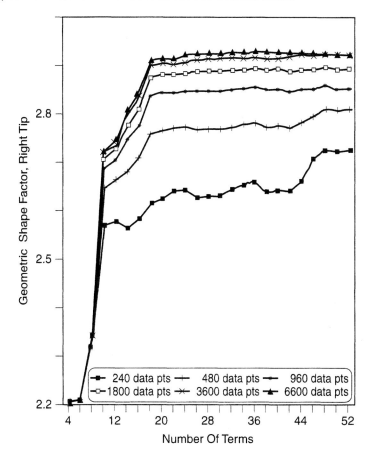

Figure 4.9 Influence of the number of boundary stations on convergence for the internal-crack problem.

in Appendix C). The function $P(z)$, the boundary influence function, is a series of the general form

$$P(z) = \sum_{j=0}^{J} A_j h_j(z) \qquad (4.14)$$

where A_j are real constants.

The functions, $h_j(z)$, are subject to the requirements that on the crack plane, $h_j(x)$ are real functions, so as to ensure that $Re\, Z(z) = 0$ on the traction-free portions of the crack surfaces. (Recall that all acceptable choices for $Z_\infty(z)$ have this property.) In addition, $h_j(z)$ should reflect any symmetries associated with the problem of interest.

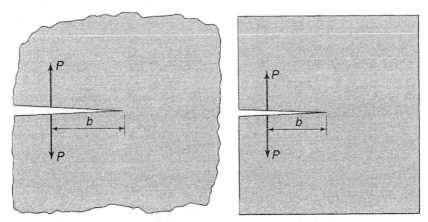

Figure 4.10   A corresponding pair of infinite-body (left) and finite-body (right) problems.

As an example of the boundary influence function approach, consider the pair of problems shown in Figure 4.10. For the infinite-body problem, the Westergaard stress function, $Z_\infty(z)$, is given in Appendix C as

$$Z_\infty(z) = C\frac{1}{z + b}\sqrt{\frac{b}{z}} \tag{4.15}$$

where C is a real constant determined from the requirement of static equilibrium along the crack plane. We note that the two key features that strongly affect the stresses in the body, namely the singularities at the crack tip and at the point of load application, are modeled by this stress function. As a consequence, the boundary influence function for this problem has the particularly simple form

$$P(z) = \sum_{j=0}^{J} A_j z^j \tag{4.16}$$

Accordingly, we can write the general series-solution for this crack-line loaded, single-edge notched geometry as

$$Z(z) = \frac{1}{z + b}\sum_{j=0}^{J} A_j z^{j-1/2} \tag{4.17a}$$

where the redundant constant $C\sqrt{b}$ has been merged into A_j. Alternatively, we could have written $Z(z)$ in normalized form as

$$Z(z) = [(z + b)/W]^{-1} \sum_{j=0}^{J} A'_j (z/W)^{j-1/2} \qquad (4.17b)$$

where W is a characteristic in-plane dimension, $A'_j = A_j W^{j-3/2}$ and the stress intensity factor, K, is given by

$$K = A'_0 (W/b)\sqrt{2\pi W} \qquad (4.17c)$$

The normalized form chosen for $Y(z)$ for the traction-free, single-edge crack problem [Eq. (4.7b)] is suitable for this problem as well.

The development of the equations for the Cartesian stress components in real coordinates closely follows that of the previous examples and is not repeated here.[4] The resulting expressions are somewhat more complicated than in our previous examples, since they need to retain the singular behavior at both points along the crack line. Nonetheless, the final equations for the boundary stresses are of the form of Eqs. (4.2). However, in contrast to our previous cases, wherein remote loading was applied, all of the boundary stresses are zero, and the boundary-station matrix $[\sigma]$ is null. As a result, Eq. (4.5) is homogeneous and the coefficient matrix, $[C]$, has no unique inverse. To circumvent this problem, the system of boundary equations must be augmented by the additional requirement that the collocated result satisfy static equilibrium, that is,

$$\int_0^{W-b} \sigma_y |_{\theta=0} dx = P \qquad (4.18)$$

With this additional equation the matrix $[\sigma]$ is no longer null, and the matrix of coefficients can be uniquely determined.

Sanford and Berger [1990] applied the boundary influence function approach to a crack-line loaded SEN geometry of standard ASTM proportions, as shown in Figure 4.11. A total of 102 equally spaced boundary stations were used in the solution. The convergence of the K value as the number of terms in the series expansion increases is shown in Figure 4.12. It is of some interest to note that the stress intensity factor has reached a stable value after only 13 terms from each of the Z and Y series. This phenomenon is in marked contrast to the results reported by Newman [1974] for a similar geometry using a modified Williams-type formulation. He reported that as many as 150 terms were needed for convergence. The rapid convergence observed for the boundary influence function method is the result of including the point load into the series solution rather than treating it as a boundary condition. Despite the small number of terms required to obtain convergence, the accuracy of the results

---

[4]Explicit expressions for the Cartesian stress components are presented in Sanford and Berger [1990].

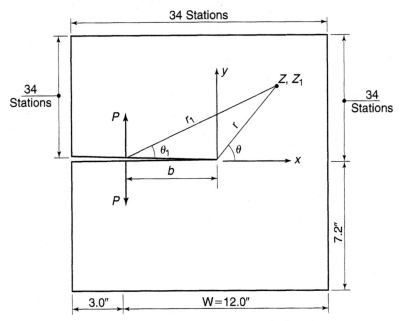

Figure 4.11   Geometry and coordinate description for a crack-line-loaded single-edge notched specimen used for boundary collocation by the influence-function approach.

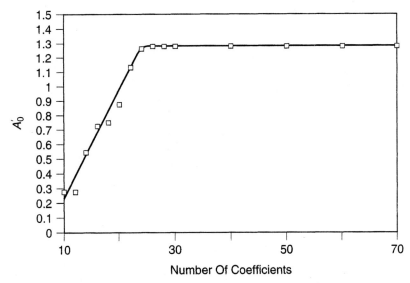

Figure 4.12   Convergence of the stress intensity factor as a function of the number of retained terms in the series expansion for the geometry shown in Figure 4.11.

Table 4.1 Boundary Collocation of the SEN Specimen

b/W	$KB\sqrt{W}/P$			Ebv/P		
	collocation analysis	ASTM 561	% difference	collocation analysis	ASTM 561	% difference
0.35	6.45	6.39	0.9	22.8	22.8	0.0
0.40	7.32	7.28	0.5	28.8	28.7	0.3
0.45	8.36	8.34	0.2	36.2	36.1	0.3
0.50	9.61	9.66	−0.5	45.9	45.7	0.4
0.55	11.34	11.36	−0.1	59.0	58.8	0.3
0.60	13.60	13.65	−0.4	77.2	77.1	0.1

is extremely high. A comparison of the nondimensional geometric stress intensity factor and the nondimensional compliance from this study with the reference values reported in ASTM Standard E 561 is given in Table 4.1. Over a wide range of crack lengths, the difference between the two results is less than one percent.

Up to this point, all of the problems we have cataloged share a common characteristic, namely that the only geometric discontinuity within the boundary of the material was the crack(s). We have taken into account load-point discontinuities through proper choice of stress functions, but have not permitted other stress raisers, such as fillets and holes. There are several ways in which we can incorporate these additional geometric features into the formulation of the problem. The optimal and most direct method would be to choose a pair of stress functions that accounts for the crack(s), the load points, and the additional stress raisers without violating the defining properties of the stress functions given by Eqs. (3.28) and (3.30).

Fortunately, for some classes of problems, it is possible to construct such augmented general solutions. For example, consider the class of problems illustrated in Figure 4.13 (i.e., a single-ended crack approaching a hole). Following the form suggested by Newman [1969], we can write

$$Z(z) = \sum_{j=0}^{J} A'_j \left(\frac{z}{W}\right)^{j-1/2} + \sum_{u=1}^{U} C'_u \frac{W^{u+1/2}}{\sqrt{z}(z-z_0)^u} \qquad (4.19a)$$

$$Y(z) = \sum_{m=0}^{M} B'_m \left(\frac{z}{W}\right)^{m} + \sum_{v=1}^{V} D'_u \frac{W^v}{(z-z_0)^v} \qquad (4.19b)$$

Note that the first summation for both $Z(z)$ and $Y(z)$ is the same as Eqs. (4.7) and models the behavior around the crack tip. As previously noted, negative powers of z below $z^{-1/2}$ lead to unbounded displacements at the crack tip ($z \to 0$) and therefore are excluded from the summations. On the other hand the second summation in both series models the stress-raising effect of the cutout and specifically includes the

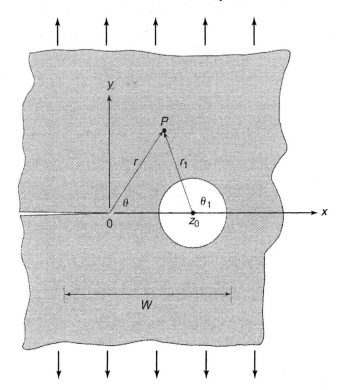

Figure 4.13 Coordinate description for a crack approaching a hole.

negative powers of z. By placing a pole, z_o, at the center of the circular boundary, this summation models the rapid rise in stress as the boundary is approached. Because the pole is located outside the boundary of the material, all negative powers of z can be included while still maintaining finite displacements within the boundary of the problem.

Kirk and Sanford [1990] used the series representations given by Eqs. (4.19) to determine the geometric stress intensity factor for the modified single-edge notched tension [SE(T)] specimen shown in Figure 4.14. This geometry was investigated as an alternative to the conventional SE(T) geometry in order to produce a more rapid rise in the K value with crack length for crack-arrest applications. From Eq. (4.19a), the σ_y stress component along the crack line becomes

$$\sigma_y = \sum_{j=0}^{J} A_j' \left(\frac{r}{W}\right)^{j-1/2} + \sum_{u=1}^{U} (-1)^u C_u' \frac{W^{u+1/2}}{r_1^u \sqrt{r}} \tag{4.20}$$

From which it follows that

$$K = \left[A_o' + \sum_{u=0}^{U} (-1)^u C_u' \left(\frac{W}{b}\right)^u \right] \sqrt{2\pi W} \tag{4.21}$$

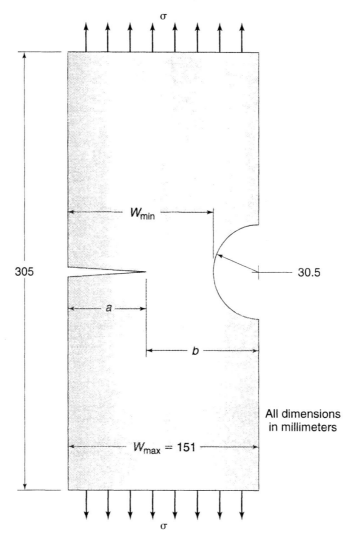

Figure 4.14   Dimensions of a model of a modified single-edge notched tension [SE(T)] specimen.

The results of the boundary-collocation analyses of this geometry for a range of crack lengths are shown in Figure 4.15. For most of the range investigated, the average boundary-stress error was of the order of 0.5 percent or less of the remotely applied stress. The results of the boundary-collocation analysis were cast into the standard form for K [Eq. (3.56)] and the dimensionless shape-function, $Y(a/W)$, was fit to a common form of polynomial for this type of problem. The results are tabulated in Table 4.2.

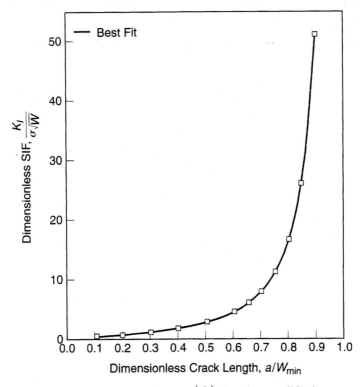

Figure 4.15 Shape function $Y\left(\frac{a}{W}\right)$ for the modified SE(T) geometry obtained by boundary collocation.

Unfortunately, it is not always possible to augment simple series solutions without violating the requisite boundary conditions on the crack plane. In such cases it may be possible to partition the geometry into several regions for which there are independent series solutions. These solutions can be mated, or "stitched," together

Table 4.2 Shape Function Coefficients for the Modified SE(T) Specimen

$$Y\left(\frac{a}{W}\right) = \frac{\beta_0 + \beta_1\alpha + \beta_2\alpha^2 + \beta_3\alpha^3 + \beta_4\alpha^4}{(1 - \alpha)^2}$$

i	β_i
0	0.558435
1	0.576951
2	−1.89903
3	2.01364
4	−1.06565

where: $\alpha = a/W_{min}$

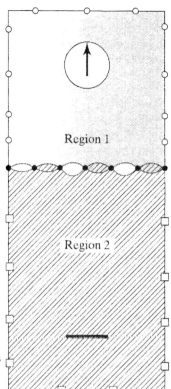

Figure 4.16 Example of the application of the partitioning method for boundary-collocation analysis. Different stress functions are used for Region 1 and for Region 2.

along the common boundary through their common boundary conditions. As an illustration of the partitioning method, let us consider the hypothetical example shown in Figure 4.16 of a pin-loaded tension strip with an internal crack. Each region can be represented by its own series solution containing unknown constants $A_j^{(i)}$ and $B_m^{(i)}$, where $i = 1$ or 2 denotes region 1 or 2, respectively. For each region we can write boundary-station equations in the form of Eqs. (4.2).

On the independent portions of each boundary, the applied boundary conditions are known explicitly. But along the common boundary, the surface tractions and displacements are not known. What is known is that they are the same for the two regions, or, more precisely, the difference between the boundary conditions at each common boundary station must be zero. As a result, the two regions are coupled together through these common boundary conditions. After assembling all of the boundary-station equations, we can write the resultant system of linear equations as a partitioned matrix equation of the general form

$$
\begin{pmatrix}
f^{(1)}\big|g^{(1)} & \bigg| & 0 \\[1em]
\hline
f^{(1)}\big|g^{(1)} & \bigg| & -f^{(2)}\big|g^{(2)} \\[1em]
\hline
0 & \bigg| & f^{(2)}\big|g^{(2)}
\end{pmatrix}
\begin{pmatrix}
A^{(1)} \\ \hline B^{(1)} \\ \hline A^{(2)} \\ \hline B^{(2)}
\end{pmatrix}
=
\begin{pmatrix}
\sigma^{(1)} \\ \hline 0 \\ \hline \sigma^{(2)}
\end{pmatrix}
\tag{4.22}
$$

Because of the unknown boundary conditions along the interface, we are required to collocate both regions simultaneously. Consequently, the resulting system of equations is quite large. Moreover, at best, the boundary conditions along the interface will be matched only at the collocation points. We are forced to accept the fact that there will be gaps or overlaps in the deformed interfacial boundary. We can minimize the adverse affects of the imperfect stitching by increasing the degree of redundancy along this boundary and by ensuring that the location of the boundary is sufficiently removed from critical areas within both regions.

There are similarities between the partitioning method and the alternating method described at the beginning of this chapter. Both methods attempt to match the boundary conditions at the interface between two independent problems. Whereas the partitioning method matches the boundary conditions on all of the boundaries (including the stitched boundary) at the expense of a large system of equations to be solved, the alternating method requires fewer equations, but needs multiple steps to reduce the residual error to zero. On the other hand, the alternating method produces at least one boundary in which the boundary conditions are matched exactly (the last one corrected). In contrast, the partitioning method will contain residual errors in the least-squares sense on all boundaries (including the stitched boundary), except where the boundary conditions are satisfied automatically.

By its nature the boundary collocation method extracts the geometric stress intensity factor by detecting the perturbations of the boundary stresses due to the presence of a crack tip. But, since the singular stresses at the crack tip decay as $1/\sqrt{r}$, their contribution to the boundary value is negligible if the boundary is far removed from the crack tip. Put another way, if the stress-field disturbance due to a crack is confined to a small region compared with the overall scale of the problem, most of the boundary will be unaware of the existence of the crack—i.e., St. Venant's principle comes into play. In these cases, boundary collocation fails to converge to accurate values of K. An example of a situation in which this problem occurs is demonstrated by the photoelastic pattern of a deep crack in a remotely loaded tension strip, shown in Figure 4.17. As can be seen in the figure, all of the variation in the stress field due to the crack (as evidenced by the density of fringes) is confined to a small region near the net ligament. The remainder of the fringe field is quite uniform. Attempts to extract the stress intensity factor using boundary collocation produce

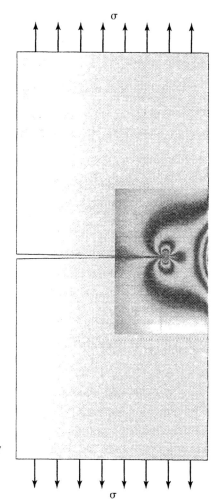

Figure 4.17   Example of a problem for which the boundary collocation method tends to fail—a deep crack in a large sheet.

the erroneous results shown in Figure 4.18. Recently, Sanford and Kirk [1991] have developed a method that combines boundary collocation with experimentally measured interior data to assist in extracting accurate geometric stress intensity factors for which boundary collocation alone is insufficient. This method, named *global collocation*, will be described in the next chapter, on experimental methods.

When properly applied, all of the boundary collocation approaches are capable of yielding highly accurate results not only for the stress intensity factor, but also for other stress-field parameters (*e.g.*, σ_{ox}, crack opening displacement, system compliance), provided that enough terms are retained in the analysis and sufficient numerical precision is available. If an exact analytical solution is not available, the boundary collocation method (or one of its variants) is generally preferred over other numerical methods (such as the finite element method), since the interior of the body is still represented by continuous functions. In addition, the stress inten-

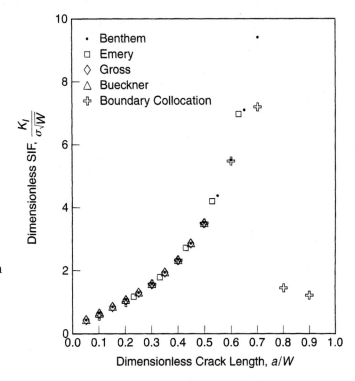

Figure 4.18 Comparison of a boundary-collocation analysis with established solutions for the SEN geometry, illustrating the loss of convergence for deep cracks.

sity factor is obtained as a consequence of the analysis from the best-fit coefficients. [See, for example, Eqs. (4.8d) or (4.12).] However, the method is not without its shortcomings. Unfortunately, the crack length as an independent parameter is lost in the formulation of the boundary equations, and the problem must be re-solved for each crack length of interest. In addition, only those problems for which an analytical solution in general series form is known, or a trial function constructed, are amenable to solution by the boundary collocation method. Finally, there are no general-purpose "boundary collocation solvers" commercially available (and supported by expert staff). Therefore, it is generally necessary to write and debug your own customized program in order to use the method. It is the combined effect of these last two drawbacks of the boundary collocation method that has lead to the widespread application of the finite element method (FEM) to determine geometric stress intensity factors. In the next section, we will explore methods to extract K values (and other parameters of interest) from FEM results.

4.3 THE FINITE ELEMENT METHOD

The limitations on practical implementation of the boundary collocation method—namely, *a priori* knowledge of a suitable general series solution and the need for customized software—are the primary motivators for pursuing the finite element method. While it is generally agreed that the discretitization of the entire body combined with the high stress-gradients at the crack tip result in loss of accuracy compared to that of the other numerical techniques, the simplicity and wide availability of FEM software have made this method of fracture mechanics analysis the method of choice for most practical applications. From a practicing engineer's perspective, the primary drawback of the approach is that, except for special formulations, general FEM algorithms do not include the stress intensity factor(s) as output parameters from the analysis. In nearly all cases, the *K* value(s) is(are) obtained by further analysis and interpretation of the postprocessor output. In this section we will focus our attention on a variety of techniques that have been proposed to extract the geometric stress intensity factor(s) from the output files of commercial FEM software.

In simplistic terms the finite element method is a systematic procedure for solving elasticity problems in which the body is divided into a large number of smaller regions (elements). Within each element the form of the displacements is *predescribed* to be fixed-order polynomials. As a consequence, we can describe the relative displacement of the corners of the element and certain locations along the edges (collectively called the *nodes*) in terms of quantities called *shape functions*. These functions describe the manner in which the displacements are approximated over the element.[5] In turn, their form depends uniquely on the order of the predefined polynomial approximation. Once the permitted variation of displacement of the element has been defined, the variation of the strain and the stress follow directly from the strain–displacement relations [Eq. (2.30)] and a suitable constitutive model, respectively.

In the earliest formulation of the finite element method, the displacements were assumed to vary linearly over the element, giving rise to constant-strain triangles as the fundamental building blocks for modeling the geometry. In current practice, the constant-strain triangle has been abandoned in favor of quadrilateral elements that permit, as a minimum, a quadratic variation in displacement. Regardless of the choice of element displacement chosen, the defining characteristic of the method is that the solution obtained depends on the form and number of elements selected to represent the body. Whether or not the output displacements, strains, and stresses realistically model the behavior of the actual structure is problematic and is best evaluated by sound engineering judgment.

In most cases, FEM programs do not include in their formulation the geometric stress intensity factor(s). Except for a special formulation that we will discuss later, we are required to extract the stress intensity factor from a fracture mechanics interpretation of the basic outputs of the program. Conceptually, we could obtain the opening mode value of *K* from the σ_y stress ahead of the crack through the defining

---

[5]The details of the construction of shape functions and the resultant assembly of the global stiffness matrix are beyond the scope of this book. Moreover, they are not needed for an understanding of the discussions that follow. The interested reader is referred to any introductory book on the finite element method for specific information.

relation

$$K_I = \lim_{x^+ \to 0} \sigma_y|_{\theta=0} \cdot \sqrt{2\pi x} \tag{4.23}$$

where x is the crack-tip coordinate along the crack plane and the limit is taken from the material side. Alternatively, we could use the crack-opening displacement behind the crack tip, v, to compute K from the relation

$$K_I = \lim_{x^- \to 0} v|_{\theta=\pi} \cdot \frac{1}{\sqrt{2\pi x}} \cdot \frac{\pi}{2} E' \tag{4.24}$$

where $E' = E$ for plane stress and $E' = E/(1-v^2)$ when plane strain conditions prevail. Recall that these definitions of the limit follow directly from the near-field equations [Eqs. (3.46)] and therefore apply only to stress and displacements computed to be within the singularity-dominated-zone. In this near-field region, Eqs. (4.23) and (4.24) predict that the products $\sigma_y \cdot \sqrt{2\pi x}$ and $v/\sqrt{2\pi x}$ will have constant values. We have previously concluded that the size of the singularity-dominated-zone is of the order of two percent of the distance to the nearest boundary in typical geometries. As a consequence, it is not practical to place enough elements within this small region in order to establish the required invariance.

On the other hand, if we examine the behavior of the stress and displacement in a region large enough to include the first higher order term in the crack-tip expansion [Eq. (3.53a)]—i.e., terms of $O(r^{1/2})$, we can write

$$\sigma_y|_{\theta=0} = \frac{K}{\sqrt{2\pi x}} + A_1 x^{1/2} \tag{4.25a}$$

and

$$E'v|_{\theta=\pi} = \frac{2}{\pi}K \cdot \sqrt{2\pi|x|} - \frac{4}{3}A_1|x|^{3/2} \tag{4.25b}$$

Equations (4.25) can be rewritten in the form

$$\sigma_y|_{\theta=0} \cdot \sqrt{2\pi x} \equiv K_{\text{apparent}} = K + \sqrt{2\pi}A_1 x \tag{4.26a}$$

and

$$\frac{\pi}{2}E' \cdot v|_{\theta=\pi} \cdot \frac{1}{\sqrt{2\pi|x|}} \equiv K_{\text{apparent}} = K - \frac{\sqrt{2\pi}}{3}A_1|x| \tag{4.26b}$$

In this form, it becomes clear that, if the modified finite element outputs (the leftmost sides of Eqs. (4.26)) are plotted as a function of the distance from the crack tip,

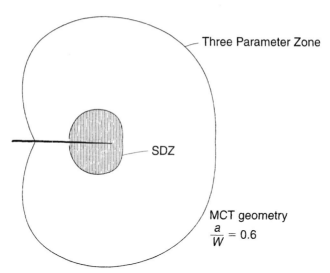

Figure 4.19 Comparison of the relative sizes of the regions for which two and three parameters accurately represent the state of stress near the crack tip in the compact-tension geometry.

$| x |$, the behavior will be linear within the region of validity of the three-parameter[6] representation of the stress and displacement field. It follows that the stress intensity factor, K, is the y-axis intercept of a best-fit line through the linear portion of the data when plotted as described, and the slope of the line is proportional to the higher order parameter A_1.

Fortunately, the size of the three-parameter region is significantly larger than that of the near-field region. For example, a comparison of the size of these two regions for the modified compact-tension geometry of Figure 3.10 is shown in Figure 4.19. Accordingly, it is relatively easy to construct a mesh with sufficient density to clearly define the linear zone by using simple elements.

Typical results from a finite element analysis of a single-edge notched specimen using the mesh shown in Figure 4.20 consisting entirely of eight-node isoparametric elements are presented in Figure 4.21. The graphs can be divided into three distinct zones. In the central zone, the data fall on a line, indicating that a three-parameter representation of the stress–displacement field is adequate over this domain. In the zone to the right, a higher order model would be required to represent the state of stress and displacement. On the other hand, the data points on the left that fail to fit on the line indicate that the elements very near the crack tip are not modeling the inverse-square-root-r stress–strain behavior and should be ignored in constructing the best-fit line to the three-parameter model. An extension of this concept to n-

---

[6]Recall that the near-field equations require two parameters, K and σ_{ox}, but σ_{ox} does not explicitly appear in either σ_y or v along the crack plane.

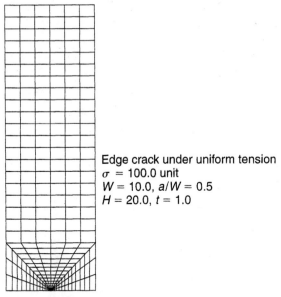

Edge crack under uniform tension
$\sigma = 100.0$ unit
$W = 10.0$, $a/W = 0.5$
$H = 20.0$, $t = 1.0$

Figure 4.20   Finite-element mesh for the SEN geometry consisting entirely of eight-node isoparametric elements.

parameter regions will be presented in the next chapter in conjunction with the use of experimental full-field data to extract K values.

Inherent in the use of the finite element method to analyze cracked geometries with conventional elements is the fact that elements near the crack tip always under-

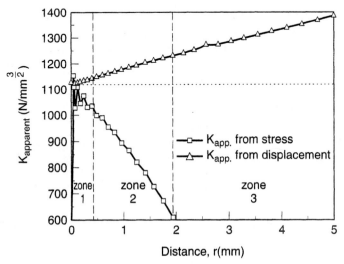

Figure 4.21 $K_{apparent}$ computed from both stress and displacement results, using the mesh of Figure 4.20.

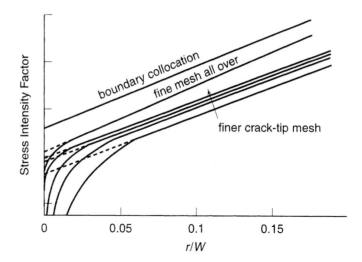

Figure 4.22 Effect of mesh refinement on the calculation of *K* from finite-element results.

estimate the sharply rising stress–displacement gradients. As a result, the discretized model always appears stiffer, that is, less compliant than the actual structure and consequently underestimates the stress intensity factor. Of the various ways to increase the accuracy of the *K* estimate the most obvious is the use of mesh refinement. As shown schematically in Figure 4.22, increasing the number of elements in the region surrounding the crack tip at the expense of increased computation time will improve the accuracy. However, Newman and Raju [1983] have demonstrated that, at best, a finite element analysis underestimates the correct *K* value by nearly two percent, and they proposed inclusion of a factor of 1.015 into the estimating scheme to correct for the increased stiffness effect.

Rather than attempt to match the inverse-square-root-*r* nature of the stress-field gradient with smaller and smaller polynomial elements, it is more efficient to model the known gradient by more direct methods. Henshell and Shaw [1975] and Barsoum [1976] independently observed that, by collapsing one side of an eight-node isoparametric quadrilateral and relocating the midside nodes to the one-quarter point, as shown in Figure 4.23, the strain distribution along the now radial sides of the element takes on the desired $1/\sqrt{r}$ characteristic. A typical result from the Henshell and Shaw paper is shown in Figure 4.24.

Later, Barsoum [1977] showed that this distribution was present along all radial lines ($\theta = $ constant) emanating from the collapsed node. He showed that the Cartesian strains could be written, in his notation, as

$$\epsilon_x = \frac{C_o}{\sqrt{r}} + \frac{d_o'}{r} + C_1 \tag{4.27a}$$

$$\epsilon_y = \frac{D_o}{\sqrt{r}} + \frac{d_o''}{r} + D_1 \tag{4.27b}$$

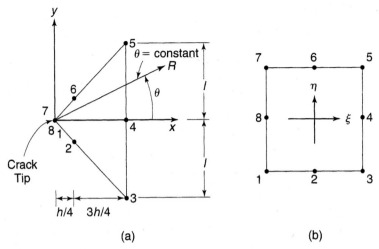

Figure 4.23 (a) Collapsed node eight-parameter quadrilateral. (b) The original form of the eight-parameter quadrilateral.

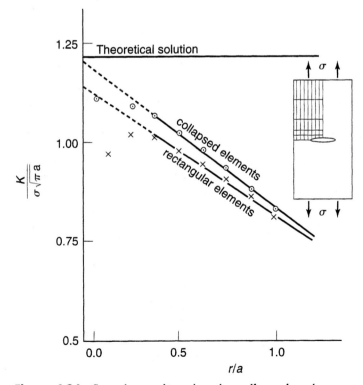

Figure 4.24 Sample results using the collapsed-node quadrilateral as compared with sample results using conventional elements [adapted from Henshell and Shaw, 1975].

He further noted that, if nodes 1, 7, and 8 in Figure 4.23 are constrained to remain equal after deformation, (i.e., $u_1 = u_7 = u_8$ and $v_1 = v_7 = v_8$), then $d'_o = d''_o = 0$, and Eqs. (4.27) reduce to

$$\epsilon_x = \frac{C_o}{\sqrt{r}} + C_1 \tag{4.28a}$$

$$\epsilon_y = \frac{D_o}{\sqrt{r}} + D_1 \tag{4.28b}$$

We note in passing that not only does the quarter-point collapsed-node element develop the required strain singularity, but it also includes a constant term that models the influence of σ_{ox} on the state of strain. Barsoum notes the significance of these constant terms for thermal analysis wherein the thermal expansion produces a pseudostrain, but makes no mention of constant strain terms in isothermal analysis.

Barsoum [1977] further noted that, if nodes 1, 7, and 8 are left unconstrained after deformation, the $1/r$ term dominates in Eqs. (4.27), and the strain-field behavior can be approximated by

$$\epsilon_x \approx \frac{d'_o}{r} + C_1 \tag{4.29a}$$

$$\epsilon_y \approx \frac{d''_o}{r} + D_1 \tag{4.29b}$$

which are characteristic of the near-tip strain distribution in perfectly-plastic materials.[7] Also, the relative motion of these initially coincident nodes emulates the observed crack-blunting behavior in elastoplastic fracture mechanics, as illustrated in Figure 4.25.

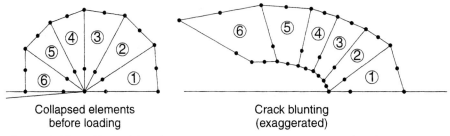

Collapsed elements
before loading

Crack blunting
(exaggerated)

Figure 4.25 Effect of leaving the crack-tip nodes unconstrained in the collapsed quadrilateral element and the resulting crack-blunting behavior.

---

[7] The nature of stress and strain fields in elastoplastic materials is described in detail in Chapter 11.

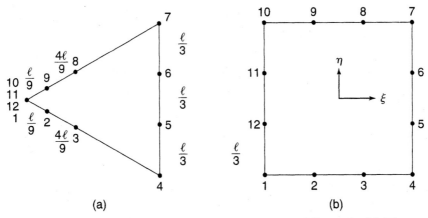

Figure 4.26 (a) Collapsed-node 12-parameter quadrilateral. (b) The original form of the 12-parameter quadrilateral.

The collapsed-node approach for developing the desired crack-tip strain-field behavior was extended to the 12-node quadrilateral isoparametric element shown in Figure 4.26 by Pu, Hussain, and Lorensen [1978]. In this case, nodes 1, 10, 11, and 12 are tied together to produce the inverse-square-root singularity, and the radial-side nodes are relocated to the $\frac{1}{9}$ and $\frac{4}{9}$ positions. This element can be used in place of the eight-node element in the best fit line method of K estimation with fewer elements required to provide the same number of data points over the three-parameter zone around the crack tip. Alternatively, Pu, Hussain, and Lorensen [1978] demonstrated a particularly simple method for extracting the stress intensity factors from collapsed-node isoparametric elements in general. They showed that the K values could be computed directly from nodal-point displacements on opposite sides of the crack plane, as shown in Figure 4.27, through the relations

$$K_I = \frac{\sqrt{2\pi}\,G(v_A - v_B)}{r_o^{1/2}(\kappa + 1)} \qquad \kappa = \frac{3 - \nu}{1 + \nu} \quad \text{plane stress} \qquad (4.30a)$$

$$K_{II} = \frac{\sqrt{2\pi}\,G(u_A - u_B)}{r_o^{1/2}(\kappa + 1)} \qquad \kappa = 3 - 4\nu \quad \text{plane strain} \qquad (4.30b)$$

In these equations, r_o is the distance from the crack tip to the first side node as shown in Figure 4.27, and G is the shear modulus. This method has the disadvantage of placing all of the emphasis on the accuracy of displacement calculations at a pair of points (or a single point in symmetric problems). On the other hand, of the methods described thus far, it is the only one that can simultaneously extract mixed-mode geometric stress intensity factors.

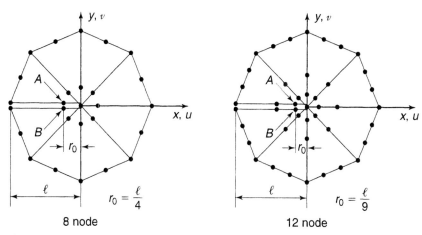

8 node   12 node

Figure 4.27   Use of nodal-point displacements to determine *K* from collapsed-node quadrilaterals.

One of the rationales for using higher order isoparametric elements over simpler elements for FEM analysis is that, since the higher order element permits a greater variation in the stress-field parameters over the element, larger elements can be used to model the problem with no loss (and possibly some gain) in accuracy. However, when the nodes are collapsed to produce the $1/\sqrt{r}$ variation, the element reflects only the near-field behavior and its size should be restricted to that of the region of validity of the near-field equations. Pu, et al. noted that the 12-node collapsed element gave good results for the problems they investigated if r_o were restricted to one to two percent of the crack length, which is comparable with the size of the singularity-dominated-zone determined by Chona, Irwin, and Sanford [1983]. Finally, in order to preserve the desired singular behavior throughout the element, it is important that the sides of the element be straight and the location of the midside nodes be precisely positioned. Otherwise, large errors of greater than 25 percent in the computed stress intensity factors are possible (Freese and Tracy [1976], Pu, et al. [1978]).

Although the collapsed-node isoparametric elements model the $1/\sqrt{r}$ characteristic of the stress–strain field in the near-field region, they do not model the known angular variations prescribed by the near-field equations. Tracy [1971] (opening mode) and Walsh [1971] (mixed mode) independently developed special crack-tip elements in which the known functional form in both r and θ of the near-field equations was used to formulate the nodal forces and displacements of the crack-tip element. However, neither included the constant stress term, σ_{ox}, in their formulation of the crack-tip element's shape function. Since the stress intensity factor(s) is (are) retained as a global parameter in the analysis, custom-written finite element programs were required to incorporate this special element into the global stiffness matrix. On the other hand, the stress intensity factor(s) is(are) determined directly from the solution of the global equation, and no additional postprocessing is required.

Subsequently, "enriched" elements were introduced by Benzley [1974] and others, in which the known singular behavior and polynomial approximations were combined within the same element. These special crack-tip elements and their associated finite element programs reduced the number of degrees of freedom required to obtain accurate results in an era during which FEM analyses were performed on large mainframe machines at relatively high cost. Today, however, the widespread use of local computing and the wide availability of general purpose FEM software, both inexpensive compared with engineering manpower costs, have all but eliminated the use of crack-tip elements for determining stress intensity factors for most applications.

Most of the methods for extracting K from two-dimensional FEM results can be extended to three-dimensional analysis if some precautions are taken. Recall that we justified two-dimensional analysis for many curved-crack problems in Chapter 2 on the basis of the marked difference in stress–strain field gradients normal to the crack versus those tangential to the crack and established the local coordinate system shown in Figure 2.7 to isolate these differences. In this coordinate system the mathematical descriptions of the stresses, strains and displacements in the ny-plane are the same as their two-dimensional counterparts in the xy-plane. Accordingly, we can apply the same three-parameter approximating scheme in Eqs. (4.26) to determine the geometric stress intensity factor as before, provided that all of the nodal points used to construct the linear region lie along a ray in the direction of the local normal to the crack front. In addition, Barsoum [1977] has shown that the twenty-node brick element shown in Figure 4.28 has the desired $1/\sqrt{r}$ singularity behavior along radial lines. As previously noted, the use of the collapsed brick element is subject to the caveat that the size of the near-tip brick is limited to the singularity-dominated-zone size. With these restrictions the analysis of three-dimensional problems involving cracks is, conceptually at least, the same as in two dimensions. The increased complexity results primarily from the more cumbersome task of constructing acceptable

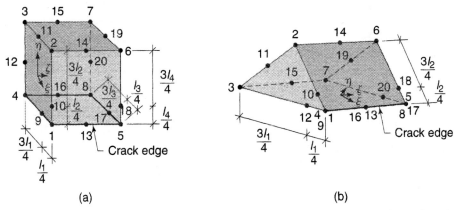

(a)   (b)

Figure 4.28 Extension of the collapsed-node approach to three-dimensional problems with the use of the 20-node brick element.

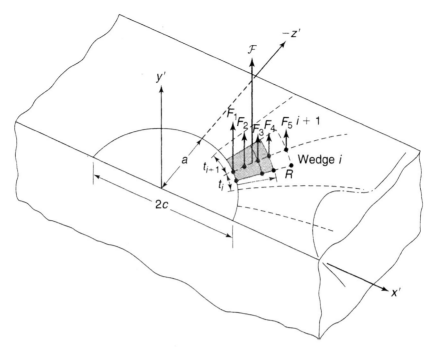

Figure 4.29 Calculation of the stress intensity factor for three-dimensional crack problems with the nodal-force method.

meshes, longer solution times, and larger output data fields to be scanned for the required nodal-point values.

Raju and Newman [1979] have pointed out that the popular crack-opening displacement approach [i.e., Eq. (4.26b)] has the disadvantage in three-dimensional applications that a state of plane strain or plane stress must be assumed. It is reasonable to approximate the local crack-tip behavior in the interior of the body as plane strain (even if the body at large does not meet the required criteria) and at the surface-breaking crack tip as plane stress. However, in the transition region, it is inappropriate to preassign the behavior to be planar when, in fact, a true three-dimensional stress state exists. To circumvent this problem, Raju and Newman proposed a variation of the three-parameter-zone method that uses nodal forces on the material side of the crack border.

Consider the finite element mesh in the neighborhood of a crack border shown in Figure 4.29. The mesh consists of thin wedges of isoparametric elements radiating from the crack border. Assuming equal contributions from both adjacent wedges, the total force, $\mathcal{F}$, normal to the crack plane acting over the shaded area bridging the wedges of elements, as shown in the figure, can be written as

$$\mathcal{F} = t_a \int_0^R \sigma_y \, dx \qquad (4.31)$$

where R is the distance from the crack front to the last nodal point included in the shaded area and t_a is the average width of the two adjacent wedges. Substituting Eq. (4.25a) into Eq. (4.31), we find that

$$\mathcal{F} = \frac{2}{\pi} K \sqrt{2\pi R} + \frac{2}{3} A_1 R^{3/2} \tag{4.32a}$$

which can be rewritten as

$$\frac{\pi}{2} \mathcal{F} \frac{1}{\sqrt{2\pi R}} \equiv K_{\text{apparent}} = K + \frac{1}{3} \sqrt{\frac{\pi}{2}} A_1 R \tag{4.32b}$$

This result is similar in form to that of Eq. (4.26b) except that, as Raju and Newman observed, no *a priori* plane conditions need be assumed. The total force, $\mathcal{F}$, is the sum of the nodal forces within the shaded area. For the shaded area shown in Figure 4.29 (i.e., four nodes within the integration region), they defined the total nodal force as

$$\mathcal{F} = F_1 + F_2 + F_3 + F_4' \tag{4.33}$$

where F_4' is only that portion of the nodal force at node 4 contributed by elements within the shaded area and connected to node 4. The remaining nodal forces (1–3) are obtained directly from the FEM output.

Raju and Newman [1979] present, as an example of the nodal-force method, the results shown in Figure 4.30 for a semicircular surface crack in remote tension. As previously noted, the data points nearest the crack tip are generally excluded from the best fit line over the three-parameter zone. In this example we can conclude that not only does the geometric stress intensity factor vary along the crack border (as determined by the intercepts of the best-fit lines) but also the A_1 parameter appears to be nearly constant all along the crack border within the body. (The slope of all of the lines is constant.) However, as the free surface is approached ($\phi = \pi/128$ in Figure 4.30), A_1 changes in both magnitude and sign.

There are numerous ways other than those described in this chapter to determine the geometric stress intensity factor from an FEM analysis. We have focused our attention on stress–strain or displacement approaches because these methods use the fracture-mechanics variables we have explored so far. The strain-energy release rate, G, introduced in Chapter 1 will be examined in detail in Chapter 7. There is a related parameter, called the *J-integral*, developed within the context of elastoplastic fracture mechanics (see Chapter 11) that has important properties which can be exploited to determine the K value. Since these concepts have not yet been introduced, we will defer until later their application to the numerical determination of stress intensity factors.

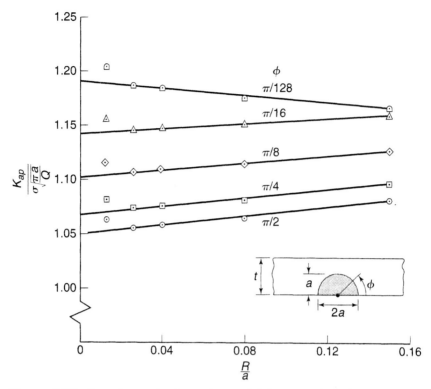

Figure 4.30 Sample results from the nodal-force method for a semicircular surface flaw [Raju and Newman, 1979].

REFERENCES

Barsoum, R. S., 1976, "On the Use of Isoparametric Finite Elements in Linear Fracture Mechanics," *Int. J. Numerical Methods in Engineering*, 10, pp. 25–37.

Barsoum R. S., 1977, "Triangular Quarter-Point Elements as Elastic and Perfectly-Plastic Crack Tip Elements," *Int. J. Numerical Methods in Engineering*, 11, pp. 85–98.

Benzley, S. E., 1974, "Representation of Singularities with Isoparametric Finite Elements," *Int. J. Numerical Methods in Engineering*, 8, pp. 537–545.

Cartwright, D. J., and Rooke, D. P., 1974, "Approximate Stress Intensity Factors Compounded From Known Solutions," *Eng. Fracture Mechanics*, 6, pp. 563–571.

Chona, R., Irwin, G. R., and Sanford, R. J., 1983, "The Influence of Specimen Size and Shape on the Singularity-Dominated Zone," *Proceedings, 14th National Symposium on Fracture Mechanics*, STP 791, Vol. 1, American Soc. for Testing and Materials, Philadelphia, pp. I1–I23.

Drude, B. T., 1993, *The General Solution to the Internal Plane Crack Problem in Westergaard Form*, M.S. Thesis, University of Maryland, College Park, MD.

Freese, C. E., and Tracy, D. M., 1976, "The Natural Isoparametric Triangle Versus Collapsed Quadrilateral for Elastic Crack Analysis," *Int. J. Fracture*, 12, pp. 767–770.

Grandt, A. F., Harter, J. A., and Heath, B. J., 1984, "Transition of Part-Through Cracks at Holes into Through-the-Thickness Flaws," *Fracture Mechanics*, STP 833, American Soc. for Testing and Materials, Philadelphia, pp. 7–23.

Gross, B., Srawley, J. E., and Brown, W. F., 1964, "Stress-Intensity Factors for a Single Edge-Notch Tension Specimen by Boundary Collocation of a Stress Function," NASA TN D–239, National Aeronautics and Space Administration, Washington, DC.

Henshell, R. D., and Shaw, K. G., 1975, "Crack Tip Finite Elements Are Unnecessary," *Int. J. Numerical Methods in Engineering*, 9, pp. 495–507.

Hulbert, L. E., 1963, *The Numerical Solution of Two-Dimensional Problems of the Theory of Elasticity*, Bulletin 198, Engineering Experiment Station, Ohio State University, Columbus, OH.

Isida, M., 1973, "Method of Laurent Series Expansion for Internal Crack Problems," *Methods of Analysis and Solutions of Crack Problems, Mechanics of Fracture*, vol. 1, G. C. Sih, ed., Noordhoff Int. Pub., Netherlands, pp. 56–130.

Kirk, M. T., and Sanford R. J., 1990, "A K_I Calibration for a Modified Single Edge Notched Tension Specimen," *J. Testing and Evaluation*, 18(5), pp. 344–351.

Kobayashi, A. S., Cherepy, R. D., and Kinsel, W. C., 1964, "A Numerical Procedure for Estimating the Stress Intensity Factor of a Crack in a Finite Plate," *J. Basic Engineering*, ASME, Series D, 86, pp. 681–684.

Longley, J. W., 1984, *Least Squares Computations Using Orthogonalization Methods*, Marcel Dekker, New York.

Newman, J. C., Jr., 1969, *Stress Analysis of Simply and Multiply Connected Regions Containing Cracks by the Method of Boundary Collocation*, MS Thesis, Virginia Polytechnic Institute and State University, Blacksburg. (See also NASA TN D–6376, National Aeronautics and Space Administration, 1971.)

Newman, J. C., Jr., 1974, "Stress Analysis of the Compact Tension Specimen Including the Effects of Pin Loading," *Fracture Analysis*, ASTM STP 560, American Soc. for Testing and Materials, Philadelphia, pp. 105–121.

Newman, J. C., Jr., and Raju, I. S., 1983, "Stress-Intensity Factor Equations for Cracks in Three-Dimensional Finite Bodies," *Fracture Mechanics: Fourteenth Symposium—Volume I: Theory and Analysis*, ASTM STP 791, J. C. Lewis and G. Sines, Eds., American Society for Testing and Materials, pp. 238–265.

Pu, S. L., Hussain, M. A., and Lorensen, W. E., 1978. "The Collapsed Cubic Isoparametric Element as a Singular Element for Crack Problems," *Int. J. Numerical Methods in Engineering*, 12, pp. 1727–1742.

Raju, I. S., and Newman, J. C., Jr., 1979, "Stress-Intensity Factors for a Wide Range of Semi-Elliptical Surface Cracks in Finite-Thickness Plates," *Eng. Fracture Mechanics*, 11, pp. 817–829.

Rooke, D. P., 1977, "Stress Intensity Factors for Cracked Holes in the Presence of Other Boundaries," *Fracture Mechanics in Engineering Practice*, Applied Science Publishers, Barking, pp. 149–163.

Rooke, D. P., 1986, "An Improved Compounding Method for Calculating Stress Intensity Factors," *Eng. Fracture Mechanics*, 23, pp. 783–792.

Sanford, R. J., and Berger, J. R., 1990, "An Improved Method of Boundary Collocation for the Analysis of Finite Body Opening Mode Fracture Problems," *Eng. Fracture Mechanics*, 37(3), pp. 461–470.

Sanford, R. J., and Kirk, M., 1991, "A Global Collocation Technique for the Solution of Opening Mode Crack Problems," *Int. J. Fracture*, 49, pp. 273–289.

Sneddon, I. N., and Lowengrub, M., 1969, *Crack Problems in the Classical Theory of Elasticity*, John Wiley, New York.

Tada, H., Paris, P. C., and Irwin, G. R., 1973, *The Stress Analysis of Cracks Handbook*, Del Research Corp., Hellertown, PA.

Tracy, D. M., 1971, "Finite Elements for Determination of Crack Tip Elastic Stress Intensity Factors," *Eng. Fracture Mechanics*, 3, pp. 255–265.

Walsh, P. F., 1971, "The Computation of Stress Intensity Factors by a Special Finite Element Technique," *Int. J. Solids Structures*, 7, pp. 1333–1342.

Williams, M. L., 1957, "On the Stress Distribution at the Base of a Stationary Crack," *J. Applied Mechanics*, 24, pp. 109–114.

EXERCISES

4.1 A wide thin plate subjected to remote uniform stress, σ, contains a keyway edge slot in the form of a rounded-end notch, as shown in Figure E4.1. Assume that the plate is very wide compared with any crack that may be present.

   (a) In the *absence* of a crack, what is the state of stress (i.e., provide specific expressions for the Cartesian stress components σ_x, σ_y, and τ_{xy}) at the root of the notch?

   (b) If there is a small crack of length, a, growing from the root of the notch, what would be a suitable *approximate* expression for the geometric stress intensity factor?

   (c) If the crack in part (b) becomes long (compared with the zone of influence of the notch), what would be a suitable *approximate* expression for the geometric stress intensity factor?

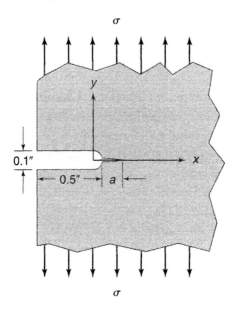

Figure E4.1

(d) Construct a finite element model of this problem, and compare your results with those of parts (a), (b), and (c) for $a = 0, 0.05$, and 0.30. Make whatever assumptions you deem appropriate.

4.2 Repeat Exercise 4.1(d) for $a = 0, 0.07$, and 0.50.

4.3 Repeat Exercise 4.1(d) for $a = 0, 0.03$, and 0.80.

4.4 Repeat Exercise 4.1(d) for $a = 0, 0.09$, and 0.90.

4.5 A finite plate with four crack-line point loads is shown in Figure E4.2.

(a) What would be a suitable set of Westergaard stress functions for this combination of geometry and loading?

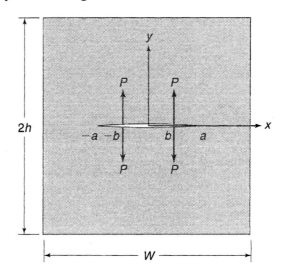

Figure E4.2

   (b) For the stress functions you selected in part (a), derive an expression for the geometric stress intensity factor at the right crack tip in terms of the unknown constants in your stress functions.

   (c) Describe the boundary conditions that must be imposed in order to solve this problem by the boundary-collocation method.

   (d) How do the required boundary conditions change if the problem is to be solved by the finite element method?

4.6 For $2h = W$ and $2a/W = 0.5$, determine the geometric stress intensity factor(s) for the problem posed in Exercise 4.5 by the boundary collocation method. Let $P/t = 500$ lb/in.

4.7 For $2h = W$ and $2a/W = 0.5$, determine the geometric stress intensity factor(s) for the problem posed in Exercise 4.5 by the finite-element method. Let $P/t = 500$ lb/in.

4.8 For $2h = 0.8W$ and $2a/W = 0.4$, determine the geometric stress intensity factor(s) for the problem posed in Exercise 4.5 by the boundary collocation method. Let $P/t = 500$ lb/in.

4.9 For $2h = 0.8W$ and $2a/W = 0.4$, determine the geometric stress intensity factor(s) for the problem posed in Excrcise 4.5 by the finite-element method. Let $P/t = 500$ lb/in.

CHAPTER 5

Experimental Methods for *K* Determination

5.1 OVERVIEW

Experimental techniques for determining the geometric stress intensity factor provide an attractive, cost-effective alternative to the approaches described in the previous chapters. Analytical methods, although highly desirable, most often require a significant investment of intellectual effort if the geometry and/or boundary conditions vary significantly from known solutions. Moreover, a great many practical problems are simply not amenable to analytical solution within the confines of the existing body of knowledge in solid mechanics.

The use of numerical techniques, particularly finite element methods, has vastly broadened the range of problems that can be solved by computational approaches. With the wide availability of general-purpose finite element analysis (FEA) codes on stand-alone computer workstations or via local area networks, the means for calculating the stress intensity factor of two-dimensional elastic bodies with well-defined boundary conditions is (or soon will be) available directly at the engineer's desk. For this class of problems the methods for extracting the stress intensity factor presented in the previous chapter, as well as other approaches to be introduced later in the book, can be employed by engineers with only a modest background in fracture mechanics, provided that reasonable precautions are taken and common-sense arguments prevail. For more complex problems, such as those involving three-dimensional geometries or poorly defined boundary conditions (e.g., contact problems), more sophisticated techniques that require special codes and significantly more computational power are needed. For these cases, the extraction of fracture mechanics parameters from the numerical codes should be attempted only by those with the specialized training and experience to properly interpret the computer results. Clearly, this situation is one case in which the adage "A little knowledge is a dangerous thing" should be taken quite literally.

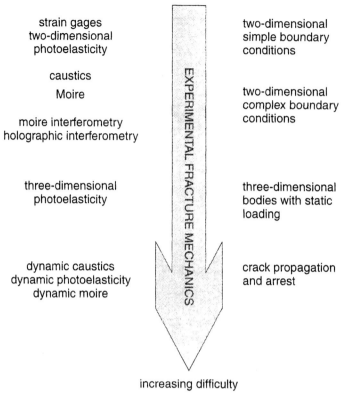

strain gages
two-dimensional
photoelasticity

caustics

Moire

moire interferometry
holographic interferometry

three-dimensional
photoelasticity

dynamic caustics
dynamic photoelasticity
dynamic moire

EXPERIMENTAL FRACTURE MECHANICS

two-dimensional
simple boundary
conditions

two-dimensional
complex boundary
conditions

three-dimensional
bodies with static
loading

crack propagation
and arrest

increasing difficulty

Figure 5.1 The spectrum of experimental methods applied to fracture problems.

Experimental techniques, ranging from the very elementary to the highly sophisticated, provide the vehicle for determining the stress intensity factor for a wide spectrum of problems, as illustrated in Figure 5.1. At the low end of the scale, two-dimensional photoelasticity or strain gages can be cost-effective alternatives to the other two approaches. They might also be used to verify results obtained by other methods. As the complexity of the problem increases, the need for (or at least the consideration of) an experimental solution increases. Finally, at the far end of the spectrum, there are some problems that can be addressed only by experimental methods. Included in this class are such problems as dynamic crack propagation and arrest behavior and in-situ measurements on flawed structures ranging from bridges to high-performance aircraft.

Although experimental techniques provide new alternatives and opportunities for solving fracture-mechanics problems, they are not a panacea. As with the other techniques and tools previously described, specialized training and experience are required in order to understand their capabilities and, more importantly, their limitations. In this chapter, we will describe a variety of experimental techniques that have been applied to fracture mechanics problems. The focus of the discussion will be on approaches for extracting the stress intensity factor from data provided by experimental techniques. As such, we will confine our attention to the mathematical relations between the experimentally measured quantities and the desired fracture mechanics parameters. Any discussion of the underlying principles of the experimental methods and the practical aspects of implementing the techniques is beyond the scope of this text. For the most part, we will limit our discussion to relatively common techniques of experimental mechanics that are described amply in suitable textbooks.

5.2 CLASSICAL PHOTOELASTIC METHODS

It is one of those curious events in the history of fracture mechanics that the first experimental technique devised specifically to determine the stress intensity factor was the result of an attempt to explain one of the most difficult problems in fracture mechanics: dynamic crack propagation. In 1958, Wells and Post presented the results of their photoelastic experiments of a running crack, but offered no interpretation in terms of the crack driving force, K. Irwin [1958] observed that the isochromatic fringes (contours of constant maximum shearing stress) near the crack tip in the Wells and Post experiment formed closed loops, as predicted by the recently formulated singular crack-tip equations, but had a distinct tilt away from the vertical, in defiance of the theoretical prediction.

Irwin reasoned that since a mathematical crack has zero thickness, a uniform stress parallel to the crack plane would have no tendency to open the crack flanks and, accordingly, would not alter the singular stress state. From this argument, he proposed that the singular near-field equations needed to be modified to include an unknown uniform stress, σ_{ox}, parallel to the crack plane, resulting in the form given by Eqs. 3.44.[1] With this modification, the governing equation describing the fringe loops in the near field can be expressed as

$$(2\tau_{max})^2 = \left(\frac{Nf_\sigma}{t}\right)^2 = \left(\frac{K}{\sqrt{2\pi r}}\sin\theta - \sigma_{ox}\sin\frac{3\theta}{2}\right)^2 + \left(\sigma_{ox}\cos\frac{3\theta}{2}\right)^2 \quad (5.1)$$

where N is the photoelastic fringe number, f_σ is the photoelastic constant for the material, and t is the optical thickness of the model.

---

[1] In Irwin's 1958 work he subtracted the σ_{ox} term from the σ_x component (with the consequence that a negative value denoted tension), and much of the early follow-on work continued to use this notation. However, in the present development, the contemporary notation, in which a tensile stress, σ_{ox}, is positive, has been adopted uniformly throughout the text. The reader is cautioned to be aware of this difference when reading the historical literature.

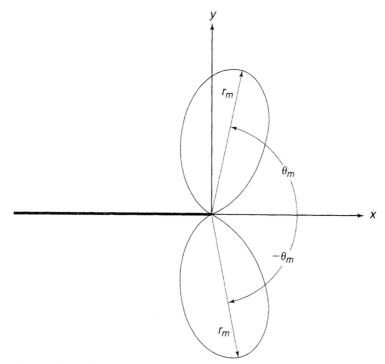

Figure 5.2 Coordinates of the apogee point in the photoelastic fringe pattern for an opening-mode crack problem.

Irwin also observed that, as shown in Figure 5.2, at the extreme point on a fringe loop (the apogee), it must be true that $\partial \tau_{\max}/\partial \theta = 0$ and

$$\sigma_{ox} = \frac{K}{\sqrt{2\pi r_m}} \frac{\sin \theta_m \cos \theta_m}{\left[\cos \theta_m (\sin 3\theta_m/2) + \frac{3}{2} \sin \theta_m \cos(3\theta_m/2) \right]} \tag{5.2}$$

where r_m, θ_m are the coordinates of the apogee point on any closed fringe loop. Evaluating Eq. (5.1) at r_m, θ_m and solving simultaneously with Eq. (5.2) provides an independent evaluation of *K* and σ_{ox} on each fringe loop. In practice, limited photoelastic sensitivity and localized three-dimensional effects very near the crack tip typically restrict the number of usable fringe loops to one or two measurements.

A more severe drawback of the apogee method is its sensitivity to small errors in θ_m. Etheridge and Dally [1977] made a systematic study of the error in the *K* measurement of due to uncertainty in θ_m. They found that for θ_m errors in the order of ±2 degrees, the variation in *K* value ranged from 3 percent at 90 degrees to 70 percent at 72 degrees. In addition, the method breaks down completely for fringe tilts of 69.4 degrees or smaller angles.

Bradley and Kobayashi [1970] proposed a differencing method that does not depend on the apogee point and its inherent uncertainty. In order to reduce the fringe equation, Eq. (5.1), to a single variable, they assumed that

$$\sigma_{ox} = -\frac{K}{\sqrt{2\pi}} \qquad (5.3)$$

We note that this is the exact value of σ_{ox} only for the case of an infinite plate with a central crack in uniaxial tension; however, for other uniaxial geometries, Bradley and Kobayashi argued that Eq. (5.3) might provide a reasonable estimate of σ_{ox}. With this assumption, Eq. (5.1) becomes

$$\tau_{max} = \frac{N f_\sigma}{2t} \approx \frac{K}{2\sqrt{2\pi r}} \cdot f\left(\frac{r}{a}, \theta\right) \qquad (5.4a)$$

where

$$f\left(\frac{r}{a}, \theta\right) = \left(\sin^2\theta + 2\sqrt{\frac{2r}{a}}\sin\theta\sin\frac{3\theta}{2} + \frac{2r}{a}\right)^{1/2} \qquad (5.4b)$$

Selecting the apogee points of two fringe loops as illustrated in Figure 5.3, the stress intensity factor can be estimated from

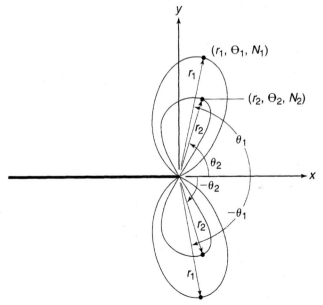

Figure 5.3   Coordinates of two apogee points for use with the differencing method to determine K.

$$K \approx \frac{\sqrt{2\pi}(N_2 - N_1)\sqrt{r_1 r_2} f_\sigma}{(f_2 \sqrt{r_1} - f_1 \sqrt{r_2})t} \tag{5.5}$$

The use of the differencing approach partially cancels the error due to the estimated value of σ_{ox}. Bradley and Kobayashi report that Eq. (5.5) provides estimates of K that compare favorably with the apogee method, but with less sensitivity (and greater freedom of choice) to angular errors.

Despite their limitations the apogee method and, to a lesser extent, the differencing method formed the basis for the analysis of dynamic photoelastic studies of polymers under a variety of conditions for almost two decades. An extensive review of these studies has been prepared by Dally [1979]. Starting in the early 1980s, the analysis of fringe patterns (both static and dynamic) based solely on the near-field equations has been replaced by more general methods to be described later in this chapter.

Concurrent with, but quite independent from, the dynamic photoelastic studies just described, C. W. Smith and coworkers [Schrodel and Smith, 1975; Smith, McGowan, and Peters, 1978; Smith, 1980] were making significant advances in the analysis of three-dimensional crack problems using the photoelastic method. Both the approach and analysis procedure were new and innovative.

Although there are several photoelastic techniques that can be applied to the three-dimensional problem, the stress-freezing method (a misnomer) is the most versatile and ideally suited to the study of surface-crack problems. In this method an epoxy model of the desired geometric proportions is fabricated and placed in an oven with a temperature programming controller. The model is slowly heated to above its transition temperature, typically 250 to 300°F, at which the material becomes rubberlike—that is, highly elastic, but low modulus. The external loads are then applied and the model slowly cooled to room temperature (1 to 2°C/hr). The unique characteristic of these materials to lock-in the birefringence (i.e., the fringe pattern) at room temperature even after the external loads are removed makes this method very useful for fracture analysis. The stress-freezing property is illustrated in Figure 5.4, which shows the through-thickness birefringence pattern in a cylindrical rod in tension with a thumbnail crack. Even more important for fracture analysis is the fact that the fringe pattern is undisturbed when the model is sliced into thin strips, provided that precautions are taken not to introduce any localized heating, which would anneal the material. Figure 5.5 shows typical slices from the model in Figure 5.4. Smith was also the first to observe that if the magnitude of the applied load is carefully controlled, slow, stable cracks would grow from a starter crack under sustained load at elevated temperature which match (or nearly so) the profile in metallic parts grown under fatigue conditions. Figure 5.6 compares the profiles of two sustained load cracks in a stress-freezing material with those of a

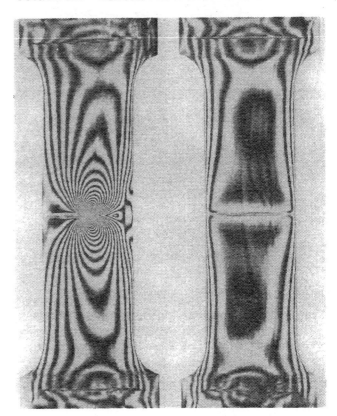

Figure 5.4 Integrated photoelasticic pattern locked into a stress-frozen circular cylinder with an edge crack after removal of the applied load.

fatigue-cracked aluminum model. Of course, only one crack profile per photoelastic model can be analyzed, due to the need to destroy the model to perform the analysis.

Once the model is sliced into thin strips along planes perpendicular to the crack front, the fringe patterns can be analyzed with the two-dimensional field equations of fracture mechanics, for the reasons discussed in Chapter 2. To simplify the analysis, Smith chose to confine his attention to the radial line $\theta = \pi/2$. This line is a logical choice, because, regardless of the fringe tilt, all of the closed fringe loops cross this line. Also, if additional data are required, classical methods of photoelasticity for point-by-point measurement of fractional fringe orders are easily obtained along a line.

Schroedl and Smith [1975] called the analysis procedure they adopted to extract the stress intensity factor from the stress-frozen patterns the *Taylor Series Correction Method* (TSCM). Although derived by a somewhat different argument, the approach is based on the observation that since the maximum shearing stress can be obtained from the difference of the principal stresses, and the principal stresses can be expressed as linear combinations of the Cartesian stresses through suitable coordinate transformation equations, the general form of τ_{max} must be expressible in power-series form, similar to the form for the Cartesian stress components. Accordingly,

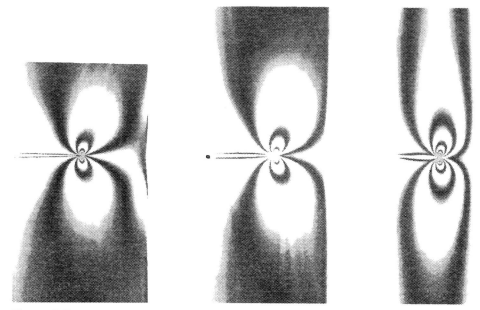

Figure 5.5 Three thin slices perpendicular to the crack front taken from the model shown in Figure 5.4.

along $\theta = \pi/2$, the maximum shearing stress must be of the form

$$\tau_{\max} = \frac{K}{\sqrt{8\pi r}} + \sum_{n=0}^{N} G_n r^{n/2} \tag{5.6}$$

where G_n are constants. Rearranging Eq. (5.6), we find that

$$\sqrt{8\pi r}\,\tau_{\max} \equiv K_{\text{apparent}} = K + \sqrt{8\pi}\,G_0\sqrt{r} + \text{H.O.T.} \tag{5.7}$$

Figure 5.6 Comparison of the sustained-load crack profiles in epoxy models with the fatigue-crack profiles in aluminum.

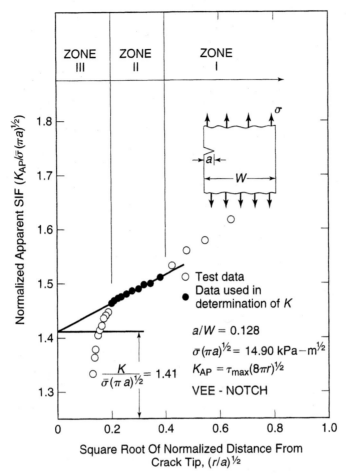

Figure 5.7 K_{apparent} versus $\left(\frac{r}{a}\right)^{1/2}$ from a typical
stress-frozen fringe pattern (courtesy C. W. Smith,
VPI&SU). Copyright (1975) Elsevier Science, reprinted
with permission.

As we observed in Chapter 4, the form of this equation suggests that the stress
intensity factor can be obtained as the y-axis intercept of a plot of the modified
maximum shearing stress (i.e., $\sqrt{8\pi r}\,\tau_{\text{max}}$) versus $\sqrt{r}$ [not r, as was the case for
Eqs. (4.26)]. A typical result, due to Smith [1975], is shown in Figure 5.7. As in the
numerical analysis case, the data can be divided into three regions. Over a limited
portion of the field, the data follow a straight line, indicating that over this region the
stress field can be represented by a two-parameter model. Beyond this region, the
additional higher order terms make significant contributions to the state of stress and
cannot be ignored. At the other end of the plot the data also deviate from linearity.
Since the stress-freezing material is truly elastic and three-dimensional effects are
suppressed by the thin slices, the behavior in this region cannot be attributed to either

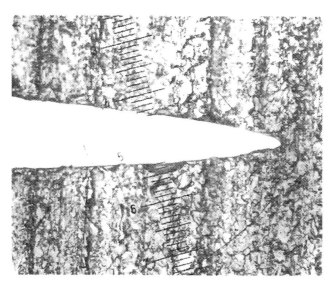

Figure 5.8 Micrograph of the natural crack profile of a stress-frozen epoxy model.

localized plasticity or antiplane variations in the stress field. Smith, McGowan, and Peters [1978] attribute the failure of the data to follow the mathematical model to the crack-blunting effect under load, due to the very low elastic modulus under stress-freezing conditions (2000 psi to 6000 psi is typical). The micrograph shown in Figure 5.8 demonstrates the crack blunting effect on a naturally grown crack under sustained load for the cylindrical rod shown in Figure 5.4.

5.3 THE METHOD OF CAUSTICS

While the focus of experimental stress analysts in the United States was on photo-elasticity for determining the stress intensity factor throughout the 1960s and 1970s, a different optical approach was being developed in Germany. First proposed by Mannog in 1964, the method of caustics, or the shadow-spot method, relies on the deflection of light rays due to stress-field gradients. Although the method is applicable to stress concentrations of any type, its primary application has been to fracture mechanics and, in particular, to dynamic fracture.

Conceptually, the method of caustics is easily understood. Consider the cross-section of a plate containing a crack depicted in Figure 5.9. (The plane of the crack is perpendicular to the plane of the page.) Since the in-plane stresses near the crack tip are both tensile, the Poisson effect produces a localized contraction (or thinning) of the material in the out-of-plane direction. This change in thickness behaves like a divergent lens and deflects the light away from its initial path, as illustrated in the figure. If we were to follow the path of light rays impinging on the face of a fracture

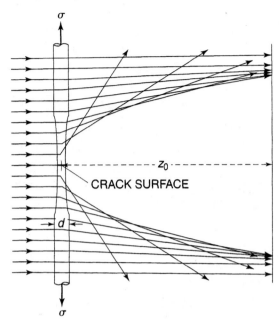

Figure 5.9 Divergent lens effect on light
rays near the tip of a crack (adapted from
Kalthoff, [1987]).

specimen as the ray moved along a radial line toward the crack tip, as shown in
Figure 5.10, we would observe that, at distances far removed from the crack tip, the
light ray is hardly affected, and the image point is the projection of the corresponding
object point. As we select rays closer to the crack tip, the path of the ray is influenced
by the surface gradient and is increasingly deflected in accordance with Snell's law. At
some point the light rays cease moving toward the crack tip and, for rays even closer to
the crack tip, the image points move away, as illustrated by the arrows on the accented
path in Figure 5.10. The concentration of light just at the points where the path of
the light ray reverses direction gives rise to a bright halo (a caustic) surrounding
a dark region around the crack tip (the shadow spot). In the mathematical sense
we observe that there is not a one-to-one correspondence between points on the
specimen plane and the image plane. We will use this observation to formulate the
governing equations for the method.

The caustic curve which is localized on the image plane is highly visible and
easily measured. There is another curve called the *initial curve* that is localized on
the specimen plane. This curve is the locus of all points on the specimen plane that
map onto the caustic curve in the image plane. Points on either side of the initial
curve map outside of the caustic curve. The initial curve cannot be seen, but its
location is critical for extraction of the stress intensity factor from the caustic image.
We will pursue this point further after we develop the theory of the caustic method.

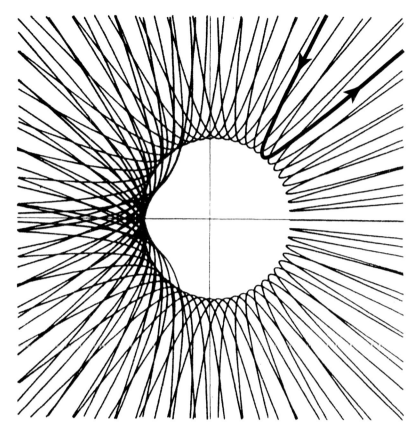

Figure 5.10 Ray mapping around a crack tip, illustrating the formation of the caustic curve (adapted from Kalthoff [1987]).

The optical system used to generate caustics is deceptively simple. Consider the optical arrangement shown in Figure 5.11 for the projection of a transparent object onto an image plane located at a distance z_o. From Fermat's principle,[2] points, P, on the object plane (the fracture specimen in this case) are mapped onto points on the image plane, P', through the relation

$$\vec{r}'(x', y') = M\vec{r}(r, \theta) - z_o \mathbf{grad}(\Delta s) \tag{5.8}$$

where M is the magnification factor for the optical arrangement ($>$, $=$, or < 1 for divergent, parallel, or convergent light, respectively) and Δs is the change in optical path length as the light traverses the object. For an optically isotropic material (for example, Plexiglas), the change in path length at normal (or near normal) incidence

---

[2]Fermat's principle states that the path a ray chooses between an entry point and an exit point is that which minimizes the transit time.

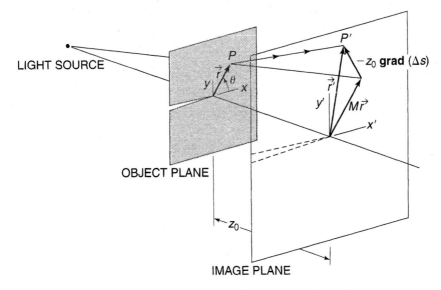

Figure 5.11 Optical system for analyzing caustic formation.

consists of two parts: the change in thickness due to Poisson contraction, and a change in refractive index, n, as the stretched material becomes optically less dense, i.e.,

$$\Delta s = (n - 1)\Delta t + \Delta n t \tag{5.9}$$

For a thin specimen in plane stress, the change in thickness is proportional to the sum of the in-plane normal stresses [Eq. (2.27)]. Similarly, from the Maxwell–Neumann stress–optic law [Coker and Filon, 1957], the change in refractive index is likewise proportional to the sum of the in-plane normal stresses. Consequently, we can write

$$\Delta s = ct(\sigma_x + \sigma_y) \tag{5.10}$$

where c is the "shadow optical constant" and incorporates both the optical and mechanical properties of the specimen material [Kalthoff, 1987]. Also, in the near field of an opening mode crack,

$$\sigma_x + \sigma_y = \frac{2K}{\sqrt{2\pi r}} \cos\frac{\theta}{2} + \sigma_{ox} \tag{5.11}$$

Finally, we can express the mapping of points in the near field of the crack tip on the specimen plane onto points in the image plane in terms of the geometric stress intensity factor, K, and the distance between the specimen and image planes, z_o, as

$$x' = Mr\cos\theta + \frac{K}{\sqrt{2\pi}} z_o c t r^{-3/2} \cos\frac{3\theta}{2} \tag{5.12a}$$

$$y' = Mr \sin \theta + \frac{K}{\sqrt{2\pi}} z_0 c t r^{-3/2} \sin \frac{3\theta}{2} \qquad (5.12b)$$

From our earlier observation that the mapping is not one-to-one, a necessary and sufficient condition for the existence of the caustic curve is that the Jacobian of the transformation, Eqs. (5.12), vanish—that is,

$$\frac{\partial x'}{\partial r} \frac{\partial y'}{\partial \theta} - \frac{\partial x'}{\partial \theta} \frac{\partial y'}{\partial r} = 0 \qquad (5.13)$$

Equations (5.12) describe generalized epicycloids in the image plane and the points in *P* that satisfy Eq. (5.13) form the initial curve on the specimen plane. This curve is a circle of radius, r_0, given by

$$r_0 = \left[\frac{3}{2} \frac{K}{\sqrt{2\pi}} z_0 c t \right]^{2/5} \qquad (5.14)$$

The shape of the caustic is characterized by its "diameters" in the longitudinal and transverse directions, as illustrated in Figure 5.12, and is uniquely related to the size of the initial curve. In general, the longitudinal diameter, D_l, is less clearly defined

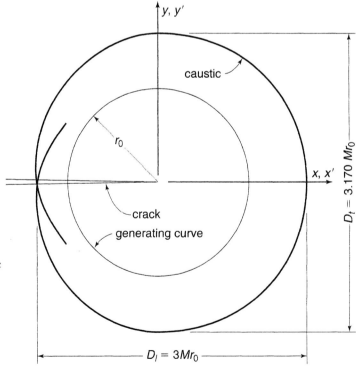

Figure 5.12 Characteristic dimensions of the initial and caustic curves for an isotropic material in transmission or an opaque material in reflection.

than is the transverse diameter, D_t, and consequently this latter dimension is usually measured to determine the stress intensity factor through the relation

$$K = \frac{2}{3}\frac{\sqrt{2\pi}}{ct}\left(\frac{D_t}{3.170M}\right)^{5/2} \tag{5.15}$$

From the foregoing analysis, it would appear that the method of caustics offers significant advantages over the photoelastic methods described earlier. In particular, we note that the constant stress term, σ_{ox}, does not contribute to the caustic curve, since only stress gradients influence the mapping. As a result, only a single measurement is required to obtain the stress intensity factor while still retaining a two-parameter representation of the near-field stress state. On the other hand, there are several factors that complicate the analysis. First, we note that the K value is related to the transverse diameter raised to the $\frac{5}{2}$ power. As a result, even small errors in measurement are amplified. This problem is compounded by the fact that the location of the caustic boundary is often blurred by imperfections in the optical system and the photographic recording process. A less obvious problem is related to the size of the initial curve relative to the dimensions of the specimen. On one hand, the shadow optical constant, c, is of the order of 10^{-10} m^2/N and, from Eq. (5.14), would result in very small initial circles relative to the specimen thickness unless suitable precautions are taken to compensate by choice of larger imaging distances, z_o. An experimental study of the influence of the optical parameters on the computed value of K has been performed by Rosakis and Ravi-Chandar [1986]. Their results, reproduced in Figure 5.13, demonstrate that for radii of the initial curve less than one half of the specimen thickness, the method of caustics underestimates the stress intensity factor. Rosakis and Ravi-Chandar attribute attribute this result to the three-dimensional stress state that exists in the immediate neighborhood of the crack tip in specimens of finite thickness (in contrast to the infinitely thin assumption of plane stress used to develop the governing equations).

On the other hand, if the size of the initial curve is allowed to be too large, the size and shape of the caustic curve becomes influenced by higher order terms in the stress-field representation ignored in Eq. (5.14). A systematic study of the distortion of the caustic curve due to the higher-order terms of up to order r^2 has been performed by Phillips and Sanford [1981]. Their results indicate that significant distortions of the caustic curve do occur, and any deviation in the ratio of the measured diameters from the classical aspect ratio of 3.170/3.000 predicted by the elementary theory should be treated as suspect.

Finally, for the case of optically anisotropic materials (i.e., photoelastic materials), the polarization effect results in a pair of caustics, such as the ones shown in Figure 5.14, rather than a single caustic. In this case the σ_{ox} term does influence the caustic diameters. Fortunately, it has been demonstrated [Phillips and Sanford, 1981]

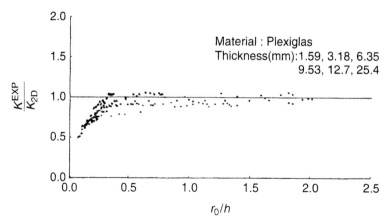

Figure 5.13 Effect of the initial curve radius as a fraction of the plate thickness on the accuracy of the determination of K by the method of caustics (courtesy A. Rosakis, California Institute of Technology). Copyright 1986 Elsevier Science, reprinted with permission.

that the average of the inner and outer diameters (both the transverse and longitudinal) is remarkably close to that of the single caustic for optically isotropic materials. As a consequence, the stress intensity factor can be determined from observations with photoelastic materials with sufficient accuracy by using the average transverse diameter in Eq. (5.15).

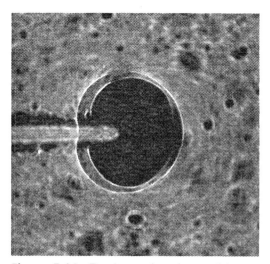

Figure 5.14 Dual caustics observed in birefringent materials.

5.4 STRAIN GAGES

Of all of the techniques of experimental mechanics, by far the most commonly used and highly developed is the use of the electrical-resistance strain gage. However, until recently, this popular measurement technique had not been employed in fracture mechanics research. The most often cited reason is the high strain gradient in the immediate neighborhood of the crack tip. Also, as a surface measurement technique, the three-dimensional effects at the crack tip were believed to be more detrimental with strain gage measurements than with optical methods, such as photoelasticity, which integrate through the thickness. Both of these problems can be eliminated by moving the measurement point(s) further from the crack tip; however, in most cases, the gage(s) would now be located outside the domain of the near-field equations. Dally and Sanford [1987] were able to overcome this obstacle by examining the state of strain in a region modestly beyond the singularity-dominated-zone wherein the strains could be described by a low-order series expansion.

Consider the strain gage placement shown in Figure 5.15—namely, a single-element strain gage located at an arbitrary point (r, θ) relative to a crack tip, with the gage oriented at an arbitrary angle, α. The state of strain in the local coordinate system, (x', y'), that is aligned along the direction of strain gage grid and the state of strain measured in the global coordinate system, (x, y), are related by

$$\epsilon_{x'x'} + \epsilon_{y'y'} = \epsilon_{xx} + \epsilon_{yy} \tag{5.16a}$$

$$\epsilon_{y'y'} - \epsilon_{x'x'} + i\gamma_{x'y'} = (\epsilon_{yy} - \epsilon_{xx} + i\gamma_{xy})e^{2i\alpha} \tag{5.16b}$$

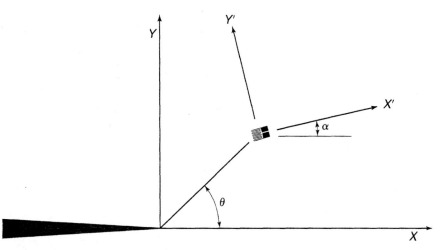

Figure 5.15 Location and orientation of a single strain gage relative to the crack tip.

Combining these equations with Eqs. (3.25), we find that exact expressions for the normal strain components in the rotated coordinate system can be written in terms of the stress functions $Z(z)$ and $Y(z)$ as

$$2\mu\epsilon_{x'x'} = \frac{1-\nu}{1+\nu}(Re\,Z + Re\,Y) - (y\,Im\,Z' + y\,Im\,Y' - Re\,Y)\cos 2\alpha$$

$$- (y\,Re\,Z' + y\,Re\,Y' + Im\,Y)\sin 2\alpha \qquad (5.17a)$$

$$2\mu\epsilon_{y'y'} = \frac{1-\nu}{1+\nu}(Re\,Z + Re\,Y) + (y\,Im\,Z' + y\,Im\,Y' - Re\,Y)\cos 2\alpha$$

$$+ (y\,Re\,Z' + y\,Re\,Y' + Im\,Y)\sin 2\alpha \qquad (5.17b)$$

If we now confine our attention to a suitably-sized region around an isolated crack tip, the stress functions of Eqs. (3.50) can be approximated as

$$Z(z) \approx \frac{K_I}{\sqrt{2\pi z}} + A_1\sqrt{z} \qquad (5.18a)$$

$$Y(z) \approx B_o \qquad (5.18b)$$

With these choices for the functions, Z and Y, Eqs. (5.17) become

$$2\mu\epsilon_{x'x'} = K_I(2\pi r)^{-1/2}\left[k\cos\left(\frac{\theta}{2}\right) - \left(\frac{1}{2}\right)\sin\theta\sin\left(\frac{3\theta}{2}\right)\cos 2\alpha\right.$$

$$+ \left(\frac{1}{2}\right)\sin\theta\cos\left(\frac{3\theta}{2}\right)\sin 2\alpha\right] + B_o(k + \cos 2\alpha)$$

$$+ A_1 r^{1/2}\cos\left(\frac{\theta}{2}\right)\left[k + \sin^2\left(\frac{\theta}{2}\right)\cos 2\alpha - \left(\frac{1}{2}\right)\sin\theta\sin 2\alpha\right] \qquad (5.19a)$$

$$2\mu\epsilon_{y'y'} = K_I(2\pi r)^{-1/2}\left[k\cos\left(\frac{\theta}{2}\right) + \left(\frac{1}{2}\right)\sin\theta\sin\left(\frac{3\theta}{2}\right)\cos 2\alpha\right.$$

$$- \left(\frac{1}{2}\right)\sin\theta\cos\left(\frac{3\theta}{2}\right)\sin 2\alpha\right] + B_o(k - \cos 2\alpha)$$

$$+ A_1 r^{1/2}\cos\left(\frac{\theta}{2}\right)\left[k - \sin^2\left(\frac{\theta}{2}\right)\cos 2\alpha + \left(\frac{1}{2}\right)\sin\theta\sin 2\alpha\right] \qquad (5.19b)$$

where $k = (1-\nu)/(1+\nu)$.

Within the region surrounding the crack tip for which the state of strain can be described by a three-parameter series representation, these equations are completely general and offer various possibilities for determining the opening mode stress intensity factor. However, by noting certain particular relationships among the parameters, we find that significant simplifications result. First, note that the B_o term ($= \sigma_{ox}/2$) can be eliminated if we require that

$$\cos 2\alpha = -k = -\frac{1 - \nu}{1 + \nu} \tag{5.20}$$

With the gage rotation angle, α, fixed by Eq. (5.20), the A_1 term can be negated if the gage is radially located such that

$$\tan\left(\frac{\theta}{2}\right) = -\cot 2\alpha \tag{5.21}$$

Substituting these trigonometric requirements into Eq. (5.19a), we find that K_I can be determined from the strain gage reading through the relation

$$K_I = E\sqrt{2\pi r}\, \epsilon_{x'x'} \cdot f(\nu) \tag{5.22}$$

where $f(\nu)$ is obtained from the first bracketed term in Eq. (5.19a).

As a consequence of the restrictions imposed on the position and orientation of the gage, a single-element strain gage, strategically located, will provide an accurate measurement of K_I when it is placed within the region describable by a three-parameter approximation to the stress state. We have previously shown in Figure 4.19 that this region is significantly larger than the singularity-dominated-zone and gage placement within this region, but, simultaneously, removed from the highest strain gradient, is practical in many geometries. We note, in particular, that the orientation and location of the gage depend only on Poisson's ratio and not on the specifics of the geometry (e.g., crack length, plate width, etc.) to which the gage is attached. As a result this method can be applied to real structures for which analytical (or numerical) expressions for the geometric stress intensity factor are not known. Table 5.1 summarizes values of θ, α, and $f(\nu)$ for typical Poisson's ratios.[3]

By logical extension the method can be extended to the measurement of mixed-mode stress intensity factors by using a pair of symmetrically placed gages that satisfy Eqs. (5.20) and (5.21), as illustrated in Figure 5.16. Taking advantage of the inherent symmetry and antisymmetry of the opening and shear deformation modes, respectively, we find that the average of the two-strain gage readings is sensitive only to K_I, and Eq. (5.22) still applies. Similarly, the difference in the strain gage readings is

---

[3] Note that there are two values of α that will satisfy Eq. (5.20); however, the oblique angle produces results that are unacceptable, for a variety of reasons.

Table 5.1

Poisson's ratio, ν	θ (degrees)	α (degrees)	$f(\nu)$
0.250	73.74	63.43	0.6143
0.300	65.16	61.29	0.6443
0.333	60.00	60.00	0.6498
0.400	50.76	57.69	0.6322
0.500	38.97	54.74	0.5587

proportional only to the shear mode of deformation and provides information from which K_{II} can be determined. The details of the algebra have been presented by Dally and Berger [1993].

In some applications of laboratory testing or field measurements, the fracture specimen is subjected to a temperature gradient. If the gradient is large or the self-temperature compensation of the gage is inadequate, an alternative measurement scheme is suggested. In these instances a stacked rectangular rosette with the pair of gages in adjacent arms of the Wheatstone bridge will provide full temperature

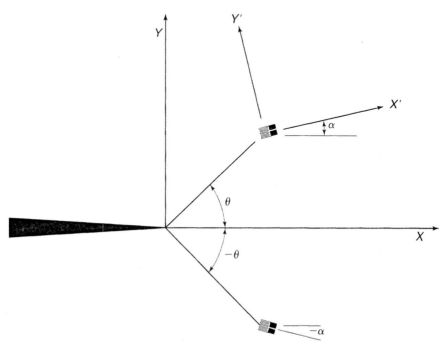

Figure 5.16   A pair of symmetrically placed strain gages for mixed-mode analysis.

compensation, and the bridge output will be proportional to $\epsilon_{y'y'} - \epsilon_{x'x'}$. From Eqs. (5.19), we then have

$$2\mu(\epsilon_{y'y'} - \epsilon_{x'x'}) = K_I(2\pi r)^{-1/2}\sin\theta\left[\sin\left(\frac{3\theta}{2}\right)\cos 2\alpha - \cos\left(\frac{3\theta}{2}\right)\sin 2\alpha\right]$$

$$- 2B_o\cos 2\alpha + A_1 r^{1/2}\sin\theta\left[\cos\left(\frac{\theta}{2}\right)\sin 2\alpha - \sin\left(\frac{\theta}{2}\right)\cos 2\alpha\right] \tag{5.23}$$

In this case the influence of B_o can be eliminated by rotating the rosette pair by $\pi/4$ with respect to the crack plane. However, unlike the previous configuration, there is no rosette orientation angle, θ, that will remove the influence of the A_1 term while retaining a meaningful expression for K_I.

Since more than one rosette will be necessary to extract an accurate estimate of the stress intensity factor, let us somewhat arbitrarily select $\theta = \pi/2$ and consider a line of gages perpendicular to the crack tip. With this choice, Eq. (5.23) becomes

$$\mu(\epsilon_{y'y'} - \epsilon_{x'x'}) = \frac{\sqrt{2}}{4}K_I(2\pi r)^{-1/2}\left[1 + \left(\frac{A_1}{A_o}\right)r\right] \tag{5.24}$$

which can be rewritten as

$$K_{\text{apparent}} = K_I\left[1 + \left(\frac{A_1}{A_o}\right)r\right] \tag{5.25}$$

where

$$K_{\text{apparent}} = \mu(\epsilon_{y'y'} - \epsilon_{x'x'}) \cdot \sqrt{2\pi r} \tag{5.26}$$

Equations (5.25) and (5.26), are similar to those used to extract an accurate estimate of the K value from the value of stress or displacement obtained with finite element models, and the procedure is the same here; namely, a plot of K_{apparent} versus r for the line of gages is constructed, and a best fit straight line is placed through the linear region of the data. The true K_I value is then the intercept of the best fit line with the origin, as shown schematically in Figure 5.17.

The results of the preceding paragraphs can be extended to measuring the propagation toughness, K_{ID}, of a running crack by following the same analysis procedure. It can readily be shown that, even when the dynamic equivalents of Eqs. (5.19) are employed, Eq. (5.20) still holds, and the influence of the constant stress term B_o can be eliminated from a strain gage's response by suitable rotation of the gage with respect to the crack propagation plane. However, unlike the static case, the gage position relative to the crack tip changes as the crack passes the fixed gage location. In this case the strain recorded by the gage rises and falls in a well-defined manner during the crack propagation event.

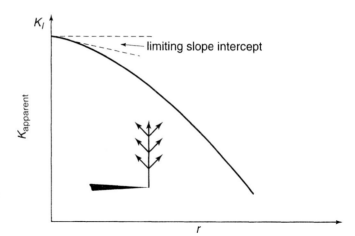

Figure 5.17 Use of a line of strain-gage rosettes to determine K in temperature-gradient fields.

To within the same three-parameter representation used earlier, the strain gage output for a properly rotated gage can be expressed as

$$2\mu\epsilon_{x'x'} \;=\; \frac{K}{\sqrt{2\pi}}\,f_0(x,c) \;+\; \Lambda_1 f_1(x,c) \tag{5.27}$$

where the functions f_0 and f_1 depend on the crack velocity, c, and the position of the crack tip, x.[4]

In this case, the value of the A_1 coefficient must be known before the dynamic stress intensity factor can be extracted from the strain–time trace. Fortunately, this information can be unambiguously obtained from the strain record itself by a straightforward procedure. Representative traces for a typical case are shown in Figure 5.18 for differing values of A_1/A_o. Note that the width of the trace changes in consort with the A_1/A_o ratio. This fact, along with the magnitude of the peak strain, is sufficient to determine the propagation toughness at the instant the crack passes under the gage location.

In a typical experiment a row of strain gages, rotated at the proper angle, is placed a fixed distance above the projected crack plane. (A height equal to one plate thickness is often used to avoid three-dimensional effects.) A sample result for a series of six equally spaced strain gages on a six-inch-wide single-edge notched plate of $\frac{1}{2}$-inch-thick 7075-T651 aluminum is shown in Figure 5.19. From these data, an average K_{ID} of 50 ksi-in$^{1/2}$ (55 Mpa-m$^{1/2}$) at a velocity of 900 m/s was obtained.

---

[4]The derivations of these equations are beyond the scope of this book; however, they are available in full detail in a 1990 paper by Sanford, Dally, and Berger.

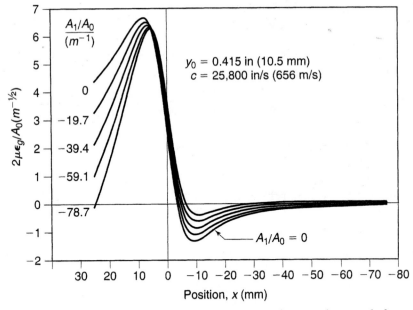

Figure 5.18 Characteristic strain-time traces of a running crack for various A_1/A_0 ratios.

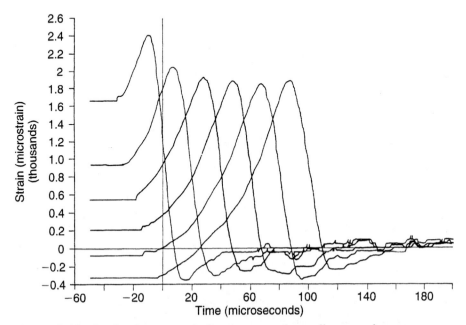

Figure 5.19 Strain-time records from a row of equally spaced gages parallel to the path of a running crack.

5.5 MULTI-PARAMETER FULL-FIELD METHODS: LOCAL COLLOCATION

All of the approaches described in the preceding sections share a common feature, namely that the analysis procedure used to extract the stress intensity factor from the experimental data relied on either the modified near-field equations or, at most, a three-parameter approximation to the state of stress or strain. In principle, the accuracy of these methods should improve as the region of data acquisition is reduced to smaller regions around the crack tip. However, as we have previously seen, there is a limit as to how close to the crack tip data can be reliably taken, because of the three-dimensional nature of the stress state in specimens of finite thickness. In addition, the near-field stress state is influenced by crack blunting and localized plasticity. Finally, the analysis procedures were based on the assumption that only a very limited number of data points are available (or have interpretive meaning). In the case of full-field optical methods (e.g., photoelasticity), this assumption means that the essentially unlimited number of potential data points are discarded either because they fail to meet a predefined criteria (as is the case for the apogee method) or they fall outside the domain of the governing equation (e.g., Smith's TSCM approach). Fortunately, there are methods to overcome these limitations, at the expense of increased algebraic complexity and data acquisition demands. With the advent of low-cost image-processing equipment interfaced to desktop computers that contain symbolic manipulation languages, these arguments against full-field algorithms cease to be relevant.

The new methods for determining the stress intensity factor from full-field optical patterns take advantage of the wealth of information contained in the pattern to improve the accuracy of the analysis and purposely exclude those regions in which the validity of the data may be suspect. In order to use this additional information, the governing equation that describes the optical behavior must be sufficiently general to include all of the factors that influence the observed experimental pattern. At a minimum, these methods abandon the near-field approximation and instead rely on a truncated form of the general series solution for the class of problem under analysis, containing unknown coefficients to describe the pattern. Determining these coefficients, which include the parameters of interest (K, σ_{ox}, etc.), is the goal of the analysis. To implement the method, we write a set of equations that relate the defining parameter of the optical pattern (usually described in terms of interference fringe numbers of either optical or mechanical origin) over an expanded region around the crack tip to the general expression involving unknown coefficients. We then solve these equations in the least-squares sense to obtain estimates of the coefficients. The number of terms retained in the series depends on the size of the region of data acquisition. Convergence of the series coefficients as the number of retained terms is increased can be used as a solution criterion. This approach is analogous to the boundary collocation method, except that the number of terms necessary to obtain an accurate estimate of the stress intensity factor is dramatically reduced, since the state of stress in the region around a crack is dominated by the lower-order coefficients. Because of the mathematical similarities between this approach and the purely numerical method of boundary collocation, the method is called the *local collocation* method.

Nearly every full-field optical stress analysis method has at one time or other been used to study stress fields in the neighborhood of a crack tip. In the upcoming paragraphs, we will develop the mathematical formulation of the local collocation method for the more prominent full-field methods. Following the theme developed in the preceding chapters, we will use the generalized Westergaard formulation to relate the observed experimental pattern to the underlying two-dimensional stress state. Also, following the earlier practice, only the opening mode of crack extension will be considered, but, by logical extension, the formulation of the mixed-mode problem can be obtained by superposition; in fact, the first application of the local collocation method was to mixed-mode photoelastic patterns [Sanford and Dally, 1979].

5.6 INTERFERENCE PATTERNS

Interferometric methods have been used for fracture studies by only a few investigators, and the analysis procedures used to extract the stress intensity factor have followed classical approaches. This situation is unfortunate, since the implementation of the local collocation method for this class of optical patterns is particularly simple and results in a system of linear equations that is easily solved. Although obtained by significantly different procedures, the governing fringe equations for classical two-beam interferometers and holographic interferometry are identical. In both cases, the fringes are contour lines of constant sum of the in-plane normal stresses, called *isopachics*. For opening mode crack behavior the shape of the fringes is similar to that of an optical caustic, except, of course, that there are many fringes in the interference pattern, as opposed to a single caustic curve. A typical double-exposure holographic interferogram of the isopachic pattern in a single-edge notched Plexiglas sample is shown in Figure 5.20.

The governing optical equation can be written in terms of Westergaard stress functions for the opening mode from Eqs. (3.24) as

$$\frac{N f_p}{t} = \sigma_x + \sigma_y = 2 Re\, Z + 2 Re\, Y + N_o \tag{5.28}$$

where N is the fringe order, f_p is the optical sensitivity constant, t is the thickness of the model, and N_o is an additional parameter that accounts for any initial phase shift in the interference pattern, particularly common in classical interferometers. Since the success of the local collocation method depends on matching the observed fringe to an analytical expression, it is vital that any non-stress-related factors that affect the experimental pattern be included in the mathematical formulation. As a

Figure 5.20 Double-exposure
holographic-interference
isopachic pattern in Plexiglas.

practical matter, the constant N_o can be interpreted as a fringe-order shift factor to compensate for the lack of an absolute fringe-order reference. Including N_o as a variable in the analysis means that only the relative order of the fringes needs to be maintained, rather than their absolute value, for the algorithm to work.

Because this method is a collocation method, the selection of suitable stress functions Z and Y is not arbitrary. The form of the stress functions must be appropriate for the class of problems being investigated. However, since the region of collocation is confined to a region surrounding the crack tip, the classical crack-tip expansion form for Z and Y given by Eqs. (3.50) is often chosen and will be used here as an example of the application of the approach. Depending on the nature of the problem, others of the stress functions described in Chapter 4 might be more appropriate. For the Williams-type expansions given by Eqs. (3.50), the governing optical equation in real-variables centered at the crack tip becomes

$$\frac{N f_p}{t} = \sum_{n=0}^{\infty} 2 A_n r^{n-1/2} \cos\left(n - \frac{1}{2}\right)\theta + \sum_{m=1}^{\infty} 2 B_m r^m \cos m\theta + N_o \qquad (5.29)$$

Note that the summation of the B_m series starts at $m = 1$ rather than $m = 0$, as in Eqs. (3.50). The inclusion of $m = 0$ results in a term of the same functional form as N_o (i.e., the constant B_o), and these two constants must be merged into one. As a consequence, the local collocation algorithm cannot distinguish between these constants and B_o (or, equivalently, σ_{ox}) cannot be uniquely determined from interferometric patterns.

Since Eq. (5.29) must be valid at every point in the field, substitution of the coordinates (r, θ) and the relative fringe order at any convenient point results in a linear equation in the unknown coefficients A_n, B_m, and N_o. For a large number of such points, an overdetermined system of equations is obtained that can be solved only in the least-squares sense. In matrix notation these equations are of the form

$$[N] = [S][C] \tag{5.30}$$

where $[N]$ is the row vector of measured relative fringe orders, [i.e., the left side of Eq. (5.29)], $[S]$ is the matrix of coefficients of the unknowns, and $[C]$ is the row vector of unknown coefficients, $[A_n, B_m, N_o]^T$. As previously shown in Eq. (4.6), the solution of this set of equations in the least squares sense is

$$[C] = [S^T S]^{-1} [S]^T [N] \tag{5.31}$$

In this case, the stress intensity factor is obtained from the first element in $[C]$, since $A_o = K\sqrt{2\pi}$.

5.7 MOIRE PATTERNS

The displacement patterns obtained with classical moire or the higher sensitivity moire interferometry can be analyzed with the linear algorithm just described once the governing optical equation is formulated. For grating lines parallel to the Cartesian axes the appropriate equations are

$$N_u = \frac{u}{f_d} + \text{R.B.M.} \tag{5.32a}$$

$$N_v = \frac{v}{f_d} + \text{R.B.M.} \tag{5.32b}$$

where N_u and N_v are the fringe counts for the u- and v-field moire patterns, respectively; f_d is the fringe sensitivity; and R.B.M. represents the additional non-strain-related motion due to rigid body motion. This term is particularly important in moire analysis, since the u and v displacement equations derived from Eqs. (3.26) assume that the forces are applied symmetrically about the crack plane and that the origin stays fixed. In practice, the fixed point is usually one of the remote load points, and the crack plane undergoes both a rigid body translation and rotation. This non-strain-related displacement can be expressed as

$$\text{R.B.M.} = Pr\cos\theta + Qr\sin\theta + R \tag{5.33}$$

where P, Q, and R are constants to be determined. As in the previous case, the constant term R can also be interpreted as a fringe shift factor, and the moire fringe numbering can again be relative.

As in the previous example, the fringe equations in real variables for a Williams-type series expansion are:

$$N_u E f_d = \sum_{n=0}^{\infty} A_n \frac{r^{n+1/2}}{n+1/2} \left[(1-v) \cos\left(n+\frac{1}{2}\right)\theta \right.$$

$$\left. - (1+v)\left(n+\frac{1}{2}\right) \sin\theta \sin\left(n-\frac{1}{2}\right)\theta \right]$$

$$+ \sum_{m=0}^{\infty} B_m \frac{r^{m+1}}{m+1} \left[2\cos(m+1)\theta - (1+v)(m+1)\sin\theta\sin(m\theta) \right]$$

$$+ Pr\cos\theta + Qr\sin\theta + R \tag{5.34a}$$

$$N_v E f_d = \sum_{n=0}^{\infty} A_n \frac{r^{n+1/2}}{n+1/2} \left[2\sin\left(n+\frac{1}{2}\right)\theta - (1+v)\left(n+\frac{1}{2}\right)\sin\theta\cos\left(n-\frac{1}{2}\right)\theta \right]$$

$$+ \sum_{m=0}^{\infty} B_m \frac{r^{m+1}}{m+1}\left[(1-v)\sin(m+1)\theta - (1+v)(m+1)\sin\theta\cos(m\theta) \right]$$

$$+ Pr\cos\theta + Qr\sin\theta + R \tag{5.34b}$$

Figure 5.21 shows an example of the application of the local collocation method to moire patterns. On the left is the experimental u-field displacement pattern for a

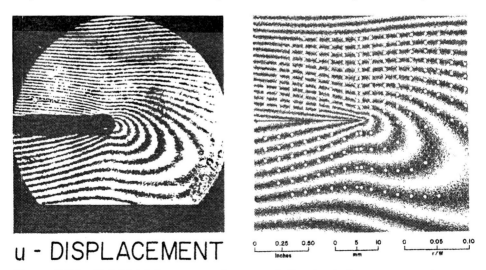

u - DISPLACEMENT

Figure 5.21 U-field Moire pattern for a crack-line-loaded compact-tension specimen (experimental on the left, and computer-generated reconstruction on the right).

crack-line-loaded, compact-tension specimen, and on the right is a computer reconstruction based on a six-term $(A_o, A_1, A_2, B_o, B_1, B_2)$ expansion of the displacement field plus the rigid body motion terms. The dark broad band in the experimental pattern was caused by damage to the grating when the simulated crack was machined into the specimen. The white dots on the reconstruction indicate the locations of the data points used in the local-collocation analysis. In this example, the experimentally measured stress intensity factor was within one percent of that calculated according to ASTM Standard E 561 (1981) for the same applied load. Further details of these results are presented in Barker, Sanford, and Chona [1985].

5.8 PHOTOELASTICITY

The photoelastic method is by far the most commonly used full-field technique for studying fracture behavior. The numerous studies of dynamic crack-propagation behavior in polymers undertaken in the 1970s and 1980s were conducted photoelastically, because of the birefringent properties of nearly all polymeric materials [Dally, 1979]. It is natural, then, that the earliest local collocation algorithms were developed for this application.

In contrast to the previously described optical methods, the local collocation solution scheme for the photoelastic method results in a set of nonlinear equations to be solved, and, accordingly, the approach is somewhat different than that previously described. On the other hand, the simplicity of a photoelastic experiment compensates for the added complexity of the analysis. Moreover, experience has shown that the characteristic features of the photoelastic fringe pattern are highly sensitive to the nonsingular terms in the series representation of the fringe field. As a consequence, the collocation algorithm is able to separate singular fields from nonsingular behavior with relative ease.

As in the previous examples, the starting point for the analysis is a proper description of the optical phenomenon in terms of the underlying stress state and other experimental variables, which, for photoelasticity, can be expressed as

$$\left(\frac{Nf_\sigma}{2t}\right)^2 = (\tau_{max})^2 = \left(\frac{\sigma_y - \sigma_x}{2}\right)^2 + \tau_{xy}^2 = D^2 + T^2 \qquad (5.35a)$$

where

$$D = yImZ' + yImY' - ReY \qquad (5.35b)$$

$$T = -yReZ' - yReY' - ImY \qquad (5.35c)$$

Unlike the other experimental methods, the optical equation for photoelasticity does not need to be modified to incorporate additional experimental effects in order to completely describe the fringe pattern.

In parallel with the earlier developments the stress functions of Eqs. (5.35b) and (5.35c) for a Williams-type series representation of the stress state in real variables become

$$D = \sum_{n=0}^{\infty} \left(n - \frac{1}{2} \right) A_n r^{n-1/2} \sin \theta \sin \left(n - \frac{3}{2} \right) \theta$$

$$+ \sum_{m=0}^{\infty} B_m r^m [m \sin \theta \sin(m\theta) + \cos(m\theta)] \qquad (5.36a)$$

$$T = -\sum_{n=0}^{\infty} \left(n - \frac{1}{2} \right) A_n r^{n-1/2} \sin \theta \cos \left(n - \frac{3}{2} \right) \theta$$

$$- \sum_{m=0}^{\infty} B_m r^m [m \sin \theta \cos(m\theta) + \sin(m\theta)] \qquad (5.36b)$$

Substituting Eqs. (5.36) into quadratic equation Eq.(5.35a) results in a governing optical equation at each selected data point that is nonlinear in the unknown constants A_n and B_m, and the solution scheme used for the previously described methods does not apply. Fortunately, there is a relatively simple Newton–Raphson method for solving systems of nonlinear algebraic equations in the least-squares sense that has proven to be effective when the number of unknowns is small and system is mathematically well behaved. This is the situation for many of the fringe patterns commonly occurring in fracture analysis. To implement the method, we need to recast the optical equation Eq. (5.35a) in the modified form

$$g_k = D_k^2(A_n, B_m) + T_k^2(A_n, B_m) - \left(\frac{N_k f_\sigma}{2t} \right)^2 = 0 \qquad (5.37)$$

where the subscript k denotes the value of the function evaluated at a point in the fringe field (r_k, θ_k, N_k). Taking a Taylor's-series expansion of Eq. (5.37) and retaining only the linear terms yields an iterative equation of the form

$$g_k^{i+1} = g_k^i + \frac{\partial g_k^i}{\partial A_o} \Delta A_o + \frac{\partial g_k^i}{\partial A_1} \Delta A_1 + \dots + \frac{\partial g_k^i}{\partial B_o} \Delta B_o + \frac{\partial g_k^i}{\partial B_1} \Delta B_1 + \dots \quad (5.38)$$

where the superscript i denotes the iteration step and ΔA_o, ΔA_1, ..., ΔB_o, ΔB_1, ... are corrections to the previous estimates of the unknowns A_o, A_1, ..., B_o, B_1, ..., respectively. Since, from Eq. (5.37), the desired result is $g_k^{i+1} = 0$, we obtain a system of linear equations (one for each data point k) in the correction terms ΔA_n and ΔB_m of the form

$$-g_k^i = \frac{\partial g_k^i}{\partial A_o}\Delta A_o + \frac{\partial g_k^i}{\partial A_1}\Delta A_1 + \cdots + \frac{\partial g_k^i}{\partial B_o}\Delta B_o + \frac{\partial g_k^i}{\partial B_1}\Delta B_1 + \cdots \quad (5.39)$$

which can be written in matrix form as

$$[-g] = \left[\frac{\partial g}{\partial A}\middle|\frac{\partial g}{\partial B}\right][\Delta] = [c][\Delta] \quad (5.40)$$

where the now redundant superscript i has been omitted. We note in passing that the partial derivatives in the matrix of derivatives, $[c]$, are not as complex to compute as might first appear. Since

$$\frac{\partial g_k}{\partial A_n} = 2D_k\frac{\partial D_k}{\partial A_n} + 2T_k\frac{\partial T_k}{\partial A_n} \quad \text{and} \quad \frac{\partial g_k}{\partial B_m} = 2D_k\frac{\partial D_k}{\partial B_m} + 2T_k\frac{\partial T_k}{\partial B_m} \quad (5.41)$$

and D_k and T_k are linear functions of the differentiation variables, A_n and B_m, the required partial derivatives can be obtained by inspection from Eqs. (5.36).

The functional form of matrix equation Eq. (5.40) is the same as that used previously, and its solution, in the least-squares sense, is

$$[\Delta] = [c^T c]^{-1}[c]^T[-g] \quad (5.42)$$

However, unlike the previous cases, the solution does not yield the desired coefficients, but rather the change in the unknown coefficients from the prior estimate. As a result, an iterative procedure must be used to obtain the best-fit set of coefficients. Experience indicates that the convergence rate is very rapid, typically with fewer than 10 steps required to obtain stable values for $K (= \sqrt{2\pi} A_o)$ and $\sigma_{ox} (= 2B_o)$. Occasionally, the solution will converge to an erroneous solution, if the initial estimates used to start the iteration process are far removed from their correct values, as shown by the computer-generated fringe patterns in Figure 5.22 [Sanford and Chona, 1981]. These false solutions can readily be identified by their high RMS error, as illustrated in Figure 5.23.

As an example of the application of the photoelastic local collocation method to determine the stress intensity factor, a series of photoelastic experiments was performed on a model of the three-point bend geometry shown in Figure 5.24. This geometry is commonly used for standardized fracture toughness testing.[5]Figure 5.25

---

[5]A thorough discussion of standardized testing in fracture mechanics is presented in Chapter 8.

RDCB SPECIMEN (a/w = 0.40)

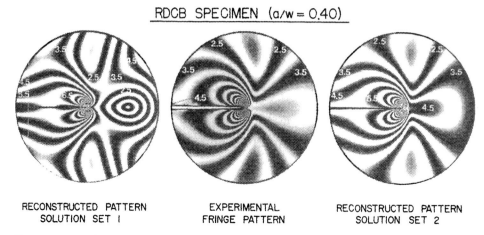

RECONSTRUCTED PATTERN
SOLUTION SET I

EXPERIMENTAL
FRINGE PATTERN

RECONSTRUCTED PATTERN
SOLUTION SET 2

Figure 5.22 Comparison of two computer-generated solutions for the same set of photoelastic data (false on the left, and true on the right).

shows the portion of the experimental pattern used for data acquisition, the data set extracted from it, and a computer-generated reconstruction of the fringe pattern based on a six-term approximation to the state of stress within the region under analysis. A comparison of the results from the local collocation analysis of this geometry with the result of a carefully performed boundary collocation analysis of the same problem [Srawley, 1976] is shown in Figure 5.26. The boundary collocation analysis

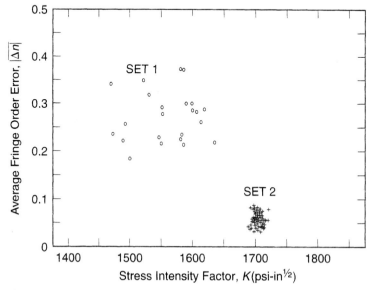

Figure 5.23 Identification of the false solution, based on high RMS error zones in sampled least-squares *K* solutions (Sanford and Chona, 1981).

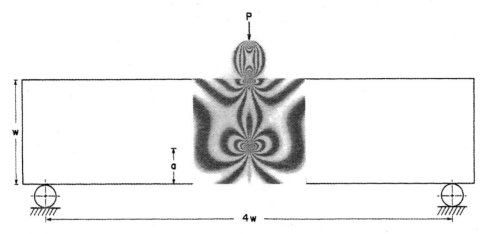

Figure 5.24 Geometry of the three-point bend specimen with photoelastic-pattern insert for one crack length.

is reported to have taken over one year to complete, whereas the local collocation analysis required less than one month. Of course, the boundary collocation analysis yielded much more information than did the photoelastic analysis, but, depending on the level of detail required, the photoelastic method provides a cost-effective alternative to the numerical analysis.

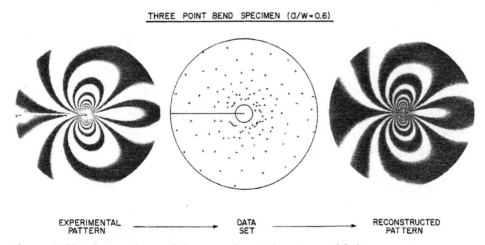

Figure 5.25 Comparison of the experimental pattern with its computer-reconstructed counterpart, based on a local collocation analysis using the data set shown for $\frac{a}{W} = 0.6$.

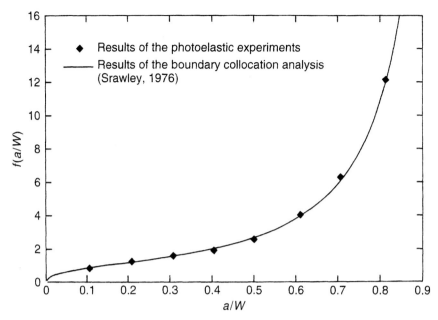

Figure 5.26 Comparison of the results from the photoelastic analysis of the three-point bend geometry with those of the boundary-collocation analysis for the full range of crack lengths.

REFERENCES

Barker, D. B., Sanford, R. J., and Chona, R., 1985, "Determining *K* and Related Stress-Field Parameters from Displacement Fields," *Experimental Mechanics*, 25(12), pp. 399–407.

Bradley, W. B., and Kobayashi, A. S., 1970, "An Investigation of Propagating Cracks by Dynamic Photoelasticity," *Experimental Mechanics*, 10(3), pp. 106–113.

Coker, E. G., and Filon, L. N. G., 1957, *A Treatise on Photoelasticity*, 2nd ed., Cambridge University Press, London, pp. 187–210.

Dally, J. W., 1979, "Dynamic Photoelastic Studies of Fracture," *Experimental Mechanics*, 19(10), pp. 349–361.

Dally, J. W., and Sanford, R. J., 1987, "Strain-Gage Methods for Measuring the Opening-Mode Stress-Intensity Factor, K_I," *Experimental Mechanics*, 27(4), pp. 381–388.

Dally, J. W., and Berger, J. R., 1993, "The Role of the Electrical Resistance Strain Gauge in Fracture Research," *Experimental Techniques in Fracture*, J. S. Epstein, Ed., Soc. Experimental Mechanics, Bethel, CT, pp. 1–37.

Etheridge, J. M., and Dally, J. W., 1977, "A Critical Review of Methods for Determining the Stress Intensity Factors From Isochromatic Fringes," *Experimental Mechanics*, 17(7), pp. 248–254.

Irwin, G. R., 1958, "Discussion of: The Dynamic Stress Distribution Surrounding a Running Crack—A Photoelastic Analysis," *Proceedings of the SESA*, 16(1), pp. 93–96.

Kalthoff, J. F., 1987, "Shadow Optical Method of Caustics," *Handbook on Experimental Mechanics*, Prentice-Hall, Englewood Cliffs, NJ, pp. 430–500.

Mannog, P., 1964, *Anwendungen der Schattenoptik zur Untersuchung des Zerreißvorgangs von Platten*, Ph.D. Dissertation, Ernst-Mach Institut, Freiburg, Germany.

Phillips, J. W., and Sanford, R. J., 1981, "Effect of Higher-Order Stress Terms on Mode-I Caustics in Birefringent Materials," *Fracture Mechanics: Thirteenth Conference*, ASTM STP 743, Richard Rogers, Ed., pp. 387–402.

Rosakis, A. J., and Ravi-Chandar, K., 1986, "On Crack-Tip Stress State: An Experimental Evaluation of Three-Dimensional Effects," *Int. J. Solids Structures*, 22(2), pp. 121–134.

Sanford, R. J., and Chona, R., 1981, "Analysis of Photoelastic Fracture Patterns with a Sampled Least Squares Method," *Proceeding, Spring Meeting*, Soc. for Experimental Mechanics, May 1981, Dearborn, MI, pp. 273–276.

Sanford, R. J., and Dally, J. W., 1979, "A General Method for Determining Mixed-Mode Stress Intensity Factors From Isochromatic Fringe Patterns," *Engineering Fracture Mechanics*, 11, pp. 621–633.

Sanford, R. J., Dally, J. W., and Berger, J. R., 1990, "An Improved Strain Gauge Method for Measuring K_{ID} for a Propagating Crack," *J. Strain Analysis*, 25(3), pp. 177–183.

Schroedl, M. A., and Smith, C. W., 1975, "A Study of Near and Far Field Effects in Photoelastic Stress Intensity Determination," *Eng. Fracture Mechanics*, 7, pp. 341–355.

Smith, C. W., McGowan, J. J., and Peters, W. H., 1978, "A Study of Non-Linearities in Frozen Stress Fields," *Exp. Mechanics*, 18, pp. 309–315.

Smith, C. W., 1980, "Stress Intensity and Flaw-Shape Variations in Surface Flaws," *Exp. Mechanics*, 20, pp. 126–133.

Srawley, J. E., 1976, "Wide Range Stress Intensity Factor Expressions for ASTM E–399 Standard Fracture Toughness Specimens," *Int. J. Fracture*, 12, pp. 475–486.

Wells, A. A., and Post, D., 1958, "The Dynamic Stress Distribution Surrounding a Running Crack—A Photoelastic Analysis," *Proceedings of the SESA*, 16(1), pp. 69–92.

EXERCISES

5.1 In addition to the contours of maximum shear stress $\tau_{\max}$, the photoelastic method provides contours of the constant principal stress direction, θ, called *isoclinics*. When used together, these two fringes determine τ_{xy}.

(a) Develop a linear local collocation method for experimentally determining the stress intensity factor from photoelastic models that uses information from both types of fringe patterns. Assume an edge-crack configuration and opening mode loading.

(b) Demonstrate the utility of your method with experimental data. Alternatively, the data can be simulated from a finite element analysis.

5.2 Shown in Figure E5.1[6] is a dark-field isochromatic pattern of a large plate with a deep crack ($a = 5.1''$) approaching the free edge. The model was made from 0.200 inch thick polycarbonate ($f_\sigma = 40$ psi/in/FR), and the scribed grid lines are 1.0 inch apart. Estimate the stress intensity factor from this pattern,

Figure E5.1

---
[6]This figure is available in high-resolution digital form at www.wam.umd.edu/~sanford.

using both the apogee and differencing methods. Compare the two methods as applied to this example.

5.3 Use the local-collocation method to determine the stress intensity factor for the model described in Exercise 5.2. Estimate the size of the singularity-dominated-zone, using a five-percent error criterion in σ_y.

5.4 The isopachic pattern shown in Figure E5.2[7] was recorded with through-transmission double-exposure (zero and full load) holographic interferometry at a wavelength of 0.514 nm. The material was 0.250 inch-thick polymethylmethacrylate (PMMA). The length of the crack is 0.54 inch.

Figure E5.2

(a) Develop a method to estimate the stress intensity factor from the gradient of the isopachic pattern.

(b) Determine the stress intensity factor by the local collocation method, and compare it with your findings for part (a). Make whatever assumptions you deem necessary.

---

[7] This figure is available in high-resolution digital form at www.wam.umd.edu/~sanford.

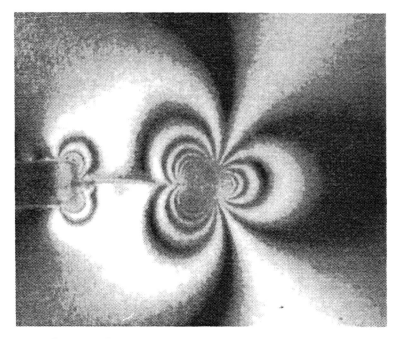

Figure E5.3

5.5 A double-exposure hologram of the the same model described in Exercise 5.4 was recorded *at fixed load* by shifting the model 0.050 inch parallel to the crack between exposures. The resulting interferogram is shown in Figure E5.3.[8]

   (a) What stress-field characteristic do these fringe contours represent?

   (b) How would you determine the fringe orders from a pattern such as this one?

   (c) Develop a method to determine the stress intensity factor from this type of pattern, and compute its value for the fringe pattern shown in Figure E5.3.

---

[8]This figure is available in high-resolution digital form at www.wam.umd.edu/~sanford.

CHAPTER 6

A Stress-Field Theory of Fracture

6.1 THE CRITICAL STRESS-STATE CRITERION

In Chapter 3 we observed that, regardless of the geometry or type of loading, all singular crack problems in elastic bodies share a common feature, namely, that the distribution of stress, strain and displacement over some region, albeit small, surrounding each crack tip has exactly the same functional form. Only the magnitude, represented by the parameter K, varies as the geometry or type of applied forces changes. In other words, we can always state that, for the opening mode of deformation, the stresses are of the form

$$\sigma_x = \frac{K}{\sqrt{2\pi r}} f_x(\theta) + \sigma_{ox}$$

$$\sigma_y = \frac{K}{\sqrt{2\pi r}} f_y(\theta) \qquad (6.1)$$

$$\tau_{xy} = \frac{K}{\sqrt{2\pi r}} f_{xy}(\theta)$$

where f_x, f_y, and f_{xy} are known functions of θ. This observation is pivotal to the formulation of the theory of brittle fracture within the framework of linear elastic fracture mechanics. In contrast to the solution of nonsingular problems in the theory of elasticity, in which both the magnitude and the distribution are different for each problem, this form invariance is unique to singularity-dominated crack problems.

With these observations as background, we are now in a position to state the fundamental law of brittle fracture within the context of LEFM:

A material body containing a crack-like defect will fail in brittle fracture when the state of stress in the immediate neighborhood of the crack tip enclosing the fracture process zone reaches a critical value, provided that the process zone is completely contained within the singularity-dominated zone.

From this law we can state that, at the instant of fracture,

$$\sigma_x|_{\text{critical}} = \frac{K_c}{\sqrt{2\pi r}} f_x(\theta) + \sigma_{ox}$$

$$\sigma_y|_{\text{critical}} = \frac{K_c}{\sqrt{2\pi r}} f_y(\theta) \tag{6.2}$$

$$\tau_{xy}|_{\text{critical}} = \frac{K_c}{\sqrt{2\pi r}} f_{xy}(\theta)$$

But, since the distribution of stress within the singularity-dominated zone is always the same to within a constant stress, σ_{ox}, parallel to the crack plane, the law of brittle fracture is equivalent to the requirement that

$$K|_{@\text{failure}} = K_c \tag{6.3}$$

where K_c is a measure of the material's resistance to fracture, a material property.

It is important to recognize that the quantities K on opposite sides of Eq. (6.3) represent fundamentally different quantities. The K on the left side represents the magnitude of the applied forces and is expressible in the form

$$K = \sigma\sqrt{\pi a} \cdot Y\left(\frac{a}{W}\right) \tag{6.4}$$

where $Y(a/W)$ is a dimensionless shape function within which all of the influences of the finite geometry are included. Since this K is determined by the geometry of the body, we shall refer to it as the "geometric stress intensity factor." Alternatively, since this parameter is linearly proportional to the magnitude of the applied forces, the geometric stress intensity factor is often referred to as the "applied K field." Our focus in the preceding three chapters was on the development of techniques for determining the geometric stress intensity factor, or, equivalently, the shape function $Y(a/W)$, for specific combinations of geometry and applied loads.

On the other hand, the parameter on the right side of Eq. (6.3), K_c, called the *critical stress intensity factor*, represents the material's resistance to fracture and, ideally, is a material property independent of the geometry used to measure its value.[1]

---

[1] We will explore this point in detail in Chapter 8.

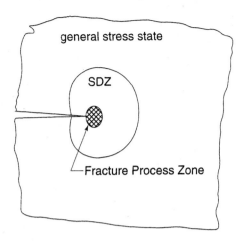

Figure 6.1 LEFM conditions apply if the fracture-process zone is fully contained in the singularity-dominated zone (SDZ).

As a material property, K_c can be determined only by experiment. Note further that K_c is a continuum mechanics parameter, which we mean that the validity of the LEFM concept requires only that the process zone (i.e., the region of material at the crack tip undergoing transformation leading to fracture) be fully contained within a region for which the near-field equations Eqs. (6.1) dominate the stress-, strain-, and displacement-field behavior. This size requirement on the process zone is illustrated in Figure 6.1. The theory does not require that we define (or even understand) the mechanisms occurring at the atomic level in order to establish the critical stress condition. As a consequence, linear elastic fracture mechanics provides a unified theory of fracture for a wide variety of materials, ranging from truly brittle materials, such as glass and refractory ceramics, to metals that, although ductile by nature, fail under certain conditions in a brittle manner. Although the theory does not require an intimate knowledge of the fracture mechanisms occurring within the process zone, it is clear that the more we understand about these mechanisms, the better we will be able to fashion materials to enhance their resistance to fracture (i.e., increase K_c).

The law of brittle fracture is illustrated schematically in Figure 6.2, in which the applied K value is plotted versus applied remote stress, σ, for a body containing a crack of varying lengths a_1, a_2, a_3, and a_4. The applied stress intensity factor for a fixed crack length increases linearly (slope $= Y(a/W) \cdot \sqrt{\pi a}$) until it reaches the limiting value of the material's resistance, K_c, at which point brittle fracture occurs. Figure 6.2 also graphically illustrates the size dependence on fracture previously described in Chapter 1. For the case of geometrically similar bodies of increasing size containing a crack that grows in proportion to the size of the body, the shape function $Y(a/W)$ is a constant. Nonetheless, the value of the critical stress decreases as the size increases, such that the product $\sigma_c \sqrt{a}$ remains constant.

This simple example illustrates the shortcomings of conventional (stress-based) design in the presence of possible brittle fracture. The common practice of "scaling up" the dimensions to redesign for larger structures fails to account for the fact that, all others factors being equal, a large structure cannot support as high a stress

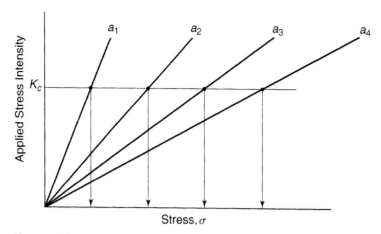

Figure 6.2 The effect of crack length on the maximum allowable stress under brittle-fracture conditions.

as can its smaller counterpart. Many of the boiler failures around the turn of the 20th century (such failures were quite common) can be attributed to the practice of scaling up the dimensions of a successful smaller boiler to accommodate the demand for increased steam capacity to support the rapidly growing industrial revolution.

The interdependence of maximum allowable stress and critical crack size can be illustrated by plotting the locus of critical points (shown as circles in Figure 6.2) for a given geometry and material, as illustrated in Figure 6.3. When presented in this form, it is clear that all combinations of stress and crack length below the fracture locus represent admissible combinations (some authors refer to this area of the graph as the safe region, but that is too subjective for our purposes), whereas points beyond

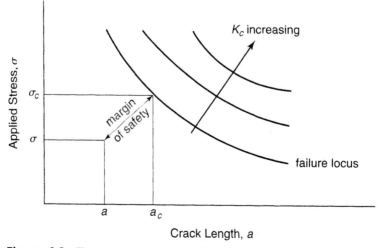

Figure 6.3 Fracture-locus curves for materials with differing fracture resistance.

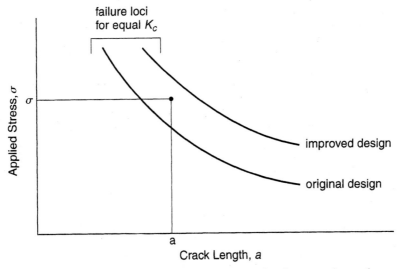

Figure 6.4 Influence of design changes on the fracture locus for the same material.

the locus line will result in sudden fracture. The margin of safety is described by the distance between (the applied load, crack length) coordinate position and the failure locus. For materials of increasing fracture toughness the fracture locus moves outward, as shown in Figure 6.3. It is often assumed (erroneously) that the only way to increase the margin of safety is to replace the material of the structure with one of higher fracture toughness (generally at higher cost), as Figure 6.3 would imply. However, the diagram shown in Figure 6.3 was constructed for a specific geometry with a prescribed shape function $Y(a/W)$. If we change the design parameters, the failure locus will be altered even for the same material, as shown schematically in Figure 6.4, and in many cases, the margin of safety can be increased economically by design changes alone. Finally, in complex structures with many potential failure sites, a separate failure locus must be constructed for each location of crack growth, and the margin of safety of the structure, taken as a whole, is governed by the minimum margin of safety among all of the potential failure sites.

Example 6.1   The center-cracked panel (W = 4.0 inches, B = 0.5 inch) containing a crack of length 1.6 inches, as shown in Figure 6.5, is to be loaded to failure. If the material has a yield strength of 60 ksi and a fracture resistance of 50 ksi-in$^{1/2}$ in plates of similar thickness, at what value of applied load will failure occur?

For this geometry the stress intensity equation due to Fedderson, Eq. (3.62), is appropriate:

$$K = \sigma\sqrt{\pi a}\sqrt{\sec\frac{\pi a}{W}}$$

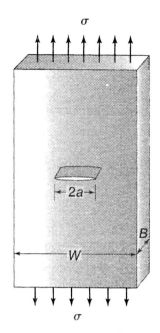

Figure 6.5 Center-cracked
plate with finite cross-section.

At the instant of brittle fracture,

$$K_c = \sigma_c \sqrt{\pi a} \sqrt{\sec \frac{\pi a}{W}}$$

Therefore, the critical stress for fracture is,

$$\sigma_c = 28.4 \text{ ksi}$$

Alternatively, the critical stress for net section yield is

$$\sigma_{\text{net}} = \frac{W - 2a}{W} \sigma_{\text{ys}} = 36 \text{ ksi}$$

Since $\sigma_c < \sigma_{\text{net}}$, failure will occur by brittle fracture at 56,800 lb. ∎

Example 6.2　For the same geometry described in Example 6.1, what is the longest crack that the panel can support without failure if the applied remote stress is 35 ksi?

In this situation the failure equation to be solved is transcendental, and there are a variety of methods that can be used to solve for the crack length. To facilitate a solution, the equation can be rewritten as

$$\left(\frac{K_c}{\sigma} \right)^2 \cos \frac{\pi a_c}{W} = \pi W \left(\frac{a_c}{W} \right)$$

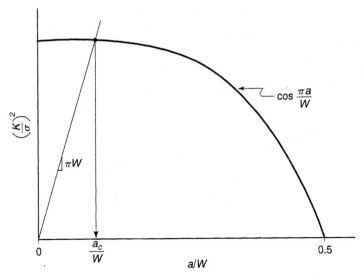

Figure 6.6   Graphical representation of the transcendental equation for determining the critical crack length.

The solution is shown graphically in Figure 6.6, and the result is $a_c = 0.583$ in. A check for possible net section yielding verifies that brittle fracture is the governing mode of failure.   ∎

Note that although the near-field equations Eqs. (6.1) explicitly include the uniform stress parallel to the crack plane, σ_{ox}, we have imposed no restrictions on its magnitude in establishing the law of failure. This does not imply that this stress is unimportant. Irwin recognized in 1957 that the stress intensity factor could not be extracted from the photoelastic data of Wells and Post [1957] without considering the impact of σ_{ox} on the inplane maximum shearing stress. Indeed, this argument was extended to all of the experimental measurement techniques discussed in Chapter 5. Likewise, in Chapter 4, the collapsed-node isoparametric element was preferred over the singularity element, because the former included a uniform strain term not included in the traditional singularity element. As we will see later in this chapter, the σ_{ox} term also influences the localized yielding behavior at a crack tip in mildly ductile materials.

All of these examples support the argument for including the constant stress term in the near-field equations; however, inclusion does not necessarily imply that our stress-based failure theory requires two parameters. In formulating the concept of the critical stress intensity factor, Irwin argued that since σ_{ox} contributes nothing to the crack opening stress $\sigma_y(x > 0, 0)$ or the crack opening displacement $u(x < 0, 0)$, it has at most a second-order effect on the critical stress at failure. Independently, we will develop in Chapter 7 the failure theory based on energy arguments. If all of the energies are properly considered [Tyson and Roy, 1989], the energy balance criterion leads to a one-parameter failure theory that is equivalent to the stress-based theory introduced in this chapter.

Since the stress-based theory of fracture depends critically on the validity of the near-field equations, it is logical to ask: how accurately do these equations model the state of stress around a crack tip in real materials? To answer this question, let us examine the predictions of the theory along the crack plane $y = 0$. Along this plane the stresses become

$$\sigma_x = \frac{K}{\sqrt{2\pi x}} + \sigma_{ox}$$

$$\sigma_y = \frac{K}{\sqrt{2\pi x}} \qquad (6.5)$$

$$\tau_{xy} = 0$$

which imply that both of the normal stresses approach infinity at the same rate as the crack tip is approached from the material side. Despite the observation of Williams [1957] that the hydrostatic stress state retards yielding, our physical insight tells us that the stress cannot be infinite for infinitesimal loads ($K \ll 1$). If it were, every structure with even the slightest imperfection would fracture under its own weight, which we know of course does not happen! To resolve this dilemma, let us examine the displacement of the crack faces for a problem for which we know the exact solution: namely, a crack of length $2a$ in an infinite sheet under biaxial tension, shown in Figure 6.7.

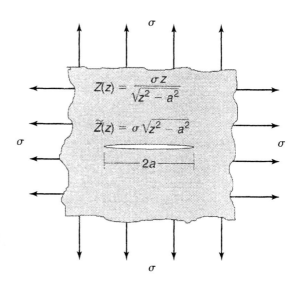

Figure 6.7 Crack profile under load for an internal crack with equal biaxial applied stress.

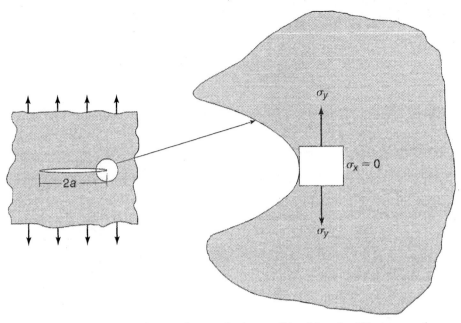

Figure 6.8 Smooth radius at the crack tip predicted by the Westergaard solution to a singular problem.

From Eq. (3.26b), the crack-face displacement for this geometry is

$$v = \frac{2\sigma}{E}\sqrt{a^2 - x^2} \qquad (6.6a)$$

which we can write as

$$\frac{v^2}{(\frac{2\sigma a}{E})^2} + \frac{x^2}{a^2} = 1 \qquad (6.6b)$$

We recognize Eq. (6.6b) as the equation of an ellipse. Consequently, under load, the crack assumes the form of an ellipse with a smooth radius at the crack tip, as shown in Figure 6.8. Thus, we see that our earlier dilemma is the result of our decision to use the undeformed coordinates as the reference frame for our formulation of the linear theory of elasticity (the Lagrangian formulation). If we had instead used a formulation based on the deformed shape (the Eulerian formulation), the elastic solution would have predicted finite stresses at the crack tip,[2] and the solution of this crack problem would have reduced to the Inglis solution for an elliptical slot. In this case the stresses ahead of the (now blunted) crack would be as shown in Figure 6.9.

[2] The use of Eulerian coordinates for problems in the linear theory of elasticity is very uncommon. In nearly all cases, this case being the exception, the difference between the Lagrangian and Eulerian solutions is negligible.

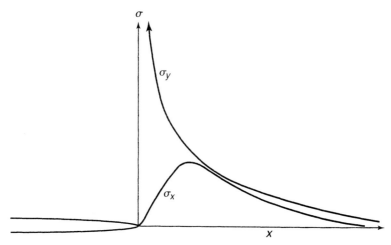

Figure 6.9 Stress profiles ahead of the blunted (Eulerian)
elastic crack tip.

Although finite, the crack opening stress is still very large and, for most practical materials, exceeds the yield stress. We can obtain an initial estimate of the extent of the plastically deformed zone, Δ_p, by limiting the stress to the plastic flow stress, σ_o, as shown in Figure 6.10. Therefore,

$$\sigma_o = \frac{K}{\sqrt{2\pi\,\Delta_p}} \tag{6.7}$$

or

$$\Delta_p = \frac{1}{2\pi}\left(\frac{K}{\sigma_o}\right)^2 \tag{6.8}$$

where K is the magnitude of the applied stress field at the crack tip.

Unfortunately, this argument fails to account for the redistribution of stress caused by yielding. That is, the contribution to equilibrium due to the shaded area in Figure 6.10 is not compensated for by modification of the elastic stress field $(x > \Delta_p)$. To rigorously determine the elastic–plastic boundary would require an elastic–perfectly plastic solution to the opening-mode problem, which is not available. Fortunately, an exact solution of the antiplane shear (mode III) problem of an isolated crack has been obtained by Hult and McClintock [1957]. The key results of their study are:

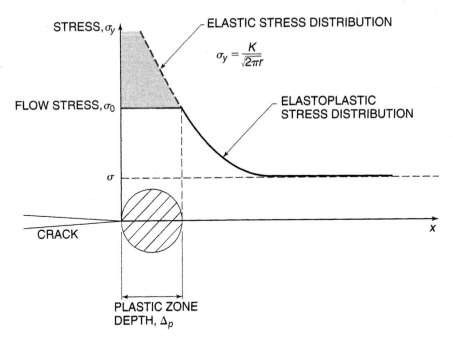

Figure 6.10 Estimate of the flow-stress limit of the plastic zone size.

(a) The elastic–plastic boundary has the same shape (a circle) as that obtained by an elastic analysis, but is displaced by an amount equal to its radius, r_y, given by

$$r_y = \frac{1}{2\pi}\left(\frac{K_{\text{III}}}{\tau_o}\right)^2 \tag{6.9}$$

where τ_o is the flow stress in shear.

(b) The elastic stress and strain distributions in the region beyond the plastic zone are identical to their elastic counterparts when calculated relative to a hypothetical crack-tip origin located at the center of the plastic zone. These results are illustrated in Figure 6.11.

If we adopt these exact elastic–plastic results for the mode-III problem as being approximately true for the mode-I problem [McClintock and Irwin, 1965], we can obtain an improved estimate of the extent of the plastic zone. Specifically, let us assume that, to the first order, the primary influence of localized plastic deformation on the elastic stress distribution for the opening-mode problem is to translate the distribution to the right by the amount necessary to restore equilibrium. As shown graphically in Figure 6.12, this effect is equivalent to requiring that the area gained by translating the elastic distribution (area B) is equal to the area lost by truncating the

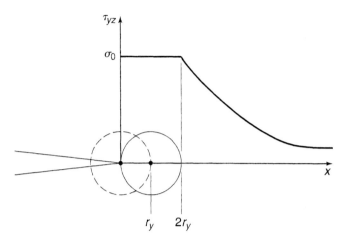

Figure 6.11 Exact elastoplastic stress distribution for mode III loading.

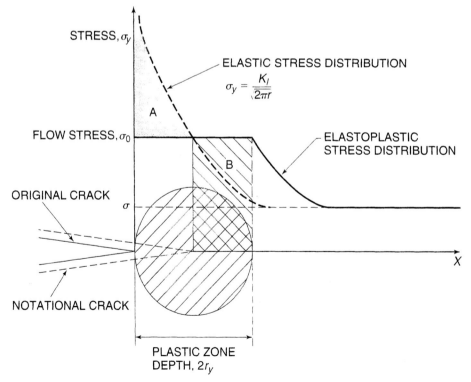

Figure 6.12 Approximate elastoplastic stress distribution for mode I loading, using the notational crack-length argument.

opening stress at the level of the flow stress (area A). Performing the force balance

$$\int_0^{\Delta p} \left(\frac{K}{\sqrt{2\pi x}} - \sigma_o \right) B\, dx = \sigma_o \left(2r_y - \Delta_p \right) B \tag{6.10}$$

where $2r_y$ is the equilibrium diameter of the plastic zone and B is the plate thickness, yields

$$r_y = \frac{1}{2\pi} \left(\frac{K}{\sigma_o} \right)^2 \tag{6.11}$$

This result and the arguments leading to it are highly significant to the extension of linear elastic fracture mechanics to materials that exhibit highly localized yielding at the crack tip, and we summarize the key points as follows:

(a) The primary influence of localized plastic deformation on the elastic stress distribution is to translate the distribution to the right by an amount equal to the plastic zone radius, r_y called the *notational crack tip* and given by Eq. (6.11).

(b) Both the size of the plastic zone and the stress distribution in the singularity-dominated region surrounding it are characterized by the single parameter K.

(c) If the region of plastically deformed material is contained within the singularity-dominated zone, the failure law of linear elastic fracture mechanics applies—provided that the physical crack length a is replaced by an effective crack length $a_{\text{eff}} = a + r_y$.

Collectively, these three arguments for the applicability of LEFM to yielding materials are called the *small-scale yielding approximations*.

Example 6.3   For the same conditions given in Example 6.1, what is the magnitude of the critical stress to cause fracture, including the small-scale yielding approximation correction. Assume that the flow stress is equal to the yield stress.

The geometric stress intensity factor, modified for localized yielding, becomes

$$K = \sigma \sqrt{\pi a_{\text{eff}}} \sqrt{\sec \frac{\pi a_{\text{eff}}}{W}}$$

where $a_{eff} = a + r_y = a + \frac{1}{2\pi} \left(\frac{K}{\sigma_o} \right)^2$.

At the instant of fracture instability,

$$K_c = \sigma_c \sqrt{\pi a_{eff}} \sqrt{\sec \frac{\pi a_{eff}}{W}}$$

where $a_{eff} = a + r_y = a + \frac{1}{2\pi} \left(\frac{K_c}{\sigma_o} \right)^2$.

Solving for the stress to cause fracture, $\sigma_c = 25.7$ ksi.   ■

Example 6.4   For the same conditions as described in Example 6.2, what is the largest crack length that the bar can support without fracture at an applied stress level of 35 ksi, including the small-scale yielding correction?

The procedure in this case is identical to that for Example 6.2, except that the unknown critical crack length is the effective crack length, $a_{eff}|_c$ at failure, that is,

$$\left(\frac{K_c}{\sigma}\right)^2 \cos\frac{\pi a_{eff}|_c}{W} = \pi W \left(\frac{a_{eff}|_c}{W}\right)$$

Therefore,

$$a_{eff}|_c = 0.583 \text{ inch.}$$

and the physical crack length is

$$a_c = a_{eff}|_c - r_y = 0.473 \text{ inch.} \qquad\blacksquare$$

Before closing this section, it is instructive to examine some of the consequences and implications of Eq. (6.11). First, we observe that the size of the plastic zone is an elastic parameter in the sense that it is determined by the magnitude of the applied stress state (i.e., K), not by the material's resistance to fracture, K_c, although K_c provides an upper bound on the diameter of the plastic zone. On the other hand, we must also view r_y as a nonlinear parameter, first in the geometric sense that the plastic zone grows as the square of the applied load. Second, from an elasticity perspective, the plastic zone is not reversible, and when a cracked body is unloaded, the effect of its prior plastic zone is to produce a residual compressive stress field at the crack tip that must be overcome in subsequent loadings of the body. This latter factor is of particular importance in some aspects of fatigue crack growth to be discussed in Chapter 9. Finally, from Eq. (6.11), we observe that the size of the plastic zone depends only on the magnitude of the local stress state (i.e., K) and is independent of specimen size. Accordingly, proportionally scaled specimens of increasing size, loaded to the same K level, will have smaller relative plastic-zone sizes (i.e., r_y/a decreases).

6.2 CRACK-TIP PLASTICITY

In the previous section we arrived at the size of the plastic zone by considering equilibrium of the truncated near-field stresses in light of Irwin's small-scale yielding assumptions. However, we should expect that this estimate of the plastic zone will apply only when the size of the zone is small compared with the crack length and when the influence of the boundaries can be ignored. If these conditions are not met, the overall stress state is more complex than that predicted by the near-field equations, and the equilibrium balance is more difficult to formulate. Moreover, Eq. (6.11) provides no information on the state of stress, strain or displacement throughout the

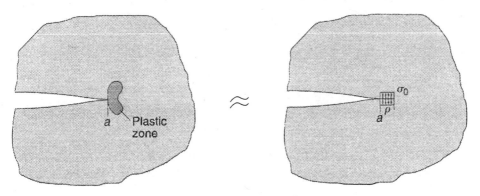

Figure 6.13 Elastoplastic solution approximated by the elastic-strip-zone model.

field. Fortunately, there is an approach short of resorting to a detailed elastoplastic solution that will provide a more complete description of the state of stress in the presence of a confined plastic zone. If we assume that all of the effects of plastic deformation can be concentrated into a narrow strip ahead of the crack tip, the plastic zone can be represented by a region of material stressed to the flow stress, σ_o, as illustrated in Figure 6.13. But, from the principle of superposition, the (elastic) strip yield model is equivalent to the sum of two other elastic solutions shown in Figure 6.14—that is, the elastic solution for the remotely loaded problem with a crack of length $a + \rho$, and the crack-line-loaded solution with a closing stress equal to the inverse of the flow stress over the length of the "strip."

In the terms of Westergaard stress functions, the principle of superposition illustrated in Figure 6.14 is equivalent to the expressions

$$Z_{\text{total}}(z, a) = Z_{\text{remote}}(z, a + \rho) + Z_{\text{strip}}(z, a + \rho) \tag{6.12a}$$

and

$$Y_{\text{total}}(z, a) = Y_{\text{remote}}(z, a + \rho) + Y_{\text{strip}}(z, a + \rho) \tag{6.12b}$$

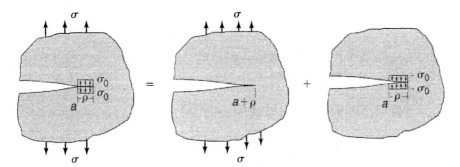

Figure 6.14 Plastic zone replaced by crack-closing stress over a limited region.

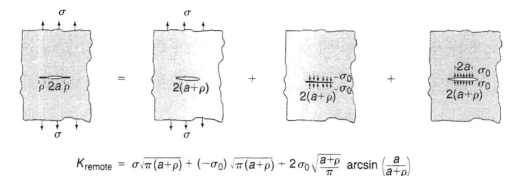

$$K_{\text{remote}} = \sigma\sqrt{\pi(a+\rho)} + (-\sigma_0)\sqrt{\pi(a+\rho)} + 2\sigma_0\sqrt{\frac{a+\rho}{\pi}}\,\text{arcsin}\left(\frac{a}{a+\rho}\right)$$

Figure 6.15 Application of the principle of superposition to the strip-yield model.

From these equations we can obtain the complete state of stress (in the elastic region) of a plastic-zone modified crack for any geometry and external loading.

Similarly, we can formally write that the geometric stress intensity factor for the strip yield model is given by

$$K_{\text{total}}(a) = K_{\text{remote}}(a + \rho) + K_{\text{strip}}(a + \rho) \tag{6.13}$$

However, since the stress at the physical crack tip $(x = a)$ is finite, the singularity must vanish. As a result,

$$K_{\text{remote}}(a + \rho) + K_{\text{strip}}(a + \rho) = 0 \tag{6.14}$$

This requirement is used to determine the equilibrium length, ρ, of the plastic (strip) zone.

Let us apply this concept to a geometry for which we know the required stress functions, namely a uniaxially remote-loaded central crack of length $2a$ containing yielded strips of length ρ. This strip yield problem, given as the superposition of two previously obtained elastic solutions, is shown in Figure 6.15.

From the results in Chapter 3, we have

$$Z_{\text{remote}} = \frac{\sigma z}{\sqrt{z^2 - (a + \rho)^2}} \tag{6.15a}$$

$$Y_{\text{remote}} = -\sigma \tag{6.15b}$$

$$Z_{\text{strip}} = \int_a^{a+\rho} \frac{2\sigma_o z}{\pi\sqrt{z^2 - (a + \rho)^2}} \frac{\sqrt{(a + \rho)^2 - \xi^2}}{(z^2 - \xi^2)}\,d\xi \tag{6.15c}$$

$$Y_{\text{strip}} = 0 \tag{6.15d}$$

After integrating and combining terms, we find that

$$Z_{\text{total}} = \frac{\left[\sigma - \frac{2\sigma_o}{\pi} \arccos\left(\frac{a}{a+\rho}\right)\right] z}{\sqrt{z^2 - (a + \rho)^2}} + \frac{2\sigma_o}{\pi} \text{arccot}\left(\frac{a}{z}\sqrt{\frac{z^2 - (a + \rho)^2}{(a + \rho)^2 - a^2}}\right) \quad (6.16a)$$

$$Y_{\text{total}} = -\sigma \quad (6.16b)$$

However, since we require that the stress field be nonsingular, the bracketed expression in the first term of Eq. (6.16a) must be zero. Therefore,

$$\cos\left(\frac{\pi\sigma}{2\sigma_o}\right) = \frac{a}{a + \rho} \quad (6.17)$$

Finally, inverting Eq. (6.17) and expanding the resultant secant function in series form, we find that, for $\sigma/\sigma_o << 1$,

$$\rho \approx \frac{\pi}{8}\left(\frac{K}{\sigma_o}\right)^2 \quad (6.18)$$

This result, first obtained by Dugdale [1960], is comparable with Irwin's [1960] estimate of the plastic zone's diameter:

$$2r_y = \frac{1}{\pi}\left(\frac{K}{\sigma_o}\right)^2 \quad (6.19)$$

We note that it was not necessary to derive Z_{total} in detail in order to obtain Eq. (6.17), as this latter result follows directly from Eq. (6.14). But, having done so, we now have the exact solution for the state of stress outside the plastic zone,

$$Z_{\text{total}} = \frac{2\sigma_o}{\pi}\text{arccot}\left(\frac{a}{z}\sqrt{\frac{z^2 - (a + \rho)^2}{(a + \rho)^2 - a^2}}\right) \quad (6.20a)$$

and

$$Y_{\text{total}} = -\sigma \quad (6.20b)$$

which we could not have obtained by considering the geometric stress intensity factor only. Moreover, we can use this result to calculate the crack opening displacement COD $(= 2v$ at $x = a)$, which is an important parameter in elastoplastic analysis.[3] From Eq. (3.26b), we have

$$COD = 2v = \frac{4Im\ \tilde{Z}(a)}{E} \tag{6.21a}$$

$$COD = \frac{8\sigma_o a}{\pi E} \ln \sec \left(\frac{\pi \sigma}{2\sigma_o} \right) \tag{6.21b}$$

Expanding the ln sec function in series form for $\sigma/\sigma_o << 1$, we find that

$$COD \approx \frac{\sigma^2 \pi a}{E\sigma_o} = \frac{K_I^2}{E\sigma_o} \tag{6.22}$$

While this analysis provides useful information on the size of the plastic zone, it provides little guidance on its shape. The strip yield model assumed *a priori* that the yielded region was confined to a narrow strip, which is a reasonable approximation for thin sheets. As the plate thickness increases, we would expect the region of plastic deformation to extend outward from the crack tip, but not necessarily assuming the circular form used by McClintock and Irwin [1965] that was modeled on mode-III behavior. An accurate assessment of the shape of the plastic zone would require a detailed elastoplastic analysis, probably best done using finite element analysis, and would, of course, need to be repeated for every geometry of interest. However, we can obtain a first-order estimate of the extent and shape of the plastic zone by examining the near-field equations.

Using the Von Mises criterion, yielding has occurred when

$$(\sigma_x - \sigma_y)^2 + (\sigma_x - \sigma_z)^2 + (\sigma_y - \sigma_z)^2 + 6\tau_{xy}^2 \geq 2\sigma_{ys}^2 \tag{6.23}$$

where σ_{ys} is the yield stress in uniaxial tension and $\sigma_z = 0$, for plane stress, or $\sigma_z = v(\sigma_x + \sigma_y)$, for plane strain. Substituting the near-field equations into Eq. (6.23), yields for plane stress (with $\sigma_{ox} = 0$)

$$\frac{K_I^2}{2\pi r} \left(1 + \frac{3}{2} \sin^2 \theta + \cos \theta \right) \geq 2\sigma_{ys}^2 \tag{6.24}$$

from which it follows that the region of plastic deformation is given by

$$r \leq \frac{1}{4\pi} \left(\frac{K_I}{\sigma_{ys}} \right)^2 \left(1 + \frac{3}{2} \sin^2 \theta + \cos \theta \right) \tag{6.25}$$

---

[3]This concept will be explored more fully in Chapter 11.

Along the crack plane ($\theta = 0$), Eq. (6.25) predicts that the elastic–plastic boundary, r_b, occurs at

$$r_b|_{\text{plane stress}} = \frac{1}{2\pi}\left(\frac{K_I}{\sigma_{ys}}\right)^2 = r_y \tag{6.26}$$

which agrees with our earlier result in Eq. (6.8), with the flow stress σ_o equal to the yield stress in uniaxial tension.

The corresponding results for plane strain (with $\sigma_{ox} = 0$) are

$$\frac{K_I^2}{2\pi r}\left[\frac{3}{2}\sin^2\theta + (1 - 2\nu)^2(1 + \cos\theta)\right] \geq 2\sigma_{ys}^2 \tag{6.27}$$

and, for $\nu = 0.33$,

$$r_b|_{\text{plane strain}} = \frac{1}{18\pi}\left(\frac{K_I}{\sigma_{ys}}\right)^2 = \frac{1}{9}r_b|_{\text{plane stress}} \tag{6.28}$$

We note that this first-order analysis violates equilibrium, and to be consistent with our earlier result, we should interpret the origin as being located at the notational crack tip. With this stipulation, Figure 6.16 shows a comparison of the plastic-zone shapes for plane stress and plane strain.

The Irwin plastic zone radius, r_y, can be defined in a consistent manner for both plane stress and plane strain by assuming that the flow stress for plane strain in Eq. (6.11) is $3\sigma_{ys}$. However, this estimate of the flow stress assumes that both normal stresses, σ_x and σ_y, approach infinity at the same rate (i.e., $1/\sqrt{r}$). In reality, the near-field stresses are more accurately represented by Figure 6.9. Irwin [1960] argued that, in light of the finite root radius at the crack tip, a reasonable estimate[4] of the flow stress in plane strain is $\sqrt{3}\sigma_{ys}$. Using this estimate of the constraint factor, we can write that

$$r_y|_{\text{plane strain}} = \frac{1}{6\pi}\left(\frac{K_I}{\sigma_{ys}}\right)^2 = \frac{1}{3}r_y|_{\text{plane stress}} \tag{6.29}$$

Since the stress field off of the crack plane is less affected by the crack blunting behavior, the general features of Figure 6.16 are unaffected by the modification to flow stress, and we conclude that the size of the plastic zone is significantly smaller in plane strain than in plane stress.

To complete the study of the plastic zone's size and shape, we can repeat the preceding analysis for nonzero values of σ_{ox} in Eq. (6.23). The algebra does not

---

[4]In fact, Irwin's first estimate of the plastic constraint factor was $\sqrt{2\sqrt{2}} \approx 1.68$, which he later replaced by $\sqrt{3} \approx 1.73$ to simplify the plastic zone calculation.

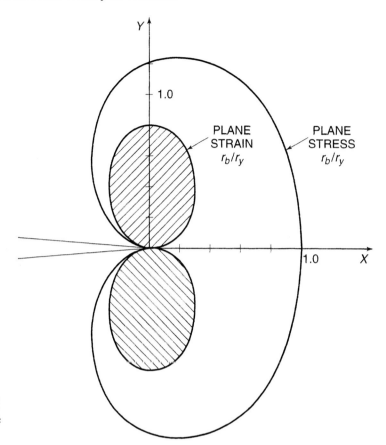

Figure 6.16 Von Mises estimates of the shape of the plastic zone for plane stress and plane strain.

lend itself to compact expressions, and the analysis is best performed numerically. Figure 6.17 shows results for representative values of $\sigma_{ox}/(K/\sqrt{a})$. These results demonstrate that the uniform stress parallel to the crack line, σ_{ox}, can significantly influence the overall shape of the plastic zone. This observation can have an impact on the choice of specimen geometries used for fracture testing, since the magnitude and sign (tension or compression) of σ_{ox} varies widely among the popular test geometries. It is interesting to note, however, that σ_{ox} has little effect on the size of the plastic zone along the crack plane. Accordingly, the r_y correction does not depend on this stress component. These results are in general agreement with an early elastoplastic FEA investigation by Larsson and Carlsson [1973].

It is highly instructive to interpret the results of the previous discussion for materials of finite thickness. For very thin sheets a state of plane stress exists throughout the sample, and a uniformly large plastic zone would be anticipated. As the thickness increases, the state of plane stress stills persists on the surface, but eventually, a plane-strain condition is approached in the interior. This situation is depicted in Figure 6.18. The conical transition region can be viewed as a surface effect, and for very thick samples, is negligible in comparison with the plane-strain region. Since the vol-

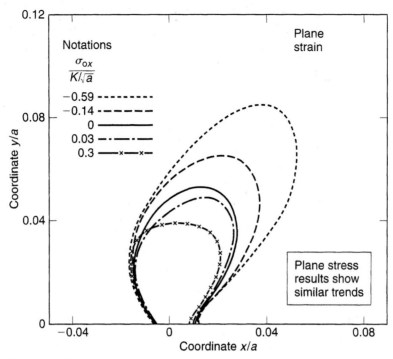

Figure 6.17   Influence of the stress parallel to the crack line (T-stress) on the shape of the plastic zone (adapted from Larrson and Carlsson, 1973). Copyright Elsevier Science, reprinted with permission.

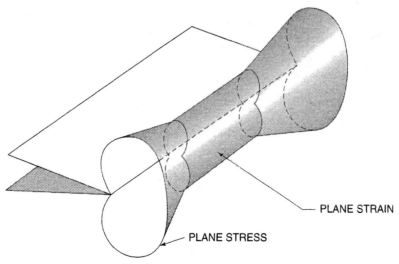

Figure 6.18   Transition of the plane-stress to plane-strain plastic zone size in plates of finite thickness.

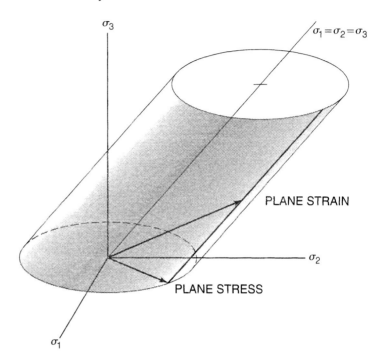

Figure 6.19 Von Mises yield surface.

ume of material undergoing plastic deformation per unit thickness is much smaller in plane strain than in plane stress, the amount of energy absorbed by plastic processes is also much smaller. As a consequence, this energy is available for brittle fracture.

We reach a similar conclusion by examining the yield surface in three dimensions. For the Von Mises yield criterion, this surface, shown in Figure 6.19, is a cylinder whose axis is along the octahedral direction (pure hydrostatic stress). For plane stress the load vector is in the $\sigma_1 \sigma_2$-plane. For plane strain there is a component of the load vector in the σ_3 direction, and the length of the load vector is increased substantially. Since the load-carrying capacity before the onset of yielding is proportional to the length of the load vector, the usual interpretation of the yield surface would lead us to speculate that the load-carrying capacity in plane strain is greater than that in plane stress. However, this interpretation is incorrect. We must view brittle fracture and plastic flow as competing mechanisms for failure. From this viewpoint, the increased length of the load vector implies that there is an *increased* probability for brittle fracture (which is controlled by a material property independent of the yield stress) in plane strain, since the probability of plastic flow is suppressed by the long load vector. In other words, even for materials that would normally fail by plastic deformation, the near-hydrostatic stress state at the crack tip, also called the *high triaxial constraint condition*, inhibits the onset of yielding, thereby providing an opportunity for failure by brittle fracture to occur.

Finally, we can reach the same conclusion from a different perspective by depicting the stress state at the crack tip on a Mohr's circle diagram. If we assume that,

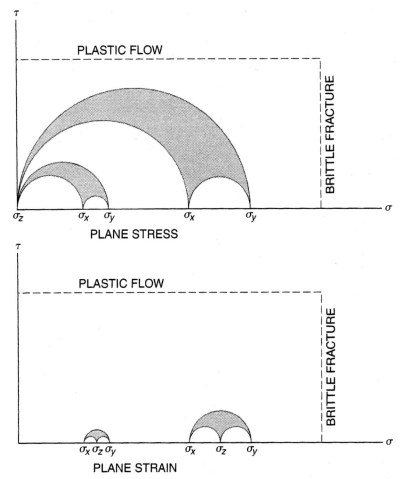

Figure 6.20 Comparison of Mohr's-circle representations for plane stress and plane strain at two levels of applied load.

due to the influence of crack blunting, $\sigma_y > \sigma_x$, the three-dimensional Mohr's-circle constructions for plane stress and plane strain at two levels of applied load are shown schematically in Figure 6.20. In the diagram, the competing mechanisms of plastic flow (Tresca yield criterion) and brittle fracture (normal stress theory) are shown as limiting values on the shear stress and normal stress axes, respectively. For plane stress, $\sigma_z = 0$, and the growth of Mohr's circles with increasing applied load must be fixed at the origin. As a result, the major diameter grows rapidly and soon reaches the limiting shear stress. On the other hand, for plane strain, the σ_z stress must always lie between the two in-plane normal stresses, and the Mohr's circles translate much more rapidly than their diameters grow. In this case the limiting normal stress (brittle fracture) is reached before the outer diameter enlarges to its critical value.

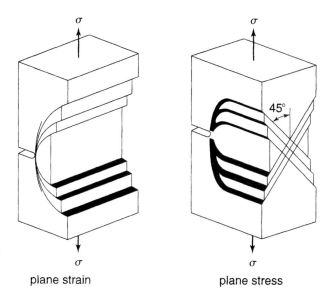

Figure 6.21 Influence of thickness on the formation of yield surfaces at the crack tip.

plane strain   plane stress

The Mohr's-circle representation provides additional information on the fracture behavior of materials. In plane stress the planes of maximum shear stress, and hence the direction of plastic flow, are in the antiplane direction, whereas these directions for plane strain are in the plane of the sheet. As a consequence, the mechanisms of plastic-zone formation are distinctly different for plane stress versus plane strain. As illustrated in Figure 6.21, plane-stress plastic flow favors slip motion toward the free surfaces. This process gives rise to the formation of fracture surfaces inclined to the free surface at 45°. This fracture feature is called a *shear lip* and its extent is closely related to the conical transition region shown in Figure 6.18. Also, in thin specimens, the height of the plastically deformed region is comparable with the thickness and in the limit behaves like the strip yield model.

The combination of all of these observations predict that the critical fracture toughness, K_c, is thickness dependent. In fact, experiments on fracture of mildly ductile materials exhibit the characteristics shown in Figure 6.22. The data suggest three distinct regions of differing behavior. For very thin specimens, the size of the plastic zone is limited to the thickness, and the fracture toughness increases with thickness. In this region the material can be classified as pure plane stress dominated. At some thickness, which depends on the yield stress, the plane-stress plastic zone of Figure 6.16 is fully developed, and any further increase in thickness promotes the formation of the transition region shown in Figure 6.18, with its corresponding decay of plastic-zone volume. Finally, at some still larger thickness, the size of the conical region is insignificant compared with the portion dominated by plane-strain behavior, and a lower bound fracture toughness, called the *plane-strain fracture toughness*, K_{Ic}, prevails. Once this thickness is reached, the fracture toughness no longer depends

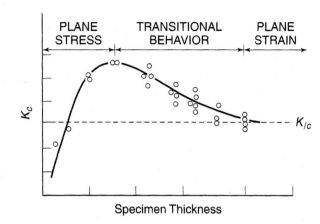

Figure 6.22 The dependence of the thickness on the fracture resistance of metallic materials.

on thickness, and the resulting fracture parameter K_{Ic} can be considered a true bulk material property.

From Figure 6.22 we can draw several conclusions. First, K_{Ic} is a conservative estimate of a material's resistance to fracture. It is reasonable to assume that the behavior is (almost) completely elastic and that failure will occur immediately when the critical stress state is reached. Second, the fact that we can design a test to measure the plane-strain fracture toughness K_{Ic} does not mean that we should design parts so that the conditions of plane strain can be met. In fact, in most cases, plane strain is the worst condition. In practice we would like to have some plastic-zone growth prior to fracture such that $K_c > K_{Ic}$. And finally, although K_{Ic} is a conservative estimate of the fracture resistance of a material, in many cases it is far too conservative for design use. An aircraft designed to plane-strain fracture toughness standards simply will not fly!

6.3 THE EFFECT OF VARIABLES ON FRACTURE TOUGHNESS

In the last section it was demonstrated that once sufficient thickness is obtained, the plane strain fracture toughness can be regarded as a material property. However, this does not mean that it is a single number. Unlike material properties such as the elastic modulus and Poisson's ratio, whose values are determined primarily by atomic variables, K_{Ic} is affected by external variables, such as temperature, loading rate, and metallurgical processing. Consequently, values of K_{Ic} listed in tables (as in Appendix D) should be regarded as representative of that class of material only. In fact, batch-to-batch variations in alloy content and heat treatment can produce significant variation in the value of K_{Ic} for a material, particularly in the transition region. For this reason, critical design decisions based on fracture toughness should always include a testing requirement to verify that a valid K_c value has been used.

Long before fracture mechanics came into being, a standardized test had been employed to assess a material's resistance to fracture. This test, the Charpy V-notch impact test (CVN), is still widely used within the metallurgical community to

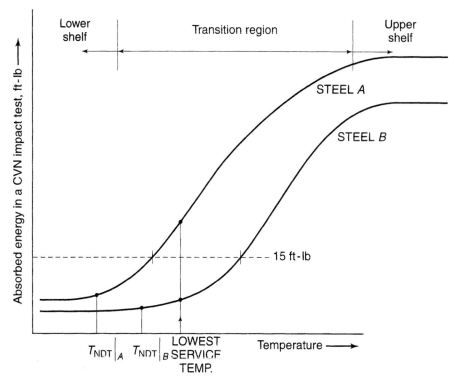

Figure 6.23   Charpy V-notch curves for two hypothetical low- to medium-strength steels over a wide temperature range.

determine the effect of compositional changes on fracture and for quality control.[5] In this test, a rectangular specimen 10 mm × 10 mm × 55 mm containing a carefully controlled sharp notch is impacted by an anvil attached to a pendulum arm. From the difference in height of the pendulum before and after impact, the energy absorbed in the fracture process is recorded as the Charpy energy (in ft-lb). Most often, Charpy tests are conducted at a range of temperatures and the results presented in graphical form, as shown schematically in Figure 6.23 for two low- to medium-strength steels ($\sigma_{ys} < 150$ ksi).

There are several features of these curves that deserve special mention. The absorbed energy of fracture from the low-temperature regime to the upper shelf at high temperatures varies by a factor of 10 or more. This large increase is accompanied by dramatic changes in the fracture appearance. At low temperatures the fracture surface is characterized by cleavage-dominated mechanisms, with their flat, shiny appearance to the unaided eye and a complete absence of any signs of ductility, as evidenced by the absence of shear lips on the faces of the specimen. Above

---

[5] The particulars of the Charpy V-notch test are described in ASTM Standard E 23.

some temperature, called the null-ductility transition temperature, T_{NDT}, the fracture surface just begins to show evidence of plastic deformation. The amount of gross deformation increases throughout the transition region until, at the upper shelf, the fracture mechanism is pure ductile tearing.

Charpy results formed the basis for the first fracture-based design criterion. In the 1940s, the large number of failures of Liberty and tanker ships led to a systematic study of the steel hulls used in their fabrication. Comparison of results for specimens taken from crack initiation regions with those for specimens taken from crack arrest regions led to the observation that Charpy V-notch energies of 10 ft-lb at the lowest service temperature mark the boundary between failure and acceptable service life. Based on these results, a conservative acceptance criterion of 15 ft-lb at the lowest expected service temperature was adopted for ship-hull plate material. By this standard, steel A in Figure 6.23 would pass, whereas steel B would not. Although developed for a particular class of hull steels, the 15 ft-lb criterion was widely adopted for structures of all types.

There are strong parallels between Charpy V-notch results and K_c behavior as a function of temperature, but to fully understand this behavior over the entire temperature range, we need to isolate the contributions due to several factors. Let us first confine our attention to results taken from samples thick enough to ensure plane-strain behavior as shown in Figure 6.22. As we will see in Chapter 8, plane-strain conditions are assured if the following requirement is satisfied:

$$\frac{1}{B}\left(\frac{K_{Ic}}{\sigma_{ys}}\right)^2 \equiv \beta < 0.4 \tag{6.30}$$

For samples of thickness satisfying this requirement, the size of the plastic zone is negligible compared with either the thickness or the crack length. For steels of modest ductility ($\sigma_{ys} < 150$ ksi), the variation of the plane-strain fracture toughness with temperature is shown schematically in Figure 6.24. However, since it is well known that the yield stress decreases with increasing temperature, albeit modestly, we can combine these two effects and recast Figure 6.24 into Figure 6.25, which shows an even larger rise with temperature.

Since $(K_{Ic}/\sigma_{ys})^2$ is proportional to the plastic zone size at fracture, we might be tempted to associate the increase in K_{Ic} with a transition from plane strain to plane stress, that is, we might attribute the rise to an increase in plastic zone size with its correspondingly larger absorption of energy. However, this is not the case, since the effect is observed for $\beta < 0.4$, and in this regime the plastic zone is negligible. Instead, the rise in K_{Ic} is due to a change in the fracture mechanism on the microscale.

At the low end of the temperature scale, the mechanism of fracture is separation along weak atomic planes (cleavage); for example, in iron, cleavage occurs along the $<100>$ plane. As each adjacent grain has a slightly different orientation, the macrocrack adjusts to the preferred plane within each grain in the

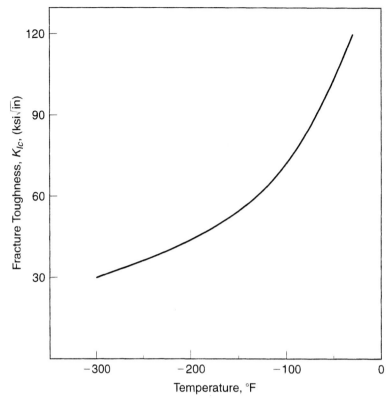

Figure 6.24 Plane-strain fracture toughness K_{Ic} as a function of temperature for a typical structural steel.

path of the advancing crack, as illustrated in Figure 6.26. These cleavage facets are highly reflective and give this type of fracture its characteristic shiny appearance. Cleavage, or transgranular, fracture is associated with truly brittle fractures at low absorbed energy.   As the temperature increases, there is a transition away from pure cleavage to a fracture mechanism that requires significantly more energy.

During normal processing of alloy steels the more brittle precipitates tend to migrate to the grain boundaries as the material undergoes phase changes upon cooling from the processing temperature. These weak particles will fracture in the high stress region ahead of the crack, creating microvoids, as shown in Figure 6.27a. As the crack approaches, the stress increases and the voids open further, creating isolated regions of very ductile material. These regions are not unlike rows of tensile specimens on the microscale, as shown in Figure 6.27b. Finally, as each ligament fails, it leaves behind a microscopic cup-and-cone fracture, called a *dimple*. This process, called *dimple rupture* or *void coalescence*, results in a rough and dull appearance of the fracture surface. Figure 6.28 shows a comparison of the fracture appearance of a quenched and tempered steel in cleavage (left) and dimple-rupture (right) modes.

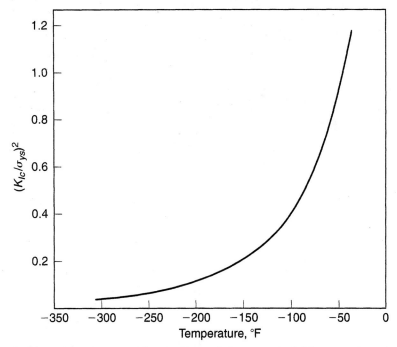

Figure 6.25 The combined effect of decreased yield strength and increased fracture toughness for steels with modest ductility (< 150 ksi) with increasing temperature.

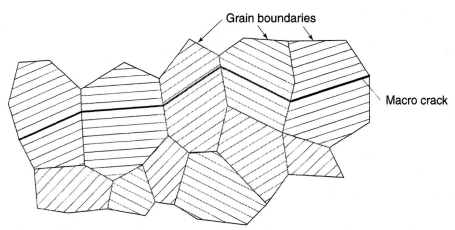

Figure 6.26 Schematic diagram of cleavage or transgranular fracture—characteristic of low-energy-absorption fractures.

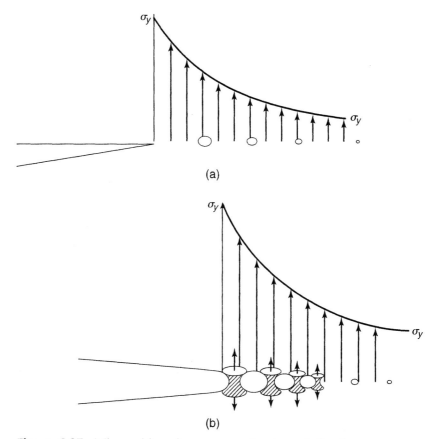

Figure 6.27 Microvoid coalescence mechanism of dimple rupture—characteristic of high-energy-absorption fractures.

Within the plane-strain region, the transition from cleavage to dimple rupture is associated with an increase in energy of fracture, due to plastic deformation on the microscale. This transition can increase the plane-strain fracture toughness K_{Ic} by a factor of two.

With a further increase in temperature, the fracture toughness continues to increase, and eventually, $\beta > 0.4$. When this occurs, plastic deformation is no longer confined to the tensile ligaments ahead of the crack, but also occurs in the bulk material. The result is a breakdown of the triaxial constraint condition of plane strain, with its associated high flow stress (refer to Figure 6.19), and the plastic zone size [Eq. (6.11)] rises rapidly, due to the combined effect of rising K in the numerator and decreasing flow stress in the denominator. In addition, the increasing size of the plastic zone ahead of the crack reduces the sharpness at its tip. This crack-blunting behavior reduces the strain gradient ahead of the crack, resulting in a further increase of energy input to sustain the void coalescence process. Because these effects are synergistic, the shift from plane strain to plane stress contributes an increase in the value

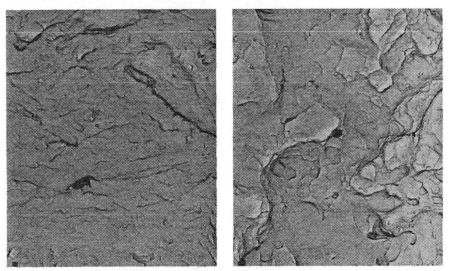

Figure 6.28 Comparison of the fracture appearances for a quenched and tempered steel under cleavage (left) and dimple rupture (right) fracture modes [courtesy C. D. Beachem, P.E.].

of K_c of the order of a factor of 10. Combining the transition from cleavage to void coalescence at the low end of the temperature scale with the transition from plane strain to plane stress at higher temperatures, we arrive at the K_c-versus-temperature behavior over the full temperature range shown in Figure 6.29.

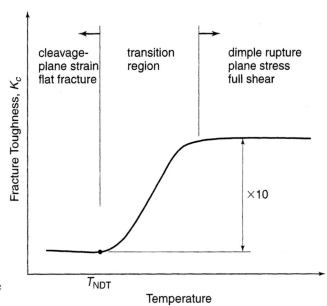

Figure 6.29 Fracture-toughness behavior over a wide temperature range.

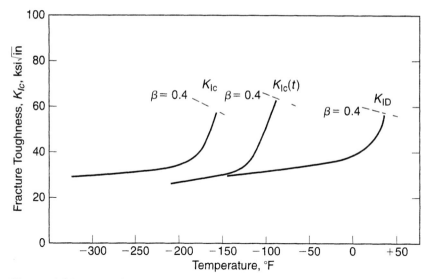

Figure 6.30 The effect of the loading rate on the fracture toughness transition.

The similarity of K_c versus T in Figure 6.29 to CVN versus T in Figure 6.23 is to be expected, since the crack-growth behavior on the microscale is the same. However, there is one difference to be noted: the K_c results were obtained from quasistatic fracture tests, whereas the Charpy results were obtained from rapid loading. If we add the effect of loading rate to the temperature effects just described, the behavior becomes even more complicated.

The effect of loading rate on the plane-strain fracture toughness (i.e., $\beta < 0.4$) is illustrated in Figure 6.30. What we observe as the time to failure is decreased from static K_{Ic}, lasting several seconds or longer, to rapid-load $K_{Ic}(t)$, of the order of milliseconds, and, finally, to dynamic K_{ID}, occurring in microseconds, is a systematic shift in null-ductility temperature. As a result, the material behaves as if it were a more brittle steel, and the cleavage-dominated mechanism persists to higher temperature. From a physics perspective the increased loading rate is suppressing the formation of the microvoid coalescence mechanism, because of the rate dependent nature of plastic deformation. In simplistic terms, as the strain rate is increased, there is not sufficient time available to accelerate slip planes to form the tensile ligaments. Finally, as the temperature increases further, the increase in available energy is sufficient to permit the void coalescence mechanism to dominate, and K_{Ic} rises sharply. Curiously, this trend is reversed in the plane-stress region, and at the upper shelf the dynamic K_c values are higher than their static counterparts. All of these effects are combined in Figure 6.31.

From a design perspective, the implications of the temperature shift due to increased loading rate can be profound. A steel with sufficient fracture resistance to static loading conditions may prove unsatisfactory in the event of unexpected dy-

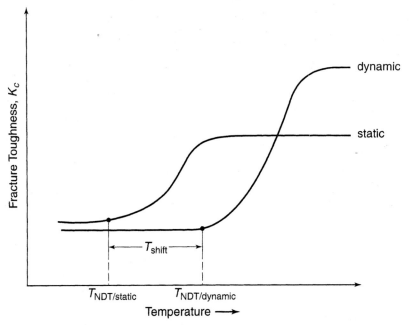

Figure 6.31 Apparent temperature shift in T_{NDT} due to the loading rate.

namic loading. (The R.M.S. Titanic comes to mind.) Unfortunately, dynamic K_c versus temperature is difficult to measure and generally is not available at the material selection phase of the design process. To overcome this problem, empirical correlations have been developed based on extensive data collected by the steel industry. Barsom and Rolfe [1987] have proposed that the shift in K_{Ic} due to loading rate can be approximated by

$$T_{\text{shift}}(°F) \approx 215 - 1.5 \ \sigma_{ys} \qquad \text{for 36 ksi} < \sigma_{ys} < 140 \ \text{ksi} \qquad (6.31a)$$

$$T_{\text{shift}} = 0 \qquad \text{for } \sigma_{ys} > 140 \ \text{ksi} \qquad (6.31b)$$

They also have suggested that, when Charpy V-notch results are available, the upper shelf dynamic fracture toughness can be estimated by

$$\frac{K_{ID}^2}{E} \approx 5(\text{CVN}) \qquad (\text{psi-in, ft-lb}) \qquad (6.32)$$

As the Barson and Rolfe equations illustrate, the effect of loading rate on K_c is most pronounced in steels with low to moderate yield stress. For steels with yield stress greater than 150 ksi (as well as aluminum and titanium), the loading rate shift does not occur. For materials in this category there is no lower shelf transition behavior, and the increase in K_{Ic} is nearly linear with temperature.

There is one other empirical relation that is quite useful for estimating K_c. Ideally, the critical stress intensity factor should be measured in the thickness used for fabrication. However, during the design stages, when the final thickness has not been determined, K_c can be estimated with the formula proposed by Irwin [1960],

$$K_c^2 = K_{Ic}^2 \left[1 + 1.4\beta^2 \right] \tag{6.33}$$

where β is given by Eq. (6.30).

This approximation assumes that the only influence of the small, but nonneglible, plastic zone is to increase the strain energy required for fracture. We note that Eq. (6.33) has meaning only when the difference between K_{Ic} and K_c is small (i.e., plate thicknesses only slightly below the plain-strain condition). As the plate thickness becomes even thinner, the assumptions of Eq. (6.33) no longer apply, and the R-curve concept to be described next must be employed for instability analysis.

6.4 *R*-CURVES

All of the preceding discussion on a stress-based theory of fracture has assumed that unstable fracture and subsequent failure occurred when the applied stress intensity factor reached the material's resistance to fracture (i.e., the K_c appropriate for the loading and temperature under service conditions). However, in thinner sections, there is often a period of stable crack growth preceding unstable failure. In these sections the initial crack extension is confined to the center of the specimen's thickness. On the surface the effect of the larger plastic zone size (Figure 6.18) is to retard the growth of the crack through the mechanisms of energy absorption by plastic deformation and stable tearing. As the load is further increased, the crack extends by slow–stable crack growth, with the crack front leading in the center and lagging on the surface. Also, the plane of separation of the crack faces on the surface is oblique to the plane of the crack front, giving rise to a pronounced shear lip. In this case, the instability criterion is more complex.

To incorporate slow–stable crack growth into the framework of linear elastic fracture mechanics, Krafft, Sullivan, and Boyle [1962] postulated and verified the concept of the *R*-curve. They argued that each material of a given thickness at a given temperature and loading rate has a unique crack-growth resistance curve independent of initial crack length and specimen geometry (i.e., the equilibrium stress intensity factor for stable crack growth K_R depends only on the crack extension Δa and the specimen's geometry). The independence of these curves on specimen geometry was later experimentally verified by Heyer and McCabe [1972]. A typical

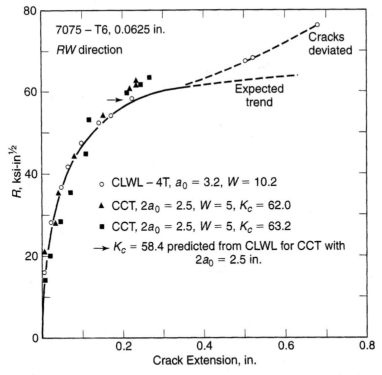

Figure 6.32 Characteristic *R*-curve shape in 7075 aluminum for several geometries (courtesy D. McCabe, Oak Ridge National Laboratory). Copyright 1972 Elsevier Science, reprinted with permission.

result from their study is shown in Figure 6.32 for 7075 aluminum. Given these observations, we can treat the *R*-curve as a material property for a given thickness of material, and, as such, the crack-growth resistance curve K_R versus a, like K_{Ic}, must be measured. Recommended procedures for measuring K_R are described in ASTM Standard Practice E 561.

The presence of stable crack extension prior to failure complicates the failure criterion, but quantitative predictions of the applied load at instability are still possible. Figure 6.33 shows an idealized *R*-curve for a specimen containing a crack of length a_o. The figure also shows a series of geometric-stress-intensity-factor curves (shown dashed) for several levels of applied load ($P_1 < P_2 < P_3$). We note that in constructing these latter curves, the total effective crack length $a_o + \Delta a + r_y$ must be used. The intersection of K_{applied} and K_R provides the measure of the amount of stable crack growth up to that level of applied load (assumed to be monotonically applied). Slow crack growth can continue until the curve of K_{applied} versus a becomes tangent to the *R*-curve, at which point instability occurs. With these observations,

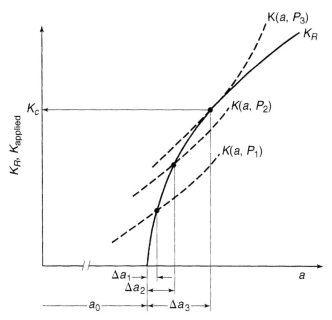

Figure 6.33 Schematic representation of the R-curve instability criteria for a given initial crack length.

we can now state the failure criteria for the onset of unstable fracture in the presence of slow crack growth prior to fracture as

$$K_R = K_c \qquad (6.34a)$$

when

$$\frac{\partial K_{\text{applied}}}{\partial a} = \frac{\partial K_R}{\partial a} \qquad (6.34b)$$

From Eqs. (6.34) we can make several important observations about unstable fracture in bodies of thinner section. The critical stress intensity factor is no longer a material property. As such, the critical stress intensity factor depends on the geometry, since K_{applied} is a geometric variable. Also, K_c depends on the initial crack length. This result is illustrated in Figure 6.34, which shows that the point of tangency moves up the K_R curve as the crack length increases for a fixed geometry. For this reason, short cracks tolerate less stable crack extension prior to fracture than do longer ones. Finally, although K_c increases for longer initial cracks, the applied load at instability decreases, since the tangency point moves to lower K_{applied} curves, as Figure 6.34 illustrates.

Because of the difficulties in measuring the R-curve for a given material and the need to construct separate R-curves for each thickness of interest, this type of analysis is rarely performed during the design stage. Instead, an R-curve analysis

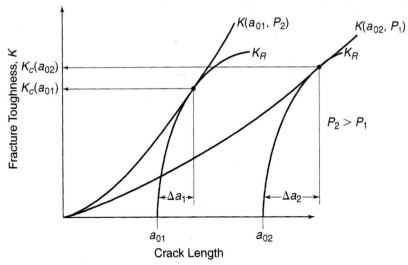

Figure 6.34 Influence of the initial crack length on the R-curve instability criteria.

is more apt to be performed after the design is finalized and all of the parameters are fixed, but only if the application is critical enough to justify the cost and time needed to complete the measurements and analysis. More often, the concepts of R-curve analysis are used to justify the assumption of a K_c larger than that predicted by Eq. (6.33), in lieu of actual measurements.

REFERENCES

Barsom, J. M., and Rolfe, S. T., 1987, *Fracture and Fatigue Control in Structures*, 2nd ed., Prentice-Hall, Englewood Cliffs, New Jersey.

Dugdale, D. S., 1960, "Yielding of Steel Sheets Containing Slits," *J. Mechanics and Physics of Solids*, 8, pp. 100–104.

Heyer, R. H., and McCabe, D. E., 1972, "Crack Growth Resistance in Plane-Stress Fracture Testing," *Engineering Fracture Mechanics*, 4, pp. 413–430.

Hult, J. A. H., and McClintock, F. A., 1957, "Elastic-Plastic Stress and Strain Distribution Around Sharp Notches Under Repeated Shear," *IXth International Congress of Applied Mechanics*, 8, Brussels, p. 51.

Irwin, G. R., 1957, "Discussion of: The Dynamic Stress Distribution Surrounding a Running Crack—A Photoelastic Analysis," *Proc. Soc. Experimental Stress Analysis*, 16(1), pp. 93–96.

Irwin, G. R., 1960, "Plastic Zone Near a Crack and Fracture Toughness," *Mechanical and Metallurgical Behavior of Sheet Materials*, Proceedings Seventh Sagamore Ordinance Materials Conference, pp. IV–63–IV–78.

Krafft, J. M., Sullivan, A. M., and Boyle, R. W., 1962, "Effect of Dimensions on Fast Fracture Instability of Notched Sheets," *Proceedings, Crack Propagation Symposium*, Vol. 1, Cranfield, United Kingdom, pp. 8–26.

Larsson, S. G., and Carlsson, A. J., 1973, "Influence of Non-Singular Stress Terms and Specimen Geometry on Small-Scale Yielding at Crack Tips in Elastic-Plastic Materials," *J. Mechanics and Physics of Solids*, 21, pp. 263–277.

McClintock, F. A., and Irwin, G. R., 1965, "Plasticity Aspects of Fracture Mechanics," *Fracture Toughness Testing and Its Applications*, ASTM STP 381, American Soc. for Testing and Materials, Philadelphia, pp. 84–113.

Tyson, W. R., and Roy, G., 1989, "Are The Foundations of Fracture Mechanics Cracked?", *Engineering Fracture Mechanics*, 33(5), pp. 827–830.

Wells, A. A., and Post, D., 1957, "The Stress Field Surrounding a Running Crack— A Photoelastic Analysis," *Proc. Soc. Experimental Stress Analysis*, 16(1), pp. 69–92.

Williams, M. L., 1957, "On the Stress Distribution at the Base of a Stationary Crack," *J. Applied Mechanics*, 24, pp. 109–114.

EXERCISES

6.1 Expand Eq. (6.17) in a suitable series, and demonstrate that Eq. (6.18) follows.

6.2 Expand Eq. (6.21b) in a suitable series, and demonstrate that Eq. (6.22) follows.

6.3 (a) Derive expressions for the Cartesian stress from the stress functions of Eqs. (6.20).

(b) Compare the stresses from part (a) to those of the corresponding elastic solution [i.e., Eqs. (3.37)].

6.4 Construct the plane-stress plastic zone size counterparts to those shown in Figure 6.17.

6.5 A thin steel plate ($\sigma_{ys} = 40$ ksi), that is 6 inches wide contains a central crack 2.0 inches in length.

(a) If the plane-stress fracture toughness of the steel is 55 ksi$\sqrt{\text{in}}$, what is the maximum stress that can be supported by the plate (including the effects of local yielding at the crack tips)?

(b) What is the longest crack the plate can support without failure at an applied stress of 20 ksi?

6.6 A very wide plate with an internal crack is subjected to a pair of opposing point loads acting at the center of the crack. If the crack starts to grow under constant load, will the fracture be stable or unstable? Justify your answer.

6.7 A $\frac{1}{4}$ inch-thick wide aluminum ($\sigma_{ys} = 40$ ksi) tension panel has an edge crack 0.80 inch long. Based on fracture tests with a similar material for the same thickness, the fracture toughness has been measured as 45 ksi$\sqrt{\text{in}}$ for slowly applied loads.

(a) Compute the diameter of the plastic zone for both plane-stress and plane-strain conditions.

(b) Based on your engineering judgment and the information available to you, which of the two plane conditions prevails? Why?

(c) What is the magnitude of the largest remotely applied steady-state stress that the panel can support without failure?

(d) What would be the consequence of an impact load on your answer to part (c)?

CHAPTER 7

The Energy of Fracture

7.1 INTRODUCTION

In the last chapter we developed a theory of failure for structures containing cracks, based on the stress field in the immediate neighborhood of the crack tip. The influence of the body taken as a whole and its remotely applied forces was incorporated into the theory through the geometric stress intensity factor. In this chapter we will take the opposite point of view: We look at the body from a global perspective and deduce failure theories from energy-balance arguments. In this case the local stress field at the crack tip enters the formulation indirectly through its influence on the strain energy.

Historically, an energy-based theory of failure predates the stress intensity factor theory by over 30 years. However, for a variety of reasons, the stress-field approach has become the most widely used method in structural applications. Nonetheless, there are two areas in which the energy approach has found application. For truly brittle materials, primarily including glass and ceramics, the theory of fracture proposed by Griffith [1921, 1925] is just as applicable today as it was when it was first proposed and continues to have a broad appeal. At the other end of the spectrum, nonlinear and elastoplastic fracture mechanics, the use of energy balance is the primary mathematical tool available for the development of suitable crack-extension theories. We will discuss elastoplastic fracture mechanics in Chapter 11.

From a purely idealistic point of view, the tensile strength of materials composed of atoms within a space lattice is governed by interatomic forces. At equilibrium, the cohesive forces trying to bind the material together are balanced by the repulsive forces keeping it apart. When a remote tensile force is applied, both the cohesive and repulsive forces decay as the atoms are strained from their equilibrium position; however, the repulsive forces decay at a faster rate. As shown in Figure 7.1, the

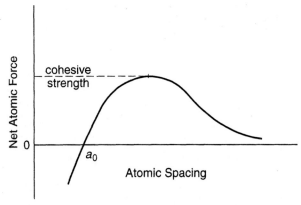

Figure 7.1 Change in the net cohesive force with increasing applied stress.

net tensile force increases on further loading until a maximum is reached, at which the equilibrium becomes unstable and failure occurs. Based on these concepts, the various theories of failure that have been proposed all lead to the same general conclusion that the cohesive tensile strength is of the order of $E/10$, where E is the elastic modulus. Glass, for example, has an elastic modulus of 10 million psi and a theoretical strength of approximately 1 million psi. We recognize that a more typical figure for the tensile strength of bulk glass is two or more orders of magnitude lower than this theoretical value. However, under carefully controlled conditions, measurements of the strength of glass that approach the theoretical strength have been reported.

Imagine the following conceptual experiment: A fine thread of freshly drawn glass is given a half twist to form a loop, as shown in Figure 7.2. Being careful not to let the thread touch itself and thereby abrade the surface, we can apply a tensile load

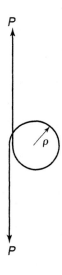

Figure 7.2 Method for determining the tensile strength of thin fibers.

and observe the radius of curvature, ρ, of the loop. Since the curvature is directly related to the extreme fiber stress in bending, the tensile strength can be estimated from the radius of the loop at the instant of fracture. Griffith [1921] performed experiments not unlike the conceptual experiment described here and made two important observations. First, by testing glass threads of varying diameters, he found that the strength was inversely related to the diameter, and in the limit the threads of smallest diameter had strengths as high as 900,000 psi. Second, he observed that these high strengths persisted for only a short period of time. The strength of "old" glass threads and rods was dramatically lower. It was these observations that led to the formulation of his theory of rupture of brittle materials.

7.2 GRIFFITH'S THEORY OF BRITTLE FRACTURE

Griffith considered various causes for the marked decrease in the strength of glass and other brittle materials from their pristine values and ultimately postulated that the origin was due to the presence of crack-like defects on the microscale. These cracks were introduced into the pristine material through the attack on the surface by exposure to the atmosphere (we now know that water vapor is the attacking agent) or by abrasion due to casual contact or handling of the samples. In his 1925 paper,[1] Griffith formulated his theory of brittle fracture based on six essential elements:

1. *All materials contain a population of fine cracks.* This condition provided the distinction between a real material with typically low strength and the near-perfect material form having tensile properties approaching the theoretical strength.

2. *Some of these cracks are oriented in the most unfavorable direction relative to the applied loads so as to have the maximum stress concentration factor.* By this argument, Griffith was able to avoid detailed stress analysis of cracks at arbitrary angles. It was necessary to assume only that the cracks were far enough from each other that there was no interaction.

3. *At one of these cracks, the theoretical strength is reached at the crack tip, and the crack grows.* This element of the theory allowed Griffith to treat the problem as an extreme-value problem rather than a statistical one.

4. *The source of the energy for crack propagation is the strain energy released as the crack extends.* Although Griffith considered only infinite bodies, he was able to show that the change in strain energy as the crack extended was finite. Since the strain energy decreased as the crack extended, this energy became available to the crack.

5. *The growth of the crack results in an increase in surface energy.* Griffith considered the fixed grip condition (see Chapter 1) and needed a receptor for the loss

---

[1] The paper was presented at the First International Congress for Applied Mechanics at Delft in 1924, but the proceedings were not published until the following year.

of strain energy. Since new surfaces were created as the crack extended, the associated surface energy became a likely candidate. This had the additional advantage that surface energy was a well-defined property of these materials and measured values were available.

6. *The crack growth process continues as long as the rate of released strain energy exceeds the energy required to form a new surface.* This final element of the theory provided the criterion for crack growth.

Given these elements, Griffith was able to formulate a mathematical expression for the critical stress to cause failure in terms of a measurable material property, the surface energy. In order to determine the strain energy available for the formation of new surfaces, he used the only complete solution for a crack-like defect available at the time, namely the Inglis [1913] solution discussed in Chapter 2.

Griffith compared the strain energy in an infinite plate of unit thickness subjected to a fixed displacement sufficient to cause a tensile stress, σ, in the absence of a crack with that of a similar plate containing a crack of length $2a$. For plane stress the decrease in strain energy, U, was

$$U = -\frac{\pi a^2 \sigma^2}{E} \tag{7.1}$$

Since there are two surfaces associated with a crack, the surface energy, W_s, required to form a crack of length $2a$ in a body of unit thickness is

$$W_s = 2 \cdot \gamma_s \cdot 2a(1) = 4a\gamma_s \tag{7.2}$$

where γ_s is the surface energy per unit area for the bulk material.

With these energies the Griffith criterion for sustained crack growth can be expressed in terms of the change in energy $\Delta U = U + W_s$ as

$$\frac{d(\Delta U)}{da} \geq 0 \tag{7.3}$$

However, the criterion for the critical stress for the onset of fracture, σ_c, is given by the lower bound of Eq. (7.3)—that is, the rate (areal) of strain energy supplied is just equal to the rate of energy required to form new surfaces:

$$\frac{d}{da}\left(4a\gamma_s - \frac{\pi a^2 \sigma_c^2}{E}\right) = 0 \tag{7.4}$$

Therefore,

$$\sigma_c = \sqrt{\frac{2E\gamma_s}{\pi a}} \tag{7.5a}$$

for plane stress and

$$\sigma_c = \sqrt{\frac{2E\gamma_s}{(1 - \nu^2)\pi a}} \qquad (7.5b)$$

for plane strain.

In the interests of historical accuracy, it should be noted that Eqs. (7.5) are not the same as those originally published by Griffith in his 1921 paper. In fact, a footnote at the end of that paper makes the observation that the author was himself aware of errors in the equations he derived, but that the order-of-magnitude arguments, which were the primary focus of the paper, were unaffected. In his 1925 paper, Griffith explained the origin of his earlier error and gave, without details, the corrected forms used here. The details of the derivation were later provided by Sih and Liebowitz in 1967.

The Griffith theory of brittle fracture as presented here is subject to two severe restrictions. First, since the change in strain energy was computed for the case of an infinite body containing an isolated crack of length $2a$, the theory is valid only for real problems that approximate this condition. Second, the theory applies only to truly brittle materials for which there is no mechanism of energy dissipation (e.g., plastic deformation) other than the formation of surface energy as the crack extends. These restrictions are of little concern to the glass and ceramic industries, and Griffith's theory in its original form is still widely used within these communities. On the other hand, these restrictions become significant when the size of critical flaws that are more than microscopic or plastic deformation at the crack tip (even highly localized) is involved. As a result the traditional structural materials community has tended to ignore the Griffith approach, favoring instead the stress intensity factor approach. Nonetheless, the energy balance concepts introduced by Griffith have broad implications in modern fracture mechanics theory, as we will see soon.

7.3 A UNIFIED THEORY OF FRACTURE

Outside the glass-fabrication industry, the consequences of the Griffith theory of fracture were largely ignored. It is easy to demonstrate that predictions of the fracture strength of metals containing measurable cracks via the Griffith's surface-energy criterion result in absurdly low values. Orowan [1955] demonstrated that even highly brittle fractures of low-carbon steels exhibited significant plastic deformation along the fracture surface.[2] As a result the plastic work of fracture exceeded the surface energy by many orders of magnitude. Orowan argued that if the plastic deformation is confined to a thin layer at the fracture surface, then the plastic work is proportional to the crack area and can be described by γ_p, the surface plastic work. Then Griffith's

---

[2] Irwin [1948] made similar observations in the context of dynamic fracture of brittle steels.

instability equations [Eqs. (7.5)], could be extended to metals by replacing the surface energy γ_s with $\gamma_s + \gamma_p$. So,

$$\sigma_c = \sqrt{\frac{2E(\gamma_s + \gamma_p)}{\pi a}} \tag{7.6a}$$

for plane stress and

$$\sigma_c = \sqrt{\frac{2E(\gamma_s + \gamma_p)}{(1 - v^2)\pi a}} \tag{7.6b}$$

for plane strain. By the measurements of Felbeck and Orowan [1955], γ_p was about 1000 times greater than γ_s, and for all practical purposes, γ_s could be ignored in Eqs. (7.6).

Orowan's extension of the Griffith theory to metals was later generalized by Irwin [1957], who collected together all sources of resistance to crack extension into a single term G_c, which he called the *strain energy release rate*. (In Irwin's earliest works, the term *crack extension force* was used.) Accordingly, the Irwin form of the Griffith criterion can be written as

$$\sigma_c = \sqrt{\frac{E G_c}{\pi a}} \tag{7.7a}$$

for plane stress and

$$\sigma_c = \sqrt{\frac{E G_c}{(1 - v^2)\pi a}} \tag{7.7b}$$

for plane strain, where G_c is a new bulk material property that can be determined only from fracture measurements on the specific material being considered.

With this generalization, the Orowan–Irwin fracture criterion [Eqs. (7.7)] has formed the basis for an energy-based theory of failure for *all materials* failing in a quasibrittle manner. However, we must observe that these equations were developed as extensions of the Griffith condition and apply only to the same geometry used by Griffith—namely, an isolated crack of length 2a in an infinite sheet under remote tension in either plane stress or plane strain. But for the same problem, we have already determined from Eqs. (3.43) that the geometric stress intensity factor is

$$K = \sigma\sqrt{\pi a} \tag{7.8a}$$

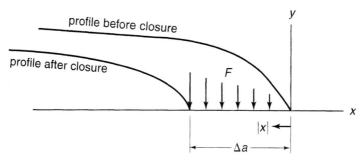

Figure 7.3 Crack-closing forces within the singularity-dominated-zone.

Therefore, from the theory developed in the previous chapter, at the instant of instability, we have

$$K_c = \sigma_c\sqrt{\pi a} \tag{7.8b}$$

Comparing Eqs. (7.7) and (7.8) for the same failure event reveals that

$$K_c^2 = E\,G_c \tag{7.9a}$$

for plane stress

$$K_c^2 = \frac{E\,G_c}{1 - v^2} \tag{7.9b}$$

for plane strain. Hence, we observe that, at least for this problem, the critical-energy-release rate theory of failure, based on global energy concepts, and the critical stress-intensity-factor criterion, based on the local stress field, are equivalent and can be used interchangeably.

We can generalize the relation between G and K to any geometry or crack configuration by examining the work required to close a crack by a small increment Δa, as illustrated in Figure 7.3. Since we have assumed elastic behavior, the energy released during a small increment of crack extension, $G\Delta a$, must be equal to the closure work over the same interval of crack length. If we confine our attention to a small increment of crack growth wholly within the singularity-dominated-zone, then we can describe the complete state of stress, strain and displacement by the near-field equations characterized by the germane stress intensity factor K. Therefore, for any geometry, the crack-opening displacement in plane stress is given by

$$Ev = \frac{2K}{\pi}\sqrt{2\pi|x|} \qquad \text{for } x \leq 0 \tag{7.10}$$

After closure, the force acting at a point in the closed region is

$$F = \sigma_y dx = \frac{K\,dx}{2\pi(\Delta a - |x|)} \tag{7.11}$$

where the quantity $\Delta a - |x|$ is the distance measured from the tip of the closed crack. Therefore, the total closure work over the interval Δa is

$$2\int_0^{\Delta a} \frac{1}{2} F v\,dx = \mathcal{G}\Delta a \tag{7.12}$$

Substituting from Eqs. (7.10) and (7.11) yields:

$$\frac{2K^2}{\pi E}\int_0^{\Delta a}\sqrt{\frac{|x|}{\Delta a - |x|}}dx = \mathcal{G}\Delta a \tag{7.13}$$

Letting $|x| = \Delta a \sin^2 u$ permits Eq. (7.13) to be integrated directly. Provided that the interval of crack extension is nonzero, the final result is

$$K_c^2 = E\mathcal{G}_c \tag{7.14}$$

for plane stress.

By this argument we have just demonstrated that the results of Eqs.(7.9), derived for the special case of an internal crack under remote loading, are, in fact, applicable for all geometries. Consequently, the stress-field theory of fracture described in Chapter 6, and the energy-release-rate theory described in this chapter are equivalent and interchangeable. For the case of pure mode-II loading, the equalities have the same form as their opening-mode counterparts, and under mixed-mode conditions for plane stress,

$$E\mathcal{G}_c = K_{I_c}^2 + K_{IIc}^2 \tag{7.15}$$

where $\mathcal{G}_c$ is interpreted as the total strain energy release rate.

7.4 COMPLIANCE

In Chapter 1 we defined the strain-energy release rate as the rate of change of stored strain energy with respect to the crack area under system-isolated (fixed-grip) conditions, that is,

$$\mathcal{G} = -\frac{\partial U}{\partial A}\bigg|_v \tag{7.16}$$

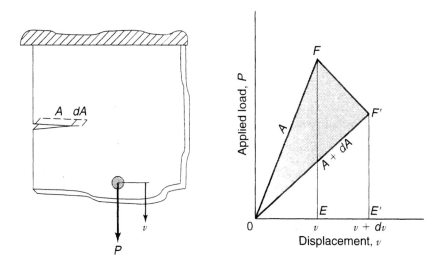

Figure 7.4 Load-deflection diagram (right) for a cracked body subjected to a force with a load-point displacement (left).

from which it followed that, for linear elastic bodies, G could be obtained from the change in compliance through the relation

$$G = \frac{P^2}{2} \frac{\partial C}{\partial A} \tag{7.17}$$

However, the system-isolated condition is unique in that all of the change in strain energy goes into the formation of new crack surfaces. In the more general case, some energy is consumed by other causes.

Consider the situation shown in Figure 7.4a of a cracked body subjected to an external load. As the crack extends by an amount dA, the load point moves by a corresponding amount dv along some path in compliance space, as shown in Figure 7.4b. However, unlike the fixed-grip condition previously discussed, not all of the shaded area OFF' is due solely to changes in stored strain energy. We need to resolve this area into its component parts. For this purpose we observe that the area OFF' can be obtained from

$$OFF' = OFE + EFF'E' - OF'E' \tag{7.18}$$

However, to the first order, the area $EFF'E'$ can be approximated by Pdv, which we recognize as the work done by the external load during the crack extension. Since $OF'E' - OFE$ is the total change in stored strain energy, dU, the energy available for crack extension is

$$OFF' = GdA = Pdv - dU \tag{7.19}$$

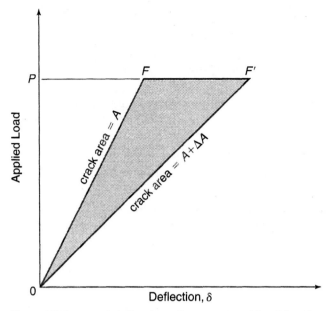

Figure 7.5   Load-deflection diagram for "dead load" fracture event. (Compare with Figure 1.8.)

or

$$G = P\frac{dv}{dA} - \frac{dU}{dA} \qquad (7.20)$$

Before proceeding further, we can make several observations about the strain-energy release rate from Eq. (7.20). First, note that if the load point is fixed, then $dv/dA = 0$, and the system-isolated result of Eq. (7.16) is recovered. Second, since there was no restriction on linear behavior imposed in the derivation, the strain-energy release rates of nonlinear elastic bodies can be determined by using Eq. (7.20), provided that the appropriate areas can be computed or measured. On the other hand, if the behavior is linear, such that $U = \frac{1}{2}Pv$, further observations can be made.

Let us now consider the other extreme-loading case—i.e., constant load (or dead load)—illustrated in Figure 7.5. In this case, Eq. (7.20) reduces to

$$G = \left.\frac{\partial U}{\partial A}\right|_P \qquad (7.21)$$

Substituting Eqs. (1.8) into Eqs. (1.7) and taking the partial derivative at constant load yields

$$G = \frac{P^2}{2}\frac{\partial C}{\partial A} \qquad (7.22)$$

Since the compliance relations, Eqs. (7.17) and (7.22) are the same for the two extreme loading cases, it must necessarily follow that the form of these equations must also hold for the more general case, because it is always possible to interpret any arbitrary unloading path as the sum of alternating infinitesimal constant-load and infinitesimal constant-displacement steps. Released from the restrictions on load-point fixity, Eq. (7.22) provides a versatile tool for determining the strain-energy release rate and, from it, the geometric stress intensity factor for geometries not otherwise amenable to analysis.

Example 7.1   Consider the double-cantilever beam (DCB) specimen shown in Figure 7.6. If the crack length, a, is short compared with the overall length of the specimen, we can treat the uncracked portion as the "rigid support" for the pair of cantilever beams formed by the crack. Then, from elementary beam theory, we have

$$v = 2 \cdot \frac{Pa^3}{3EI} \quad \text{where } I = \frac{Bh^3}{12} \tag{7.23}$$

Since the compliance is, $C = v/P$, elementary calculations yield

$$G = \frac{P^2}{2} \frac{\partial C}{\partial A} = \frac{P^2 a^2}{B_n I} \tag{7.24}$$

where the only influence of the side groove shown in Figure 7.6 used to guide the crack is its effect on the net section area of the crack (i.e., $A = aB_n$). For common specimens that are tall compared with their thickness (plane stress),

$$K^2 = EG = \frac{P^2 a}{B_n I} \tag{7.25}$$

Additional results for the double-cantilever specimen will be presented in Chapter 8. ■

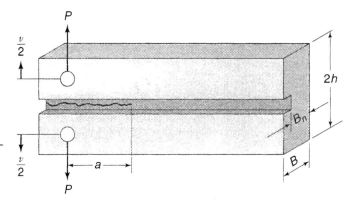

Figure 7.6 The double-cantilever-beam (DCB) geometry subjected to end-point loads.

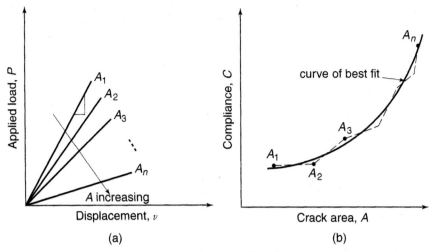

Figure 7.7 Constructing a compliance calibration curve for an arbitrary body containing a cracked area.

An important feature of the compliance formulation is its generality. For geometries that are too complex to permit direct determination of the stress intensity factor by calculation or experiment using the methods outlined in either Chapter 4 or Chapter 5, the compliance method provides an alternative procedure. This approach can be particularly important in three-dimensional bodies. The approach also finds application in failure analysis. In this latter case the examiner (you!) is asked to determine the cause of failure, given only a broken part. Examination of the broken halves usually reveals a crack that has grown to critical dimensions. The steps involved in a typical failure analysis based on compliance are listed as follows:

(a) Obtain an unbroken sample, preferably unused, of the part being investigated. Alternatively, if the part is too large or too small, a proportional replica of the part can be fabricated of a suitable material and to reasonable size. In this event, all of the results should be cast in dimensionless form for transference of the results to the actual part.

(b) Introduce a crack-like slit of area A_1 that is somewhat smaller than the critical dimensions. In the case of prior fatigue crack growth, there are often fatigue striations on the fracture surface to guide selection of the appropriate crack profile.

(c) Conduct an instrumented load test to determine the load-displacement relation.

(d) Increase the crack area to $A_2, A_3, \ldots$ and repeat step (c) at each increment until the cracked area exceeds the observed critical area by several increments.

(e) Determine the compliance from each of the $P\text{--}v$ diagrams as shown in Figure 7.7a, and plot them on a common graph, as in Figure 7.7b.

(f) Fit the compliance data to a suitable low-order fitting function in the least-squares sense. There are no rigid rules for selecting a fitting function, but some guidelines are available. For structures subjected to external force loading, the curvature (i.e., dC/dA) of the fitting function must monotonically increase with increasing crack area. For displacement-controlled structures, the converse is true.

(g) Analytically differentiate the fitting function to determine dC/dA, and plot it as a function of A. Note that this procedure is preferable to graphical differentiation of the measured compliances, as the latter are prone to experimental errors that are not averaged out in the graphical approach.

(h) Use the master curve to determine dC/dA for the crack area of interest and thereby determine either the stress intensity factor at failure, K_c or K_{Ic}, if the failure loads are accurately known, *or* the load causing failure, P_c, if the material's resistance to fracture is known, from the relation

$$\frac{K_c}{P_c} = \sqrt{\frac{E'}{2}\frac{dC}{dA}} \tag{7.26}$$

where $E' = E$ for plane stress or $E' = E/(1 - \nu^2)$ for plane strain.

In a typical failure analysis scenario, this type of analysis can be used to determine if the material's resistance to fracture, K_c or K_{Ic}, of the failed part met design minimums or if the part was "inadvertently" overloaded (i.e., did P_c exceed the design maximum?). Although the foregoing procedure was described in the context of experimental measurements of compliance, the procedure can be simulated using finite element analysis. Experience indicates that numerical estimates of compliance tend to understate the true value, due to the discretization process. However, the derivative of the compliance is less sensitive to the actual mesh, provided that it is consistent among the multiple models needed for this procedure.

REFERENCES

Felbeck, D. K., and Orowan, E., 1955, "Experiments on Brittle Fracture of Steel Plates," *Welding Journal, Research Supplement*, Vol. 34 (11), pp. 570s–575s.

Griffith, A. A., 1921, "The Phenomena of Rupture and Flow in Solids," *Philosophical Trans. Royal Society*, London, Vol. A221, pp. 163–198.

Griffith, A. A., 1925, "The Theory of Rupture," *Proceeding*, First International Congress for Applied Mechanics, C. B. Biezeno and J. M. Burgers, Eds., pp. 55–63.

Inglis, C. E., 1913, "Stresses in a Plate Due to the Presence of Cracks and Sharp Corners," *Trans. Inst. Naval Architects*, Vol. 55, pp. 219–230.

Irwin, G. R., 1948, "Fracture Dynamics," *Fracturing of Metals*, ASM International (American Society for Metals), Metals Park, OH, pp. 147–166.

Irwin, G. R., 1957, "Analysis of Stresses and Strains Near the End of a Crack Traversing a Plate," *J. Applied Mechanics*, Vol. 24, pp. 361–364.

Orowan, E., 1955, "Energy of Fracture," *Welding Journal, Research Supplement*, Vol. 34(3), pp. 157s–160s.

Sih, G. C., and Liebowitz, H., 1967, "On the Griffith Energy Criterion for Brittle Fracture," *Int. J. Solids Structures*, Vol. 3, pp. 1–22.

EXERCISES

7.1 Consider a double-cantilever beam similar to the one shown in Figure 7.6, except subjected to a constant end moment, M, rather than the point load shown in the figure. For this case the total stored strain energy in the beam is given by $U_T = \frac{1}{2}M\theta$, where θ is the angle of twist at the free end.

(a) Illustrate with a suitable diagram the energy available for crack extension for this case.

(b) Using energy principles, derive an expression for the geometric stress intensity factor K for this specimen. Assume plane-stress conditions.

(c) Design a set of test fixtures that would permit specimens of this type to be tested in a universal testing machine more commonly used for tension tests.

7.2 Based on prior experience, the $\frac{1}{4}$ in-thick steel bracket shown in Figure E7.1 is prone to develop fatigue cracks in the area of the fillet as shown. After performing a series of experiments with simulated cracks of varying lengths, we find that the compliance is fitted to an equation of the form

$$\frac{v}{P} = \left(100 + 20\sin\frac{\pi a}{4}\right) \times 10^{-6} \text{ in/lb.}$$

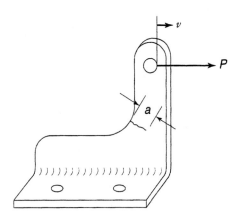

Figure E7.1

(a) Derive an expression for the geometric stress intensity factor for this bracket. Assume plane-stress conditions.

(b) Visual inspection can detect cracks as short as 0.2 inch in the bracket. If the bracket is to be used for loads up to 2000 lb, what is the minimum fracture toughness K_c of the steel that must be specified for this application? A factor of safety against fracture of 2 is to be included in your analysis.

7.3 As a failure expert, you have been called to resolve a dispute between a customer and the manufacturer of a crane hook that failed in service. The customer contends that the hook failed because of a flaw and offers as proof the fact that the fracture surface of the failed part has a obvious crack with an area of 0.35 square inch. (See Figure E7.2.) The manufacturer counters that damage-tolerant design concepts were used in the design of the hook, and a crack of this size would not lead to failure. The manufacturer counterclaims that the customer exceeded the maximum load rating of the hook of 4000 lb. To aid in the analysis, you conduct compliance tests with machined flaws of 0.30 and 0.40 square inch. The test results are shown in Figure E7.2. The hook is made of steel ($E = 30,000,000$ psi and $\nu = 0.29$) and has a plane-strain fracture toughness of $K_{Ic} = 40$ ksi$\sqrt{\text{in}}$. Using this information, who is correct? Justify your answer.

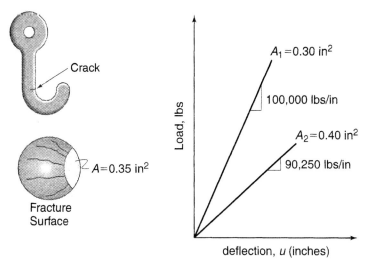

Figure E7.2

7.4 Assuming that you were to perform the experiment illustrated in Figure 7.2 with freshly drawn glass fibers, what would be the diameter of the loop at the instant of failure for a fiber with diameter (a) 0.001 inch and (b) 400 microinches, both at the theoretical strength limit for glass.

CHAPTER 8

Fracture-Toughness Testing

8.1 INTRODUCTION

The application of fracture mechanics concepts to design requires two quantities: the geometric stress intensity factor as a function of the crack length for the specific geometry, and the material's resistance to fracture (i.e., K_c). In earlier chapters we have developed a variety of techniques for determining the former, using analytical, numerical, and experimental approaches. We will now turn our attention to techniques for determining the material property K_c, or its equivalent G_c.

In principle, any geometry can be used to determine the critical fracture toughness if the geometric stress intensity factor as a function of crack length is accurately known. Then K_c can be determined from a single failure test, knowing the crack length and applied load at the instant of instability, through Eq. (6.4). The first successful application of the concept of linear elastic fracture mechanics was to the measurement of the fracture toughness of hot-stretched acrylic used as an aircraft-glazing material. Tests conducted at the Naval Research Laboratory in Washington during the period 1953–1956 demonstrated that hot-stretching greatly improved the material's resistance to fracture. Shortly thereafter, a standardized test, codified in a military specification (mil spec) was developed based on Irwin's tangent formula, Eq. (3.60). This test was then used as the acceptance criterion for processed sheets of the material.

Although the early tests on metals were for thin plates, it was determined early on that thickness had a strong influence on fracture behavior. Irwin [1960] demonstrated, as shown in Figure 8.1, that thick sections have markedly lower fracture toughness values than those of their thin counterparts. Although the explanation for this phenomenon was not yet fully understood, he correctly attributed the effect to plane strain and the decreased significance of the plastic zone. These results led to

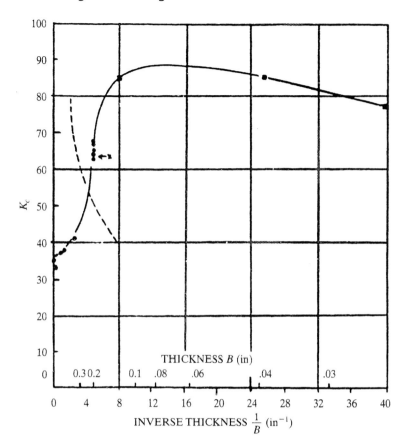

Figure 8.1 K_c vs. inverse thickness $\left(\frac{1}{B}\right)$ for 7075-T6 aluminum alloy [Irwin, 1960].

the systematic study by ASTM committee E–24 (now designated E–08) to develop a standard that measured a fundamental bulk fracture property of metallic materials, culminating in ASTM Standard E 399.

8.2 FRACTURE TOUGHNESS STANDARDS

Ideally, a standardized test should accomplish two objectives:

1. It should ensure that the quantity being measured is reproducible among independent testing laboratories; and
2. It should ensure that the quantity being measured is a physical property that can be used in later applications as a design variable.

We note that Objective 2 is much more demanding than Objective 1. The goal of the first objective is to specify the test conditions necessary to guarantee, to some acceptable degree of precision, that the results of the test will be independent of the test operator, the laboratory performing the test, and the specific brand of test equipment used. In simple terms, a standard meeting this requirement will ensure that a supplier and a receiver of the same material will obtain comparable results. From this perspective the standard provides a basis for a mutually agreeable purchase specification. In contrast, Objective 2 sets a much higher standard (no pun intended!). In the latter case, not only will the measured quantity be reproducible, but also the quantity will have its own intrinsic value. By this we mean that the quantity measured has a physical interpretation, independent of the standard used to measure it, that can be used in other physics applications. While all acceptable standards meet Objective 1, relatively few meet Objective 2. The reader should not infer from these comments that standards meeting only Objective 1 have less value. Objective accept/reject criteria based on standardized tests are vital to commerce. The correct inference is that it is the user's responsibility to ensure that the quoted value from a standard meets the more stringent criterion of Objective 2 before accepting the quantity as a design variable.

The fracture mechanics community at large is fortunate that the standards related to fracture toughness testing generated by ASTM Committee E–08 meet the more demanding criterion. This was not an accident. Committee E–24 was initially composed of academic and industry researchers, and government research laboratories actively engaged in the scientific advancement of fracture mechanics as a discipline, and this basic character has remained unchanged in 40 years of activity. Committee E–24 was an outgrowth of the ASTM Special Committee on Fracture Testing of High-Strength Sheet Materials commissioned by the federal government in 1959 to investigate a series of metal failures of national significance. The early reports of this committee (reprinted by Barsom [1987]) provide a succinct history of the fracture research that led to the initial series of standards. From its earliest reports the focus of this group of dedicated contributors has been to maintain a high degree of scientific merit in all of the standards it has issued. Table 8.1 provides a listing of many of the ASTM Standards developed by this committee. We will discuss in detail only two of the standards from this formidable list in the upcoming sections.

Terminology Relating to Fatigue and Fracture Testing: ASTM E 1823

Over the years the field of fracture mechanics has grown in parallel within the various branches of materials science and mechanics. Along the way, specialized notations and conventions have been developed that have made cross-disciplinary reading difficult. In an effort to minimize this confusion, a unified set of terminology and descriptions was formalized in ASTM Standard E 616: *Standard Terminology Related to Fracture Testing* [1989]. In 1996, Standard E 616 was replaced with Standard E 1823, which combines the terminology of both fatigue and fracture testing into a single standard. The standard has two major components: a listing of the adopted mathematical symbols associated with each specific defined term commonly occurring in fracture mechanics literature (cross-referenced to the standards in which terms appear), and a codified system for describing the specimen type, loading, and

Table 8.1 ASTM Standards Related to Fracture Mechanics

E 338–91 (Reapproved 1997)	Standard Test Method of Sharp-Notch Tension Testing of High-Strength Sheet Materials
E 399–90 (Reapproved 1997)	Standard Test Method for Plane-Strain Fracture Toughness of Metallic Materials
E 436–91 (Reapproved 1997)	Standard Test Method for Drop-Weight Tear Tests of Ferritic Steels
E 561–98	Standard Practice for R-Curve Determination (formerly a Recommended Practice 1974)
E 602–91 (Reapproved 1997)	Standard Test Method for Sharp-Notch Tension Testing with Cylindrical Specimens
E 604–83 (Reapproved 1994)	Standard Test Method for Dynamic Tear Testing of Metallic Materials
E 647–95a	Standard Test Method for Measurement of Fatigue Crack Growth Rates
E 740–88 (Reapproved 1995)	Standard Practice for Fracture Testing with Surface-Crack Tension Specimens
E 812–91 (Reapproved 1997)	Standard Test Method for Crack Strength of Slow-Bend Precracked Charpy Specimens of High-Strength Metallic Materials
E 813–89 (Discontinued 1997)	Standard Test Method for J_{Ic}, a Measure of Fracture Toughness (see Test Method E 1737)
E 1221–96	Standard Test Method for Determining Plane-Strain Crack-Arrest Fracture Toughness K_{Ia} of Ferritic Steels
E 1290–93	Standard Test Method for Crack-Tip Opening Displacement (CTOD) Fracture Toughness Measurement
E 1304–97	Standard Test Method for Plane-Strain (Chevron-Notch) Fracture Toughness of Metallic Materials
E 1737–96 (Discontinued 1998)	Standard Test Method for J-Integral Characterization of Fracture Toughness (see Test Method E 1820)
E 1820–99	Standard Test Method for Measurement of Fracture Toughness (generalization of E 399)
E 1823–96	Standard Terminology Relating to Fatigue and Fracture Testing (replaces E 616–89)
E 1921–97	Standard Test Method for Determination of Reference Temperature T_o for Ferritic Steels in the Transition Range
E 1922–97	Standard Test Method for Translaminar Fracture Toughness of Laminated Polymer Matrix Composite Materials

material orientation. Since nearly all of the defined terms in the standard appear somewhere in this book and are defined when they first occur, we need not discuss standardized notation further. The sole exception is the use of subscripts to denote the mode of deformation: The standard adopts the use of Arabic numerals 1, 2, and 3 whereas this book uses Roman numerals I, II, and III to be consistent with the early literature.

The far more useful aspect of Standard E 1823 is its standardization of the myriad of shorthand notations for specimen types that have appeared in the literature over time. The unified code consists of three basic elements in code form plus an optional prefix (spelled out in full). The coding scheme is summarized in Table 8.2.

The optional prefix is a qualifier that usually denotes some deviation from the customary configuration. A representative sampling of qualifiers is given in the first column of Table 8.2. The main designation consists of three parts all expressed in capital letters. The first letter or set of letters indicates the specimen type (col-

Table 8.2 Designation Codes for Fracture Specimens

Prefix (optional)	Code, Configuration	Applied Load	Crack Orientation
precracked	M, middle	(T), tension	L, longitudinal rolling direction
notched	DE, double edge	(B), bending	
chevron	SE, single edge	(M_x), torsion about x-axis	T, width or long transverse
contoured	C, compact single edge	(M_z), torsion about z-axis	S, short transverse or thickness
	MC, modified compact	(W), wedge	R, radial
	DC, disk-shaped compact	(W_b), wedge w/bolt	
	A, arc		
	DB, double beam		
	RDB, round double beam (formerly short rod)		C, circumferential or tangential
	R-BAR, round bar		
	PS, part-through surface		

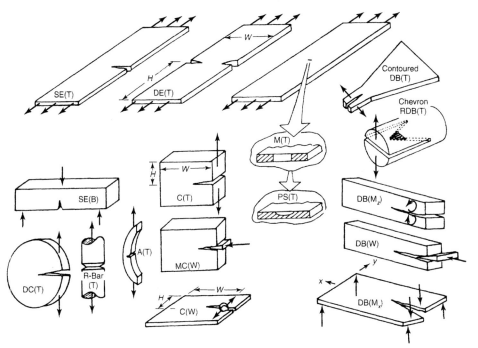

Figure 8.2 Standardized codes for specimen configurations and applied load (adapted from ASTM Standard E 1823, copyright ASTM, reprinted with permission).

umn 2 of the table). This letter or set of letters is followed in parentheses by the type of applied loading in code form (column 3 of the table). Combined, these two codes provide a complete mechanics description of the specimen. Some of the more common specimens and loading are shown in Figure 8.2. The last part of the code provides the description for the orientation of the specimen from the material blank from which it was prepared. This additional information is necessary because many metallic materials exhibit different properties for crack propagation along different planes. The orientation code, enclosed in parentheses, consists of two or more letters separated by a hyphen. The first letter or set of letters denotes the orientation of the normal to the crack plane, and the second letter or set of letters specifies the direction of crack propagation. Since this feature of fracture toughness measurement is closely associated with the form of the material blank, the designation code letters depend on the form and have separate standardized notations for plate and cylindrical stock, as shown in Figure 8.3. If the orientation of the crack normal is not along the principal directions, the designation requires two letters to describe the plane containing the normal to the crack plane; for example, the designation LT

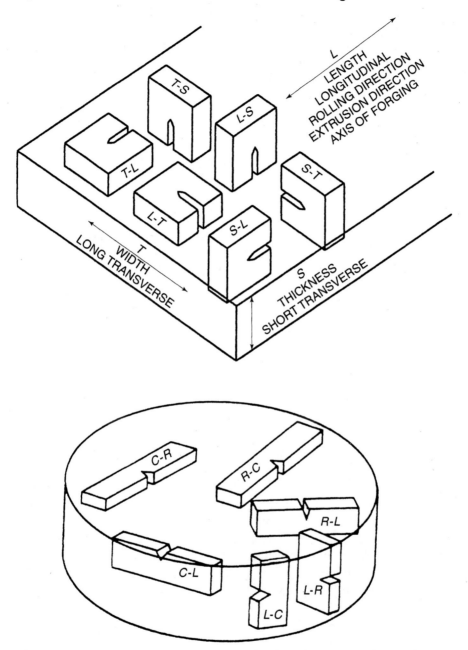

Figure 8.3 Orientation codes for sheet (top) and rod stock (bottom) (adapted from ASTM E 1823, copyright ASTM, reprinted with permission).

indicates that the normal to the crack plane lies in the plane defined by unit vectors in the L and T directions. A similar condition occurs if the direction of crack propagation is oblique to a principal material direction. In either of these cases, the use of two-letter designations is only qualitative, since the standard provides no mechanism for denoting the angle of the crack normal within the designated plane.

Plane-Strain Fracture Testing of Metallic Materials: ASTM E 399

As we have seen earlier, the fracture resistance of metals exhibits a thickness dependence on fracture resistance, due to the formation of a plastic zone around the crack tip. (See Figure 6.18.) At one extreme, in very thin specimens, plastic deformation is primarily in the antiplane direction, and the plastic zone takes the shape of a thin strip ahead of the crack. At the other extreme, the size of the plastic zone becomes negligible in thick specimens. For this latter case, the material resistance to fracture behaves as a bulk material property, and its value should be the same in specimens of any geometry, provided that the thickness meets the minimum requirement. We also have observed that K_{Ic} is a lower bound measure and represents the most conservative estimate of fracture resistance.

Under the umbrella of ASTM Standard E 399 [1997], five specimen geometries have been certified for K_{Ic} testing. The size and shape of these specimens have been selected based on needs established by the fracture-testing community and the availability of accurate geometric stress intensity factor calibrations. For the most part the geometric calibrations are based on the results of boundary collocation solutions [Gross and Srawley, 1965] and, quite recently, finite element modeling. Because of the need for extensive numerical solutions to large matrix equations, NASA and its large mainframe computers were the primary contributors of these K solutions during the 1970s. In Standard E 399, these extensive databases of numerical solutions are reduced to simple power series expressions by using conventional curve-fitting procedures. By careful choice of the fitting functions, the results are reported to be accurate to within 0.5 percent over the permitted range of a/W [Srawley, 1976]. For all of the specimen geometries, the crack tip is constrained to be near the center of the specimen by the requirement that $0.45 \leq a/W \leq 0.55$.

Of all the specimen geometries permitted by Standard E 399, the bend bar [SE(B)], shown in Figure 8.4, is the simplest to prepare and test. It consists of a single-edge notched, precracked rectangular bar of width (height), W, loaded in three-point bending. The test method requires a span, S, to width ratio, S/W of 4. The standard bend bar as defined in Standard E 399 has a width-to-thickness ratio $W/B = 2$, but a range of values $1 \leq W/B \leq 4$ is permitted. For these proportions a

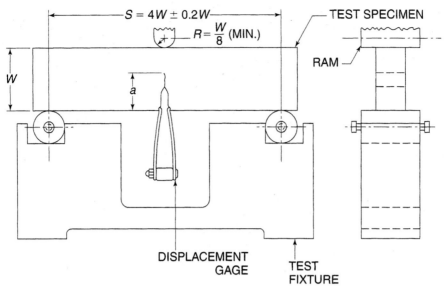

Figure 8.4 Standardized bend bar and test fixture (adapted from ASTM E 399, copyright ASTM, reprinted with permission).

provisional value of the material's plane-strain fracture toughness, K_Q, is computed from

$$K_Q = \frac{P_Q S}{B W^{3/2}} \cdot f\left(\frac{a}{W}\right) \tag{8.1}$$

where

$$f\left(\frac{a}{W}\right) = \frac{3\left(\frac{a}{W}\right)^{1/2}\left[1.99 - \left(\frac{a}{W}\right)\left(1 - \frac{a}{W}\right)\left(2.15 - 3.93\frac{a}{W} + 2.7\left(\frac{a}{W}\right)^2\right)\right]}{2\left(1 + 2\frac{a}{W}\right)\left(1 - \frac{a}{W}\right)^{3/2}} \tag{8.2}$$

and P_Q is a measure of the load at instability.

Although simple to prepare, the bend bar geometry is far from conservative in material, and the only role of most of the material is to provide a large bending moment at the plane of the crack. In applications where the availability of material from which to prepare specimens is limited or a minimum size is desirable, the compact tension geometry [C(T)] is preferred. An important example of this latter case is that of specimens subjected to irradiation degradation where storage space within an operating reactor is scarce. There are two variations of the compact geometry, as shown in Figure 8.5. The round compact-tension specimen was not in the original standard, but was added later as a convenience when the available stock was in the form of a round bar. Note that for the compact geometries, the in-plane reference dimension W is not the width of the specimen, but rather the distance from the load line to the far end of the specimen. The actual width of the specimen is $1.25W$. As

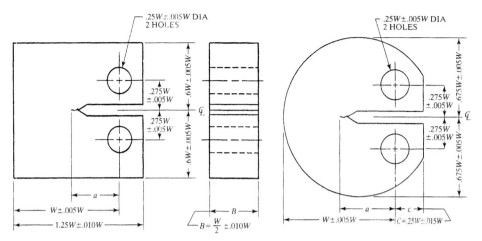

Figure 8.5 Two variations of the standardized compact tension specimen (adapted from ASTM E 399, copyright ASTM, reprinted with permission).

was the case for the previous geometry, the standard compact geometry has a width-to-thickness ratio $W/B = 2$ but now a more limited range of values $2 \leq W/B \leq 4$ is allowed. For this geometry, the provisional stress intensity factor is computed from

$$K_Q = \frac{P_Q}{B W^{1/2}} \cdot f\left(\frac{a}{W}\right) \tag{8.3}$$

where

$$f\left(\frac{a}{W}\right) = \frac{\left(2 + \frac{a}{W}\right)\left(0.886 + 4.64\frac{a}{W} - 13.32\left(\frac{a}{W}\right)^2 + 14.72\left(\frac{a}{W}\right)^3 - 5.6\left(\frac{a}{W}\right)^4\right)}{\left(1 - \frac{a}{W}\right)^{3/2}} \tag{8.4}$$

The final geometries certified in Standard E 399 are intended for use when the available form of the material is that of a hollow cylinder (the gun tube being the motivation behind these geometries). The two variations of the arc-shaped tension specimen [A(T)], are shown in Figure 8.6. The geometry shown on the right of the figure is the limiting case, in which the loading hole offset, X, relative to the inner radius is reduced to zero. As was the case for the prior geometries, the standard dimensions of this specimen are $W/B = 2$, with a range of $2 \leq W/B \leq 4$ permitted. The formula for the provisional stress intensity factor for this specimen is

$$K_Q = \frac{P_Q}{B W^{1/2}}\left[3\frac{X}{W} + 1.9 + 1.1\frac{a}{W}\right]\left[1 + 0.25\left(1 - \frac{a}{W}\right)^2\left(1 - \frac{r_1}{r_2}\right)\right] \cdot f\left(\frac{a}{W}\right) \tag{8.5}$$

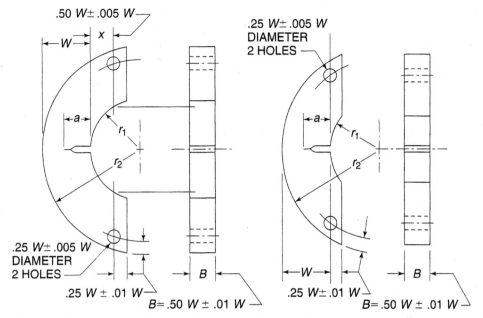

Figure 8.6 Two variations of the standardized arc-bend specimen (adapted from ASTM E 399, copyright ASTM, reprinted with permission).

where r_1 and r_2 are the inner and outer radii, respectively, and

$$f\left(\frac{a}{W}\right) = \frac{\left(\frac{a}{W}\right)^{1/2}\left[3.74 - 6.30\frac{a}{W} + 6.32\left(\frac{a}{W}\right)^2 - 2.43\left(\frac{a}{W}\right)^3\right]}{\left(1 - \frac{a}{W}\right)^{3/2}} \qquad (8.6)$$

For each of the specimen types, Standard E 399 provides very precise requirements for the machining of the starter notch and the fatigue precracking procedure. These constraints are intended to ensure that the crack tip approaches as much as is practical the mathematical requirement of infinite sharpness and freedom from compressive residual stress.

The choice of specimen shape to be tested from among the three types allowed by Standard E 399 depends on the form of the available bulk material and the need to conserve material. The bend bar is popular for a variety of reasons. It requires the least machining, and the test fixtures are less expensive to prepare; however, more importantly, the forces required to produce failure are lower in the bend bar than in either of the other two specimen shapes. In mechanics terms, the bend bar specimen is highly compliant compared with the others. By this, we mean that the increase in load is accompanied by larger displacements, δ. When viewed in the applied load–displacement plane (the $P\delta$-plane), the slope is low. However, since the stored strain energy imparted to the specimen is proportional to the product of P and δ, a sufficient amount of energy to cause fracture can be inputted to the specimen, even in low-capacity testing machines, at the expense of larger displacements. The other specimens are quite stiff by comparison with the arc-bend specimen somewhat intermediate level along the scale. We should also note that, although the specimens are scaled in the in-plane dimension by W, the critical dimension is the thickness B. Since the concept of a valid plane-strain fracture toughness is intimately connected to the minimum thickness required to ensure adequate antiplane constraint, the design of a specimen usually begins by selecting a thickness for the specimen. The other dimensions then follow, based on the range of W/B allowed for the specimen type.

ASTM Standard E 399 is an unusual standard in that a valid test cannot be assured at the outset, even if all of the procedures are carefully followed. This lack of assurance is because the minimum thickness for a valid K_{Ic} depends on the measured value of K_{Ic} itself. The requirements on minimum thickness for a valid K_{Ic} are

$$B > 2.5 \left(\frac{K_{Ic}}{\sigma_{ys}} \right)^2 \tag{8.7a}$$

and

$$a > 2.5 \left(\frac{K_{Ic}}{\sigma_{ys}} \right)^2 \tag{8.7b}$$

Since the crack tip is located at approximately $W/2$, the second condition is automatically satisfied in specimens of standard thickness. For nonstandard specimens, both conditions must be satisfied independently. The factor of 2.5 was selected based on a large number of fracture tests involving a number of metallic materials [Brown and Srawley, 1966]. However, we arrive at a similar value directly from mechanics arguments. Recall that the radius of the plastic zone is

$$r_y = \frac{1}{6\pi} \left(\frac{K_{Ic}}{\sigma_{ys}} \right)^2 \approx \frac{1}{20} \left(\frac{K_{Ic}}{\sigma_{ys}} \right)^2 \tag{8.8}$$

Therefore, $W - a \approx 50$ times the radius of the plastic zone, or the extent of the plastic zone must be less than two percent of the distance of the uncracked ligament. This is of the same order of magnitude as the size of the singularity-dominated zone in similar geometries found by Chona, Irwin, and Sanford [1983] from photoelastic measurements. From this observation, we conclude that not only must the plastic zone be small relative to the dimensions of the specimen, but it also must be well confined within the zone of dominance of the near-field equations in order for plane-strain conditions to prevail.

The procedure for conducting the test is straightforward, and the required instrumentation is widely available. Only two parameters are recorded: the remotely applied load, P, and the displacement of the crack mouth, δ, measured with a clip

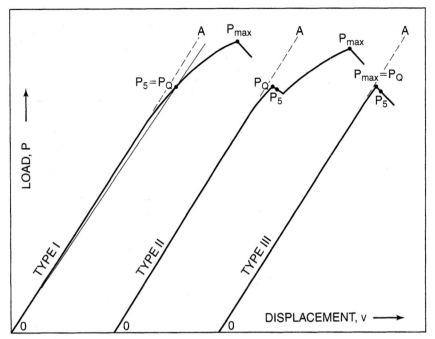

Figure 8.7 Determination of P_Q from each of the three principal load-displacement records (ASTM E 399, copyright ASTM, reprinted with permission).

gage.[1] The load–displacement record will look like one of the three curves shown in Figure 8.7. From the record, two key values are extracted: P_Q and P_{max}. In all cases, P_Q is the load value substituted into the appropriate formula above for K_Q, and P_{max} is the maximum or failure load. All specimens must fail completely in order to be considered. Each of the three curves represents a different mechanism of failure, and, accordingly, the critical load corresponding to failure, P_Q, is determined differently.

In the case of Type I behavior, the gradual deviation from linearity preceding failure indicates a period of stable crack growth. If we assume that it is desirable to limit the extent of stable crack growth, Δa, to a value less than the radius of the plastic zone, then from Eq. (8.8), we have

$$\Delta a \ \leq \ \frac{1}{20} \left(\frac{K_{Ic}}{\sigma_{ys}} \right)^2 \tag{8.9a}$$

---

[1] The standard even includes instructions for making a clip gage using strain gage sensors.

Or, in light of Eq. (8.7b),

$$\frac{\Delta a}{a} \le \frac{1}{50} \tag{8.9b}$$

Therefore, restricting the extent of stable crack growth to be within the plastic zone at the onset of crack growth is equivalent to limiting stable crack growth to two percent of the initial crack length for standard specimens. To translate this result into an equivalent reduction in slope of the load–deflection curve, we note that it is always possible to write the crack-opening displacement per unit load as

$$\frac{vEB}{P} = F\left(\frac{a}{W}\right) \tag{8.10}$$

where $F(a/W)$ is the dimensionless compliance for the geometry of interest.

Therefore, we can formally write the change of compliance as

$$\frac{\Delta vEB/P}{vEB/P} = \frac{F\left(\frac{a}{W} + \frac{\Delta a}{W}\right) - F\left(\frac{a}{W}\right)}{F\left(\frac{a}{W}\right)} \tag{8.11}$$

which, from the mean value theorem of calculus, becomes,

$$\frac{\Delta v}{v} = \frac{F'}{F} \cdot \frac{\Delta a}{a} \cdot \frac{a}{W} \tag{8.12}$$

where F' is the derivative of $F(a/W)$ with respect to its argument a/W. It can be shown that, for single-edge cracked geometries,[2] $F'/F \approx 5$ for $a/W = 0.5$. Then the change in compliance is

$$\frac{\Delta v}{v} \approx 5\% \tag{8.13}$$

We interpret this result to mean that the magnitude of the load corresponding to a two-percent stable crack growth limit can be obtained from the load–displacement record by constructing a line from the origin that has a slope five percent less than the recorded slope. The intercept of this line with the recorded data is denoted as P_5. For Type-I load–displacement records, P_5 is chosen by the standard to be the effective failure load P_Q.

In the case of the Type-II record, the slight dropoff in the load record preceding the maximum load indicates a sudden, but brief, period of rapid crack extension preceding gross failure. This condition, called a *pop-in*, corresponds to the lower bound failure strength of the material, and P_Q in this case is taken as the load at the instant of pop-in. Finally, Type-III failures are those in which failure proceeds

---

[2] See the exercises at the end of this chapter.

across the entire remaining ligament without hesitation and $P_Q = P_{max}$. For all load–displacement curve types, Standard E 399 requires that P_{max} be no more than 10 percent larger than P_Q as defined for each of the failure types. This restriction ensures that P_Q is a representative, albeit conservative, measure of the load at the onset of failure.

In addition to the foregoing requirements, Standard E 399 dictates that the minimum dimensions

$$B_{min} = 2.5 \left(\frac{K_Q}{\sigma_{ys}} \right)^2 \qquad (8.14a)$$

and

$$a_{min} = 2.5 \left(\frac{K_Q}{\sigma_{ys}} \right)^2 \qquad (8.14b)$$

must have been achieved, where K_Q is based on P_Q for the type of failure observed and the appropriate formula for the specimen geometry.

If all of the tests are satisfied, then $K_Q = K_{Ic}$, and a valid result can be reported. For tests that fail to meet all of the criteria for a valid test, the standard suggests that the test be repeated with a specimen 50 percent larger. This "after-the-fact" acceptance criterion of Standard E 399 has two unfortunate consequences. First, the standard provides no outlet for reporting the results of tests that fail to meet all of the criteria. As a result, the unsuccessful tests represent lost time and expense to the laboratory conducting the tests. Second, for specimens that nearly, but not quite, meet one or more of the acceptance criteria within the bounds of the available size of the material or the limits of the testing machine, the test engineer is placed in the unenviable position of ordering an additional test with full knowledge that, it too, may fail!

After more than 25 years since the first adoption of Standard E 399, a potential solution to the dilemma it poses has been offered in the form of a new standard, Standard E 1820, *Standard Test Method for Measurement of Fracture Toughness* [1999]. At the expense of additional instrumentation and some increased complexity in the test procedure, the new standard provides the means for reporting additional measures of fracture when plane-strain fracture toughness is not an appropriate parameter. These parameters, the crack-tip opening displacement ($CTOD$) and J-integral (J), are meaningful within the realm of elastoplastic fracture mechanics and will be discussed in that context in Chapter 11.

8.3 NONSTANDARD FRACTURE-TOUGHNESS TESTS

In metals the use of standard specimens is almost universal. Except for cases in which there is insufficient material to make one of the standard geometries, there is little justification for nonstandard test geometries considering the great effort that went into calibrating the standard geometries. On the other hand, for nonmetals, the standard geometries of ASTM Standard E 399 may not be suitable. You can imagine the difficulty in drilling holes in glass to manufacture compact-tension specimens! For these classes of materials, there has been little coordinated effort to establish accurate calibrations for suitable specimens. The reason, in part, is the diverse range of materials and applications to be considered. As a result there are a wide variety of specimens in use, all of which were developed in response to a particular need. We have already mentioned the center-cracked panel for plastics testing as an example. In the upcoming sections of this chapter, several representative nonstandard geometries that have been widely used for particular applications or materials will be described. For these specimens there are no universally accepted calibration equations, and a common feature is the use of elementary principles of mechanics to determine suitable expressions.

Double-Cantilever Beam

We first examined the double-cantilever beam (DCB) geometry in Chapter 7. This simple rectangular geometry is widely used in plastics testing, because the entire specimen (including a side groove to guide the crack along the center plane) can be made on a table saw with a suitable carbide blade. This geometry offers the opportunity for several types of fracture tests as we will soon see.

For the geometry shown in Figure 8.8, an elementary expression for the geometric stress intensity factor based on compliance calibration was shown in Chapter 7 to be of the form

$$K^2 = E\mathcal{G} = \frac{P^2 a^2}{B_n I} \tag{8.15}$$

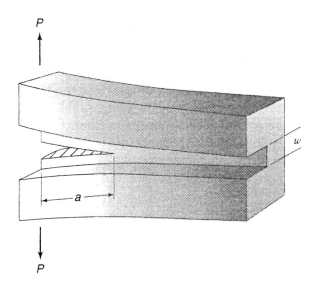

Figure 8.8 Double-cantilever-beam specimen with an end load.

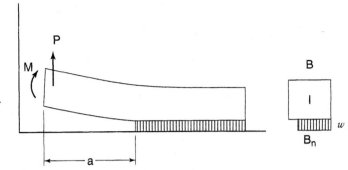

Figure 8.9 Beam-on-elastic-foundation representation of the double-cantilever-beam specimen [Freiman et al., 1973].

This result, derived from the elementary theory of beams in bending, includes the effect of the reduced cross-section on the strain-energy release rate per unit area due to the side groove, but does not consider its effect on the compliance.

A more thorough analysis of this geometry would consider the finite height, w, of the side groove. The effect of this web is to produce a larger rotation of the beam at the crack tip than the elementary theory would predict. Freiman, Mulville, and Mast [1973] reanalyzed the cantilever-beam geometry by considering the side-grooved portion of the specimen as an elastic foundation [Hetenyi, 1946] upon which the full-width portion rested, as shown in Figure 8.9. Their analysis also considered the influence of transverse shear on the energy release rate. Two loading cases, a point end load and a pure end moment, were treated.

For the case of an end load, P, Freiman, Mulville, and Mast demonstrated that the stress intensity factor can be written as

$$K^2 = EG = \frac{P^2}{B_n I}\left[(a + \delta)^2 + \frac{EI}{GA}\right] \tag{8.16}$$

where δ is the characteristic length of the beam and is given by

$$\frac{1}{\delta^4} = \frac{B_n}{4Iw} \tag{8.17}$$

The contribution due to shear is given by EI/GA (where G is the shear modulus of the material) and is generally neglected when the crack length is long compared with the beam height. The elastic foundation correction appears in Eq. (8.16) as an addition to the crack length, and its effects can be minimized by keeping the web height as small as practical.

On the other hand, for the case of pure end moment, M, the same degree of analysis leads to an equation of particularly simple form:

$$K^2 = EG = \frac{M^2}{B_n I} \tag{8.18}$$

In either case, the results remain valid only so long as the crack is sufficiently removed from the far end. The effect of finite rotation of the free end has not been considered.

Unlike the case for metals wherein the crack is most often introduced into the specimen by fatigue, the crack can be produced in brittle nonmetals by wedge, or crack-line, loading. In this type of loading, a controlled crack-line displacement produces an applied K field that decreases with increasing crack length. Cracks initiated this way will arrest when the applied K field is lower than the material's resistance to fracture. Said another way, the crack will stop when the stress field surrounding the running crack is no longer able to supply sufficient energy to maintain crack growth. This type of loading is accomplished in the DCB geometry by mechanically driving a wedge between the two arms of the specimen, as illustrated in Figure 8.10a. For plastics a tapered wedge driven by light taps with a small hammer is usually sufficient after some practice, to drive the crack to the required position. However, for very brittle materials, e.g., glass and ceramics, with extremely low fracture toughness, this approach often results in the crack traversing the whole length of the specimen unabated. To prevent this from occurring, an additional constraint in the form of a clamping force, as shown in Figure 8.10b, applied perpendicular to the crack plane has proven to be effective. In fracture mechanics terms, this superposed stress field alters the $K-a$ relationship, resulting in an applied K field that decreases much more dramatically than that of the unclamped specimen, often down to zero.

The double-cantilever beam geometry is well suited to the measurement of the critical fracture toughness of relatively brittle plastics. In these materials there is often some stable crack growth preceding unstable fracture, but the crack length at the onset of instability can usually be determined by examining the fracture surface. The fracture morphology of these materials is very distinctive and uniquely related to the crack velocity [Dally, Agrawal, and Sanford, 1990]. Another application for this

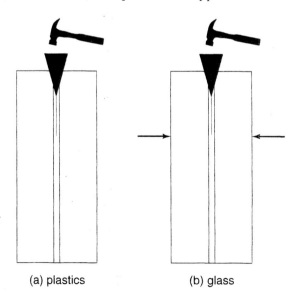

Figure 8.10   Generating starter cracks in double-cantilever-beam specimens in high (left) and very low (right) toughness materials.

(a) plastics   (b) glass

geometry is for studying the fracture resistance of adhesive joints. The adhesive is applied between the two arms of a specimen fabricated from the adherend material and allowed to cure. This application, first introduced by Ripling, Mostovoy, and Patrick [1964], has been used to study a wide range of variables that affect adhesive strength.

Even more important than its use for the measurement of fracture resistance, the DCB geometry is ideally suited to the study of subcritical crack growth. It is well known that many materials, most notably glass and plastics, support slow, stable crack growth at applied stress fields well below their critical fracture resistance. The crack in your windshield which slowly grows across your line of view is an example. This phenomenon, called *static fatigue, sustained-load crack growth* or *environmentally assisted crack growth*, is the result of some mechanism of chemical attack in the high-energy region of the crack tip. This chemical action alters the crack-growth mechanics in the fracture process zone, reducing the required strain energy for propagation [Griffith, 1921]. Ample studies have shown that there is a clear correlation between the stable crack velocity and the applied K level, and this subcritical crack growth behavior in these materials is an important aspect of their fracture mechanics characterization. Ideally, we would like to measure the crack velocity (the dependent variable) as a function of the applied stress intensity (the independent variable) under conditions in which we can change the independent variable at will. Fortunately, this geometry offers several possibilities for conducting such "constant K" experiments.

For the simple double-cantilever beam specimen with an end load, the stress intensity factor, after substituting for the moment of inertia in Eq. (8.15), can be written as

$$K = \frac{2\sqrt{3}Pa}{\sqrt{B_n B}\, h^{3/2}} \tag{8.19}$$

Clearly, for the simple rectangular geometry shown in Figure 8.8, the stress intensity factor increases as the crack grows under constant load and eventually an initially stable crack will become unstable. However, if the height of the beam were made variable, such that

$$Ch^{3/2} = a \tag{8.20}$$

then K can be written in the form

$$K = \frac{2\sqrt{3}CP}{\sqrt{B_n B}} \tag{8.21}$$

where C is a constant determined from the specimen taper.

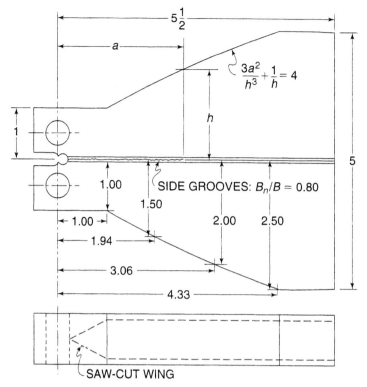

Figure 8.11 Tapered double-cantilever-beam specimen [Mostovoy et al., 1967]. Copyright ASTM, reprinted with permission.

This geometry, the tapered doubled-cantilever beam specimen, was first introduced by Mostovoy, Crosley, and Ripling [1967] for studying subcritical crack growth in adhesive joints. The geometry of their specimen is shown in Figure 8.11. In this application the increased cost of fabricating specimen halves of nonuniform height could be justified, since the arms could be remilled after each test and reused several times. While this practice is acceptable for metallic beams, most often aluminum, it becomes prohibitively expensive for materials such as glass.

For such materials, Eq. (8.18) provides a clue to an alternative double-cantilever beam specimen. Since the stress intensity factor is proportional to the applied end moment, this type of loading automatically provides a constant K specimen in beams of uniform cross-section, as long as a practical mechanism for applying such a moment can be found. For glass and other brittle materials for which the applied forces are small, Freimen, Mulville, and Mast [1973] demonstrated that the simple expedient of gluing lever arms, as shown in Figure 8.12, is all that is required.

The constant-moment DCB specimen has the advantage over other specimen types for studying environmentally assisted crack growth in glass in that large amounts

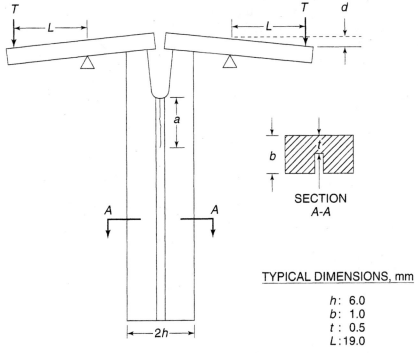

TYPICAL DIMENSIONS, mm

h: 6.0
b: 1.0
t: 0.5
L:19.0

Figure 8.12   Constant-moment double-cantilever-beam specimen for studying environmental effects in glass [Freiman et al., 1973].

of data can be obtained from a few specimens. After a fixed load has been applied to the specimen's levers, the crack position versus time can be monitored until an accurate measure of the crack velocity is obtained. Then the load is increased and the process repeated and repeated, again and again, until the crack finally gets too near the free end of the specimen for valid data. From these data a complete velocity-versus-K curve can be constructed. Figure 8.13 shows typical results due to Freiman, Mulville, and Mast [1973] for soda–lime–silica glass in a variety of liquid environments.

The constant-moment specimen described previously was widely used to obtain data on subcritical crack growth at the Naval Research Laboratory (NRL) in Washington, DC, in the early 1970s; however, it was not the first constant-moment specimen used for this purpose. In 1969, Kies and Clark, also at NRL, proposed a constant-torsion, double-cantilever beam geometry for this same purpose. This specimen is shown in Figure 8.14. By using energy-release-rate arguments, it can easily be shown that the governing equation for this type of loading is similar in form to Eq. (8.18). (The details will be left as a exercise for the reader.) Even though the double-torsion specimen is easier to prepare than the constant end-moment geometry, the latter is the geometry of choice, for several reasons. First, monitoring the position of the crack tip is more difficult with the double-torsion specimen, since the specimen is generally mounted in the horizontal plane, requiring the operator to look down on the surface. More important, the crack front in the double-torsion geometry tends to be strongly curved, as shown schematically in Figure 8.14, and the leading edge is often difficult to observe. This feature is even more pronounced in plastics for which the angle of rotation is larger.

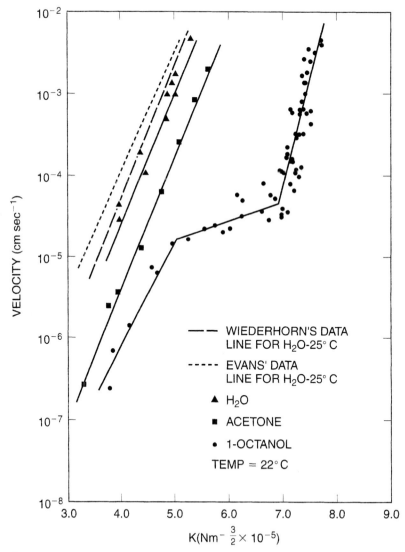

Figure 8.13 Velocity-K behavior of soda–lime–silica glass in various liquid environments [Freiman et al., 1973].

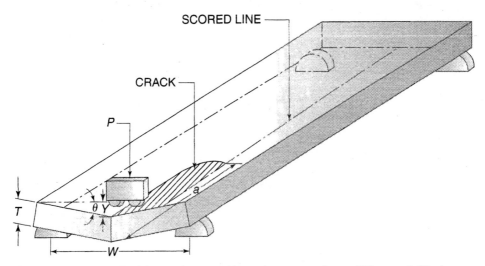

Figure 8.14 The double-torsion cantilever-beam specimen [Kies and Clark, 1969].

Chevron-Notched Specimen

In the measurement of the critical fracture toughness of some ceramic materials, a unique problem arises. Typical fracture tests require the load and crack length at failure, but in these materials, the fracture morphology often provides no clear indication of the onset of unstable crack growth. Unlike metals, for which heat-tinting[3] can be used, or plastics, for which India ink staining is commonly used, these materials do not lend themselves to simple methods for marking the critical crack length. For such materials, a test method that does not require the measurement of the crack length is highly desirable. The constant-moment tests just described could be used, but are more suitable for the measurement of subcritical fracture parameters than the onset of unstable crack growth. A specimen geometry developed to meet this goal is the chevron-notched specimen shown schematically in Figure 8.15. This specimen is similar to the compact-tension geometry, except that the load is applied along the crack line through knife edges, which are more appropriate for brittle materials than are loading holes. However, unlike the case for the C(T) geometry, the uncracked ligament is trapezoidal, with the taper angle, θ, closely controlled during the fabrication process. The isolation of the crack front from the specimen faces, combined with a very thin slot along the crack plane, produces a high degree of antiplane constraint at the crack tip. As a result, measurements of plane-strain fracture toughness can be made with specimens significantly smaller than their through-thickness cracked counterparts. For truly brittle materials, specimen widths, W, as small as 15 millimeters are not uncommon [Shannon et al., 1981]. A more

---

[3] Heat-tinting involves the heating of the specimen in a warm oven to accelerate the formation of oxides on the surface. This procedure is done before the specimen is broken completely into two parts.

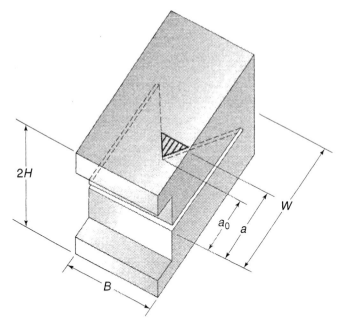

Figure 8.15 Chevron-notched short bar specimen.

important consequence of the chevron cross-section is the form of the geometric shape function $Y(a/W)$ that results.

For specimens subjected to point loads, the stress intensity factor K can be expressed as

$$K = \frac{P}{B\sqrt{W}} \cdot Y\left(\frac{a}{W}\right) \qquad (8.22)$$

where $Y(a/W)$, the dimensionless shape function, embodies all of the effects of the specimen's shape parameters. With the aid of Eq. (7.26), the change in compliance as the crack advances by an amount Δa can be written as

$$\frac{dC}{da} = \frac{dC}{dA}\frac{\Delta A}{\Delta a} = \frac{2Y^2}{E'WB^2} \cdot \frac{\Delta A}{\Delta a} \qquad (8.23)$$

For a straight-through crack, $\Delta A/\Delta a = B$, the specimen thickness, and

$$\left.\frac{dC}{da}\right|_{ST} = Y^2 \frac{2}{E'WB} \qquad (8.24)$$

On the other hand, for the chevron-notched geometry, $\Delta A/\Delta a = b$, the instantaneous width of the crack front, as illustrated in Figure 8.16, and

$$\left.\frac{dC}{da}\right|_{CN} = (Y^*)^2 \frac{2b}{E'WB^2} \qquad (8.25)$$

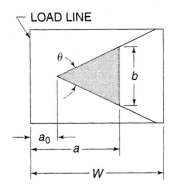

Figure 8.16   Cross-section of
the chevron-notched specimen.

The instantaneous thickness b can be expressed in terms of the crack length and the chevron angle θ as

$$b = 2(a - a_o) \tan \frac{\theta}{2} \qquad (8.26)$$

Considering the behavior of a thin slice of the material at the midplane, Munz, Bubsey, and Srawley [1980] argued that the change in compliance with crack length for the chevron-notched specimen is the same as that of a straight-through cracked specimen of comparable dimensions. Therefore, the shape function for the chevron-notched geometry can be expressed in terms of its counterpart with a straight-through crack as

$$Y^* = Y \sqrt{\frac{B}{W} \left[\frac{1}{2 \left(\frac{a}{W} - \frac{a_o}{W} \right) \tan \frac{\theta}{2}} \right]} \qquad (8.27)$$

Munz, Bubsey, and Srawley computed Y^* from well-established calibration functions for edge-cracked specimens, and their results are shown in Figure 8.17 for $H/W = 3.33$, along with independent calibrations using two-dimensional photoelastic models by Sanford and Chona [1984]. These results using the energy argument for crack growth across nonuniform cross-sections are in close agreement with detailed three-dimensional finite element analysis [Raju and Newman, 1984] and three-dimensional photoelastic studies by Nicoletto [1986]. The characteristic feature of this shape function is the presence of a minimum at a predetermined crack length. Therefore, as the load is slowly applied to an uncracked chevron specimen, a crack will initiate at a very low load level, but will quickly arrest, because the increase in surface area is greater than the energy per unit area available for crack extension. As the load continues to increase, only stable crack growth occurs as long as the crack area grows faster than the rate of energy supply—that is, the product PY^* is a constant for equilibrium crack growth. Then, at the minimum in Y^*, the crack becomes unstable, and fracture occurs from the natural crack front that preceded the instability. Since the crack length at instability depends only on the particulars of the specimen geometry, the maximum load $P_{\max}$ is all that is required to determine K_{Ic},

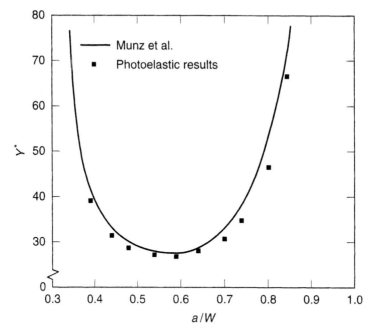

Figure 8.17   Dimensionless shape function for the chevron-notched specimen for $\frac{H}{W} = 3.33$, obtained from through-thickness geometries by two methods.

provided that plasticity effects are insignificant. Although developed initially for the ceramics community, the chevron-notched specimen, with suitable corrections, has been applied to a variety of materials, including metals [ASTM Standard E 1304].

8.4 SUMMARY

The number of specimen geometries that could be used for fracture toughness measurements is endless. Each new material and application presents new measurement challenges, and ingenuity is required to overcome problems that arise. If the purpose of a fracture test is to compare materials or maintain quality control, then there is considerable freedom in the design of a suitable and inexpensive test. However, if the value obtained from the test is to be used in design, it is necessary to ensure that the measured value be transferable to the structure for which the measurement is intended. To accomplish this more stringent requirement, two conditions need to be met. First, it is necessary to guarantee that the state of stress at the crack tip be know to a high degree of precision; that is, the geometric stress intensity factor versus the crack-length behavior must be accurately established. Second, it must be shown that there is a region around the crack tip for which the near-field equations,

Eqs. (3.46), dominate the mechanics at the crack tip. This condition is equivalent to the requirement that, in the presence of plasticity or other micromechanical damage effects controlling the onset of fracture, the fracture process zone be smaller than the singularity-dominated zone; otherwise, the precepts of LEFM do not apply.

REFERENCES

ASTM Standard E 399–97, 1997, "Standard Test Method for Plane-Strain Fracture Toughness of Metallic Materials," *Annual Book of ASTM Standards*, Vol. 03.01, American Soc. for Testing and Materials, West Conshohocken, PA.

ASTM Standard E 616–89 (now obsolete), 1989, "Standard Terminology Relating to Fracture Testing," *Annual Book of ASTM Standards*, Vol. 03.01, American Soc. for Testing and Materials, West Conshohocken, PA.

ASTM Standard E 1304–97, 1997, "Standard Test Method for Plane-Strain (Chevron-Notch) Fracture Toughness of Metallic Materials," *Annual Book of ASTM Standards*, Vol. 03.01, American Soc. for Testing and Materials, West Conshohocken, PA.

ASTM Standard E 1820–99a, 1999, "Standard Test Method for Measurement of Fracture Toughness," *Annual Book of ASTM Standards*, Vol. 03.01, American Soc. for Testing and Materials, West Conshohocken, PA.

ASTM Standard E 1823–96, 1996, "Standard Terminology Relating to Fatigue and Fracture Testing," *Annual Book of ASTM Standards*, Vol. 03.01, American Soc. for Testing and Materials, West Conshohocken, PA.

Barsom, J. M., 1987, *Fracture Mechanics Retrospective: Early Classic Papers (1913–1965)*, American Soc. for Testing and Materials, Philadelphia.

Brown, W. F., and Srawley, J. E., 1966, *Plane Strain Crack Toughness of High Strength Metallic Materials*, STP 410, American Soc. for Testing and Materials, Philadelphia.

Chona, R., Irwin, G. R., and Sanford, R. J., 1983, "The Influence of Specimen Size and Shape on the Singularity-Dominated Zone," *Proceedings, 14th National Symposium on Fracture Mechanics*, Vol I, STP 791, American Soc. for Testing and Materials, Philadelphia, pp. I–1–I–23.

Dally, J. W., Agrawal, R. K., and Sanford, R. J., 1990, "A Study of the Hysteresis in the K_{ID}–$\dot{a}$ Relation," *Experimental Mechanics*, 30(2), pp. 273–289.

Freiman, S. W., Mulville, D. R., and Mast, P. W., 1973, "Crack Propagation Studies in Brittle Materials," *J. Mat. Science*, 8, pp. 1527–1533.

Griffith, A. A. 1921, "The Phenomena of Rupture and Flow in Solids," *Philosophical Trans. Royal Society*, London, Vol. A221, pp. 163–198.

Gross, B., and Srawley J. E., 1965, "Stress-Intensity Factors for Single-Edge-Notch Specimens in Bending or Combined Bending and Tension by Boundary Collocation of a Stress Function," *NASA Technical Note D–2603*, January 1965.

Hetenyi, M., 1946, *Beams on Elastic Foundations*, Univ. Michigan Press, Ann Arbor, MI, p. 24.

Irwin, G. R., 1960, "Plastic Zone Near a Crack and Fracture Toughness," *Mechanical and Metallurgical Behavior of Sheet Materials*, Proceedings, Seventh Sagamore Ordnance Materials Conference, pp. IV–63–IV–78.

Kies, J. A., and Clark, A. B. J., 1969, "Fracture Propagation Rates and Times to Fail Following Proof Stress in Bulk Glass," *Fracture 1969*, Proc. 2nd Int. Conference on Fracture, Chapman & Hall, Ltd., United Kingdom, pp. 483–491.

Mostovoy, S., Crosley, P. B., and Ripling, E. J., 1967, "Use of Crack-Line-Loaded Specimens for Measuring Plain-Strain Fracture Toughness," *J. Materials*, 2(3), pp. 661–681.

Munz, D., Bubsey, T., and Srawley, J. E., 1980, "Compliance and Stress Intensity Coefficients for Short Bar Specimens with Chevron Notches," *Int. J. Fracture*, 16(4), pp. 359–374.

Nicoletto, G., 1986, "Three-Dimensional Photoelastic Calibration of a Chevron-Notched Short-Bar Fracture Specimen Geometry," *Eng. Fracture Mechanics*, 24(6), pp. 879–887.

Raju, I. S., and Newman, J. C., Jr., 1984, "Three-Dimensional Finite Element Analysis of Chevron-Notched Fracture Specimens," *Chevron-Notched Specimens: Testing and Analysis*, ASTM STP 855, American Soc. for Testing and Materials, Philadelphia, pp. 32–48.

Ripling, E. J., Mostovoy, S., and Patrick, R. L., 1964, "Application of Fracture Mechanics to Adhesive Joints," *Adhesion*, ASTM STP 360, American Soc. for Testing and Materials, Philadelphia, pp. 5–19.

Sanford, R. J., and Chona, R. C., 1984, "Photoelastic Calibration of the Short-Bar Chevron Notched Specimen," *Chevron-Notched Specimens: Testing and Analysis*, ASTM STP 855, American Soc. for Testing and Materials, Philadelphia, pp. 81–97.

Shannon, J. L., Bubsey, R. T., Munz, D., and Pierce, W. S., 1981, "Fracture Toughness of Brittle Materials Determined with Chevron Notch Specimens," NASA Technical Memorandum 81607, NASA Lewis Research Center, Cleveland, OH.

Srawley, J. E., 1976, "Wide Range Stress Intensity Factor Expressions for ASTM E399 Standard Fracture Toughness Specimens," *International J. Fracture*, 12(3), pp. 475–476.

EXERCISES

8.1 Starting from the ASTM E 399 equation for the provisional stress intensity factor for bend specimens [Eq. (8.1)],

(a) by suitable integration, determine the compliance v/P for this geometry;

(b) fit the results of part (a) to a power series of the form

$$A_o \left(\frac{a}{W}\right) \left[\sum_1^4 A_i \left(\frac{a}{W}\right)^{i-1} + \frac{A_5}{(1 - a/W)^2} \right]$$

(c) compare the result of part (b) with the compliance expression given in ASTM Standard E 399 for this geometry.

8.2 Starting from the ASTM E 399 equation for the provisional stress intensity factor for compact-tension specimens [Eq. (8.3)],

(a) by suitable integration, determine the compliance v/P for this geometry;

(b) fit the results of part (a) to a power series of the form

$$\frac{A_o}{\left(1 - \frac{a}{W}\right)^2} \sum_1^5 A_i \left(\frac{a}{W}\right)^{i-1}$$

(c) compare the result of part (b) with the compliance expression given in ASTM Standard E 399 for this geometry.

8.3 Derive the expression for F'/F, where F is defined in Eq. (8.10) for the compact-tension geometry, and evaluate the expression at $a/W = 0.5$.

8.4 Repeat Exercise 8.3 for the bend-bar geometry.

8.5 Using energy-release-rate arguments, derive an expression for the geometric stress intensity factor for the double-torsion geometry.

8.6 A plane strain fracture toughness test of a steel sample ($\sigma_{ys} = 57$ ksi) has been conducted in accordance with the procedures in ASTM Standard E 399. The specimen was a standard 1T compact-tension specimen ($B = 1.000$ in). The record of the load versus crack-mouth displacement is shown in Figure E8.1. After completion of the test, the crack length at the center and quarter points were 1.10, 1.06, and 1.07 in, respectively. Compute the provisional plane-strain fracture toughness K_Q, and perform all checks necessary to determine if the test yielded a valid plane-strain fracture toughness.

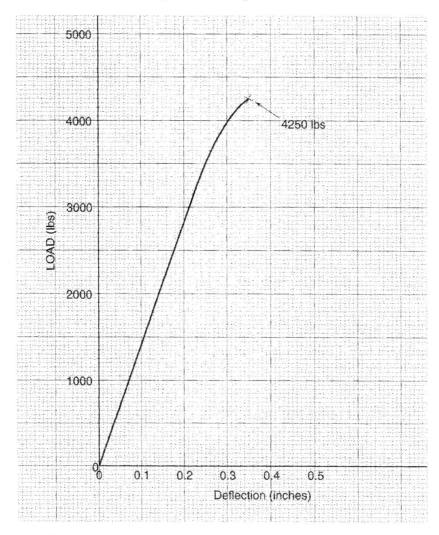

Figure E8.1

8.7 A fracture toughness test of an aluminum alloy (σ_{ys} = 65 ksi) has been conducted in accordance with the procedures in ASTM E 399. The specimen was a standard 1T compact-tension specimen, and the test was completed in 105 seconds. The load–crack-mouth deflection diagram and the results of dimensional measurements are given in Figure E8.2.

(a) From the load–deflection diagram, determine P_5, P_Q, and P_{max}.

(b) Compute K_Q.

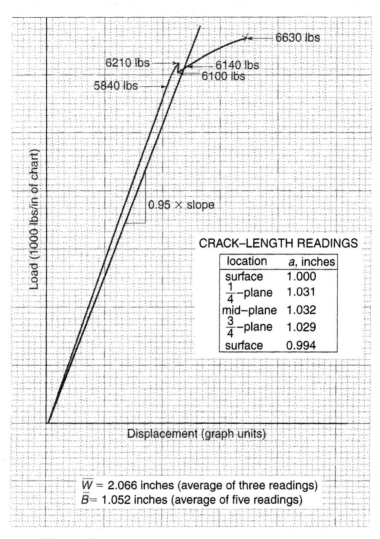

6630 lbs

6210 lbs

6140 lbs
6100 lbs

5840 lbs

Load (1000 lbs/in of chart)

0.95 × slope

CRACK–LENGTH READINGS

location	a, inches
surface	1.000
$\frac{1}{4}$–plane	1.031
mid–plane	1.032
$\frac{3}{4}$–plane	1.029
surface	0.994

Displacement (graph units)

$\overline{W}$ = 2.066 inches (average of three readings)
$\overline{B}$ = 1.052 inches (average of five readings)

Figure E8.2

(c) Was the specimen large enough for a valid K_{Ic} to be determined? Justify your answer.

(d) What is the largest fatigue load, $P_{f(max)}$, that would have been permissible for crack sharpening for this test? (Refer to ASTM Standard E 399 for the applicable requirements.)

CHAPTER 9

Fatigue

9.1 INTRODUCTION

Of all the topics in linear elastic fracture mechanics, fatigue is arguably the most important. It is estimated that greater than 80 percent of all brittle fractures are preceded by some period of crack growth in fatigue—either systemic fatigue, characteristic of rotating structures, such as a motor shaft, or random fatigue, such as that occurring in bridge structures. In either case the fatigue crack-growth portion is a major component of a structure's overall lifetime. In modern manufacturing practice, with good quality control and adequate inspection, along with sound engineering during the design phase, the existence of dangerously large initial flaws is the exception rather than the rule. This has not always been the case, for example, when the change was made from riveted pressure vessels to welded construction, failures during the initial load cycle were not uncommon. The infant science of welding combined with a lack of suitable inspection methods made such structures highly prone to failure. Over time and through bitter experiences, welding technology improved as lessons were learned from each new failure mechanism uncovered. An example of one such lesson is shown in Figure 9.1 of a large maraging-steel rocket motor that failed during its initial hydrotesting. Postfailure analysis revealed that the origin of the failure was a welder's accidental arc strike that locally embrittled the wall of the pressure vessel.

From a safety perspective the real danger of fatigue crack growth is its hidden nature. In contrast to structures that are plastically overloaded, for which the deflections become large or the permanent deformations remaining after the load is removed are obvious, failures caused by fatigue crack growth occur suddenly, with few or no precursors to warn of impending failure. Therefore, to be prudent, we should anticipate that any structure that has free surfaces (that is to say, all structures) may eventually experience fatigue crack growth starting from the surface and

Figure 9.1 Hydrotest failure of a maraging-steel solid propellant motor casing due to a weld defect.

growing to reach the critical dimension for brittle fracture, even structures that are essentially subjected to dead loads. For example, a water tower along the side of a major highway is ostensibly subjected only to hydrostatic and gravity loads. But as large trucks pass nearby, they set up ground vibrations that are transmitted to the water tower and induce small, but measurable, oscillating loads. If we adopt the Griffith argument that all structures contain a population of initial flaws, then these alternating loads will, in time, have the potential to cause fracture. While the crack growth rate may be extremely low, the important parameter is the integrated effect over the life of the structure.

As it turns out, linear elastic fracture mechanics, which has its roots in the linear theory of elasticity, is incredibly good at predicting the life of structures with initial flaws, despite the fact that the underlying mechanism that gives rise to subcritical crack growth is purely one of plastic deformation on the microscale. The primary mechanism on the microscale is that of slip-line or dislocation motion occurring within the damage process zone. On the other hand, since the loads are small, this process zone is itself well contained within the singularity-dominated zone of linear elastic fracture mechanics. Therefore, regardless of the actual mechanism that is causing the crack growth on the microscale, that mechanism is governed by the state of stress within the singularity-dominated zone and can be fully characterized by the linear elastic fracture mechanics parameter K. In other words, even though the mechanistic process giving rise to the progression of the crack through the material is a plasticity event, the behavior can be described by the equations of elasticity.

Before we proceed to develop a theory of fatigue crack growth and lifetime prediction we need to distinguish between several types of fatigue behavior. The engineering community recognizes two distinct classes of fatigue. On the one hand, there is the behavior called *low-cycle fatigue*. A common example given to describe this class is the type of fatigue observed when repeatedly bending a paper clip to failure. Low-cycle fatigue is characterized by the presence of high stress levels, most often exceeding the elastic limit on the macroscale, and a short lifetime, defined as less than 1000 cycles. Manson [1960] has demonstrated that this type of failure is governed by the total strain at the site of the fatigue damage. Low-cycle fatigue is of some concern in the metal forming industry but is rarely considered in engineering design applications. In contrast, consider the opposite extreme, high-cycle fatigue. Here, the applied loads are generally low compared with the yield stress, and the structure exhibits a long lifetime before failure, including the possibility of infinite life. Studies on smooth sided samples, the classical S-N specimen, have demonstrated that this type of fatigue is a stress dependent phenomenon. Structures undergoing high-cycle fatigue go through three identifiable stages: crack initiation, stable crack growth, and unstable failure. Each of these stages will be discussed in detail in the upcoming section. It is this type of fatigue crack growth that is amenable to analysis with fracture mechanics methods, and our attention throughout the remainder of the chapter will be limited to this mechanism of fatigue damage.

9.2 STAGES OF FATIGUE CRACK GROWTH

Previously, we indicated that the benefit of fracture mechanics for the study of fatigue is that we need not understand the actual processes ongoing within the process zone in order to make predictions of fatigue life. However, this observation assumes that a crack is already present. If there is no crack present, it is obvious that we cannot apply the principles of fracture mechanics. The question then becomes: how do we get from a structure that has no defect to a structure that fails from brittle fracture? This process is called *crack initiation* or *Stage-I crack growth*. Let us consider the fatigue lifetime of three specimens that differ only in the condition of their free surface: smooth, machined, and precracked, respectively. Figure 9.2 shows crack length–life cycle curves for the three specimens, where log scales have been used to emphasize the initial portions of the curves. For the precracked specimen, crack growth begins immediately and the lifetime is the shortest of the three cases. The machined specimen differs slightly from the precracked one in that the initial flaw

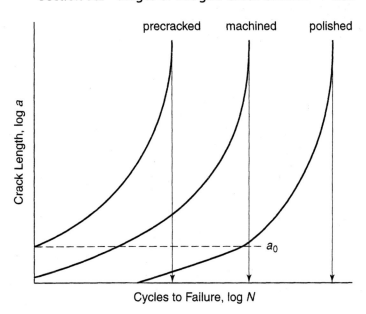

Figure 9.2   The effect of surface finish on the life cycle.

size is smaller, on the order of the RMS surface roughness, and some number of cycles are required to "sharpen" the microscopic surface notches and generate a surface crack. Again, following Griffith, we need concern ourselves only with the deepest or sharpest surface groove, as it will be the one to grow into a macrocrack. In contrast, the smooth-sided specimen exhibits a long period of zero crack growth, typically ranging from 30 percent to 80 percent of the total fatigue life. It is during this stage of crack initiation that the surface undergoes changes that lead to the formation of a macrocrack. Note that once all three specimens develop cracks of comparable dimensions, the crack growth rate is the same, as illustrated Figure 9.2.

An interesting feature of fatigue crack initiation in smooth specimens is that the crack invariably starts at the free surface. In the few cases in which a fatigue crack is shown to form in the interior, close examination reveals either an internal free surface, such as a void, or an interface (e.g., an embedded nonmetallic inclusion) as the origin of the failure. In the absence of such internal flaws the development of a measurable crack initiating at the free surface goes through several steps. Our basic understanding of the formation of initiation cracks predates the science of fracture mechanics. In 1956 the International Conference on Fatigue in Metals[1] brought together over 500 leading researchers and engineers from both sides of the Atlantic to discuss the state of the art in fatigue. It was already known prior to the conference that the formation of fatigue cracks in smooth specimens was preceded by migration of slip bands to the free surfaces, but beyond that, the mechanism that caused the slip band to transition into actual cracks was less well understood. Wood [1956] discussed

---

[1]The conference was unique in that it was held in two parts. For the first five days the meetings were held in London, and two months later, the remaining three days were held in New York.

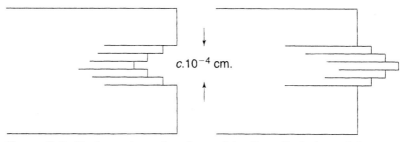

Figure 9.3 Peaks and notches formed by fine slip [adapted from Wood, 1956].

the various competing explanations and discounted all of them. Instead, he proposed that fatigue slip at low stress levels differed from the slip bands observed under an optical microscope that were formed at higher strain levels. The slip mechanism he envisioned, called *fine slip*, involved submicron motions through relatively few lattice spacings rather than the single, larger step previously assumed. He speculated that the "to" motion favored one part of the slip band and the "fro" motion another, leading to the formation of either notches or peaks, as illustrated in Figure 9.3. Finally, he proposed that when the notch deepens, a stress raiser forms with a root radius of atomic dimensions, and the ensuing very large strains result in separation of atomic planes (i.e., form cracks). In the next presentation at the conference Forsyth [1956] demonstrated that he had observed a fine dispersion of slip bands on the surface in an aluminum alloy long before the formation of coarse slip bands. Moreover, he described the eruption of thin ribbons of material, illustrated in Figure 9.4 (taken from his paper), which he called *slip-band extrusions*.

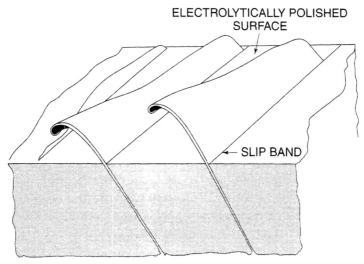

Figure 9.4 Slip-band extrusions in an aluminum alloy [Forsyth, 1956].

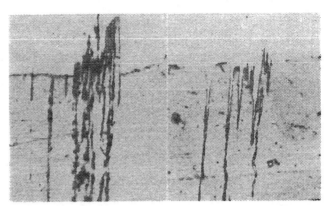

Figure 9.5 Slip-band extrusions (left, original at 12,500×) and notches (right, original at 20,000×) observed with small-angle sectioning technique [Wood, 1958]. Copyright ASTM, reprinted with permission.

Both Wood and Forsyth expanded on their brief papers in later conferences. In the 1958 ASTM Symposium on Basic Mechanisms of Fatigue, Wood [1958] used a small-angle (2° to 3°) sectioning technique that effectively magnified the disturbances on the surface by a factor of 20× to 30× and then viewed the polished sections in an optical microscope at 500× to 1000× magnification. The result was a micrograph at a total equivalent magnification of up to 30,000× —comparable with electron microscope images. With this unique technique, Wood was able to show the formation of the peaks and notches that he had described earlier. Figure 9.5 shows typical results due to Wood from his studies with brass and copper alloys. Three years later, at the Cranfield Crack Propagation Symposium,[2] Forsyth [1962] introduced the concepts of Stage I and Stage II crack growth. He stated that the criterion for slip band formation is the range of resolved shear stress and that slip band cracking is most likely to occur within those grains whose orientation places the preferred (i.e., weak) shear planes along the maximum shear directions—that is, at 45° to the applied tension. He called these grains *soft grains*. Forsyth showed that the transition from Stage I to Stage II is gradual. (See Figure 9.6.)

By 1961 the fact that the striations on the fracture surface during Stage II fatigue crack growth, such as those shown in Figure 9.7 in stainless steel, appeared to be related to the fatigue crack growth rate had been well established. In his Cranfield paper, Forsyth [1962] demonstrated that the one-to-one correspondence between striation spacing and crack growth rate did indeed correspond to one striation per cycle of crack growth, and he was the first to suggest the importance of these striations in the postmortem evaluation of service failures. However, his explanation of the origin of the striations as due to advance nucleation of the cracks and the subsequent stretching of the remaining ligaments was inaccurate. In later years, it was shown that this mechanism appears to be operative in quite brittle metals, but not in metals of moderate to large ductility. A general interpretation of the origin of striation marks in ductile materials was provided the following year by Laird and Smith [1962] and developed in more detail in 1967 by Laird. In this latter paper Laird [1967] argues

---

[2]The proceedings of the conference were not published until the following year.

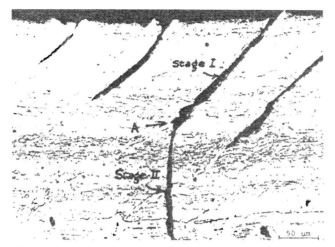

Figure 9.6 Gradual transition from Stage I to Stage II crack growth [Forsyth, 1962]. Copyright QinetiQ Limited, 2001.

that Stage I and Stage II crack growth in ductile materials are governed by the same mechanism—i.e., slip band motion along preferred shear planes—and differ only in scale. The formation of the striation and the advance of the crack are integral to

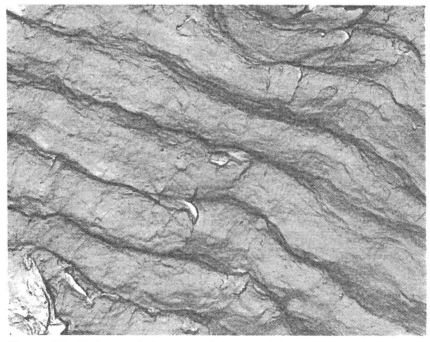

Figure 9.7 Fatigue striations in stainless steel (courtesy C. D. Beachem, P.E.).

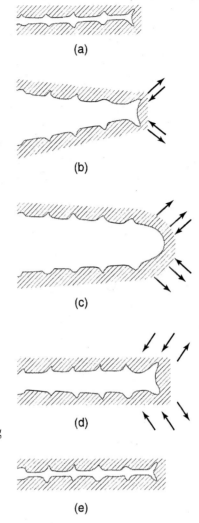

(a)

(b)

(c)

(d)

(e)

Figure 9.8   The plastic-blunting mechanism of crack extension, leading to the formation of striation markings [Laird, 1967]. Copyright ASTM, reprinted with permission.

each other, and Laird refers to the combined effect as plastic blunting. With the aid of Figure 9.8, Laird describes the process in the following manner: Starting from zero load (Figure 9.8a), the small double notches at the crack tip from the prior cycle concentrate the slip zones along 45° as the tensile load is applied (Figure 9.8b), tending to maintain the square shape. As the tensile strains approach their maximum, the slip zones at the tip broaden, and crack blunting ensues, (Figure 9.8c). When the load is reversed, the slip directions reverse, and the crack faces, including the new surface just created, are pushed together. In the process the very front of the crack buckles (Figure 9.8d), and is partly folded back on the crack to form new notches. This last step was important to Laird's explanation, as it explains the origin of the notches that serve as the slip-zone concentrators for the start of the next cycle (Figure 9.8e). Throughout the remainder of his 1967 paper Laird discusses some of the ways that the symmetry of his model is modified by the local metallurgy of the material while maintaining the one-striation-per-cycle rule.

In his 1979 review paper Laird describes the fatigue process from the perspective of the metallurgist and compares his model for Stage II crack propagation with a somewhat similar model proposed by Cottrell [1962]. He also discusses the fatigue limit in terms of a plateau strain that needs to be approached from below. Near the end of the paper he laments that there is limited communication between the fatigue community and the fracture mechanics community. Unfortunately, it took until 1992 for the fatigue community (ASTM Committee E–9 on Fatigue of Metals) and the fracture mechanics followers (ASTM Committee E–24 on Fracture of Engineering Materials) to join forces.

Finally, when the fatigue crack grows to such a length so as to approach the critical length for instability at the maximum applied stress, the crack grows rapidly. This final stage, sometimes referred to as *Stage-III growth*, is nothing more than an acknowledgement that eventually the specimen fails and the final crack jump satisfies the criterion that $K_{max} > K_c$. Taken together, the three stages of crack propagation can be represented over the full range of applied stress by a da/dN versus ΔK curve in log space, exhibiting the characteristic sigmoid shape shown schematically in Figure 9.9. In the next section we will look at the various attempts to model high-cycle fatigue behavior, that is, to provide mathematical descriptions that match the observed behavior shown in Figure 9.9.

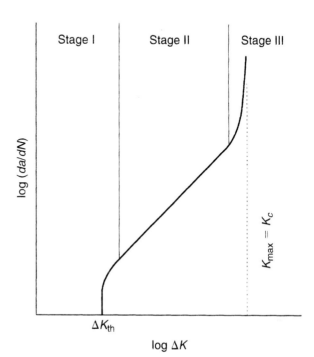

Figure 9.9   Idealized fatigue-crack growth curve over the full range of applied ΔK.

9.3 MATHEMATICAL ANALYSIS OF STAGE II CRACK GROWTH

Unlike the theory of brittle fracture developed in Chapter 6, wherein the equations governing the onset of unstable fracture were derived from first principles involving well-accepted laws of physics, the mathematical relations defining Stage II fatigue crack growth are much less precise. The formulation of a mathematical description of stable crack growth in fatigue is based on the phenomenological method of analysis, as opposed to the analytical method. For our purposes the distinction between the two is that in the former, the development is based almost entirely on observations of the physical world. In contrast, the analytical approach begins with the statement of sound physical laws, and the theory follows from a sequence of well-defined mathematical steps. Lacking a firm theoretical foundation, the equations governing fatigue crack growth have evolved over the years from behavioral observations and extensive materials testing into very sophisticated, albeit empirical, modeling tools for predicting crack growth rates. Over the last 40 years, various fatigue crack growth "laws"[3] have been proposed, and some of the more important ones will be discussed in the upcoming sections.

Before we proceed to discuss the historical development of fatigue crack growth modeling, it is instructive to see how far we can go by the dimensional analysis approach. We begin this discussion by listing as many as possible of the variables that are either *known* to influence fatigue crack growth or may *possibly* influence it. This latter, more vague, class of variables needs to be included at the outset in order to develop a comprehensive model. However, we can, and will, discount some of these variables, based on controlled experimental evidence that the results are insensitive to some of these parameters. Our model should include the following observed behaviors:

(a) Fatigue crack growth occurs in discrete increments during each load cycle and can be expressed in differential form as da/dN. As we have previously observed, optical or electron microscope examination of the fatigue crack surface in many materials shows striation marks in which the spatial period corresponds to the measured crack growth rate.

(b) The mode of crack extension corresponds to the opening mode. Once the initial surface crack has grown beyond the initiation phase, the plane of the crack invariably turns so as to be perpendicular to the applied tension. Unless constrained to remain in mixed mode—for example, by deep side grooving—the fatigue crack always turns so as to eliminate mode-II behavior. Since there is no shear motion between surfaces, surface roughness tends not to be a factor of interest.

---

[3] Of course, these relations are not laws of physics in the classical sense, but rather are empirical relations that model the observed behavior to such a high degree that they are accepted as if they were laws.

(c) The crack growth rate is sensitive to the change in applied load per cycle, $\Delta\sigma$. We have already observed that high-cycle fatigue occurs only when the applied stresses are small compared with the yield stress. Consequently, it is reasonable to describe the state of stress at the crack tip by ΔK, the change in applied stress intensity factor corresponding to $\Delta\sigma$. Moreover, since the crack extension per cycle is very small compared with the in-plane dimensions, the fracture process zone is likewise very small, and the elastic stress state in the body is unaffected by the crack growth. As a result, the principles of linear elastic fracture mechanics apply.

(d) The plastic zone size is also small compared with the thickness of the body. Because the applied stress level is low compared with the yield stress, the applied K level and its associated plastic zone [Eq. (6.11)] are correspondingly low. Accordingly, to the first order, the plastic zone size is ignored in the analysis. Thus, we have the fundamental paradox of fatigue crack growth based on principles of fracture mechanics. As we demonstrated in the introductory section, even though the mechanism of *microscopic* crack growth within the process zone which leads to the separation of atomic planes, is governed by dislocation (i.e., plasticity) behavior, we can describe the *macroscopic* crack growth behavior by purely linear elastic models. An exception to this observation is made when stress overload effects (to be discussed later) are included in the analysis.

(e) There is an effect of the maximum stress, represented by K_{max}, on the fatigue crack growth rate. As illustrated in Figure 9.10, as the mean stress increases for constant ΔK, the distance between the critical stress state K_c and K_{max}

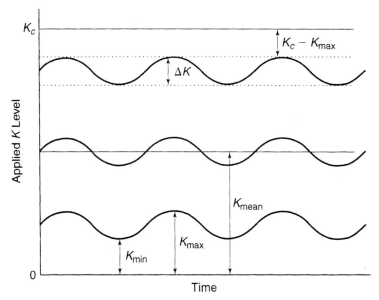

Figure 9.10 Constant-amplitude fatigue cycling at various mean stress levels.

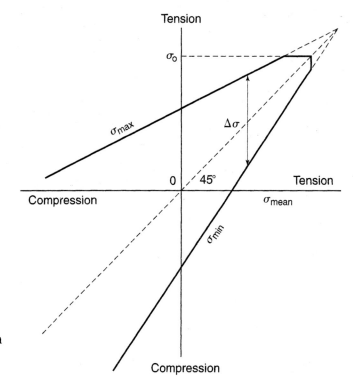

Figure 9.11 Goodman diagram corresponding to Figure 9.10.

decreases. Consequently, we would expect higher crack growth rates as $K_{\max}$ approaches K_c. This effect is analogous to the observation on a Goodman diagram in classical fatigue design analysis (see Figure 9.11) that the maximum alternating stress must decrease as the mean stress increases. (See, for example, Dieter [1986].) Traditionally, the influence of mean stress is called the R-ratio effect, where $R = K_{\min}/K_{\max}$.

(f) There appears to be a lower threshold of alternating stress magnitude, characterized by the fracture mechanics parameter ΔK_{th}, below which there is no growth. This lower threshold is comparable with the endurance limit observed in S–N diagrams of steels. From a mechanistic perspective the threshold corresponds to a crack growth rate of one atomic spacing per cycle. A more pragmatic definition of ΔK_{th} is the applied ΔK corresponding to a fatigue crack growth rate of 10^{-10} m/cycle [ASTM Standard E 1823, 1996].

(g) Fatigue crack growth rates have been shown to be very sensitive to environmental influences (temperature, humidity, salinity, etc.) and, to a lesser extent, to material processing variables (yield stress, ultimate strength, grain size, etc.) and the mechanics of load application (frequency, wave shape, etc.).

(h) Crack extension can proceed only as long as the local stress state in the process zone is tension. Initially, we will interpret this seemingly obvious statement to mean that the crack growth rate is related to ΔK only as long as $K_{\min} > 0$. In other words, crack growth cannot occur if the crack faces impinge on each other. However, the concept is much more complicated than that. We will defer until later a discussion of the concept of crack closure [Elber, 1971] to account for reversed loading and localized residual stresses due to the prior plastic zone. During that discussion, we will introduce the variable, ΔK_{eff}.

Given these observations, we propose to describe the crack propagation behavior in the most general terms by an equation of the form

$$\frac{da}{dN} = f\left(\Delta K, K_{\max}, K_{\min}, \Delta K_{\text{th}}, E, \nu, \sigma_{ys}, \sigma_{\text{ULT}}, \epsilon_i, k_i\right) \tag{9.1}$$

where, in addition to the variables defined in the above discussion, we let ϵ_i be any of the environmental variables (temperature, humidity, salinity, etc.) and k_i be any of the other material and mechanics variables (grain size, frequency, wave shape, etc.). From dimensional considerations, Eq. (9.1) can be written in the form

$$\frac{da}{dN} = \left(\frac{\Delta K}{E}\right)^2 F\left(R, \frac{K_{Ic}}{\Delta K}, \frac{\Delta K_{\text{th}}}{\Delta K}, \frac{\sigma_{ys}}{E}, \frac{\sigma_{\text{ULT}}}{E}, \alpha_i \epsilon_i, \beta_i k_i\right) \tag{9.2}$$

where α_i and β_i are conversion factors, albeit somewhat unorthodox, used to convert the environmental, material, and mechanics variables to dimensionless forms.

Equation (9.2) predicts that, for a given set of variables, the fatigue crack growth rate plotted as $\log da/dN$ versus $\log \Delta K$ should be a straight line with a slope of 2, as shown in Figure 9.12. In addition, the equation predicts that systematic changes

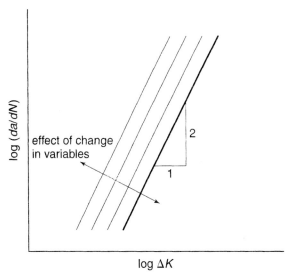

Figure 9.12 Crack growth rate $\log da/dN$ versus $\log \Delta K$, based on dimensional analysis.

in any one variable will produce shifts in the line parallel to itself. In fact, fatigue crack growth data, such as those shown in Figure 9.13, do plot as a straight line over some interval for a wide variety of metallic materials; however, unlike the prediction of the dimensional analysis, the slope, n, is generally in the range $2 < n < 4$. Paris, Gomez, and Anderson [1961] were the first to present crack growth data in this form. Two years later, Paris and Erdogan [1963] proposed the empirical relation

$$\frac{da}{dN} = C(\Delta K)^n \tag{9.3}$$

as a general equation to describe fatigue crack growth behavior and suggested that

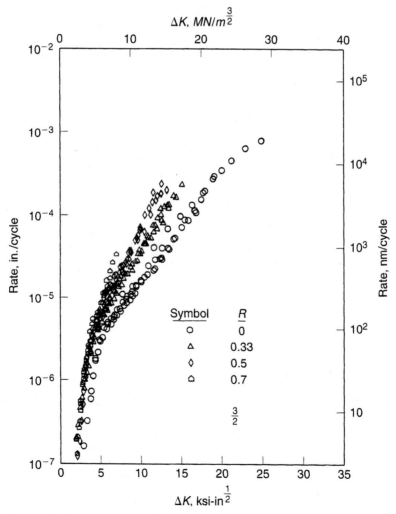

Figure 9.13 Fatigue-crack growth data for 2024-T3 aluminum [Hudson, 1969].

for 7075-T6 aluminum, a value of $n = 4$ is appropriate. Equation (9.3), now known universally as the *Paris Law*, was not widely accepted when it was first introduced. This is not surprising when we consider that linear elastic fracture mechanics was still an infant science in 1961 and fatigue was under the purview of metallurgists who had not yet accepted the stress intensity factor concept. However, mounting experimental evidence over the next few years from a number of independent laboratories validated the universal nature of Paris's relation. More importantly, the form of Eq. (9.3) lends itself to the prediction of fatigue life in the presence of a growing crack. (See Example 9.1.) It is not an exaggeration to say that this latter role of the Paris law is the seminal event that led the engineering design community to embrace the new science of linear elastic fracture mechanics.

Example 9.1   A wide thick plate ($K_{Ic} = 40$ ksi-in$^{1/2}$) contains an edge crack 0.1 inch long. The plate is subjected to alternating stresses between 0 and 20 ksi. Data on a similar material under similar environmental conditions exhibited Paris-law coefficients of n = 4 and C = 7×10^{-10} for ΔK in ksi-in$^{1/2}$. How many cycles will the plate support before failure?

It is first necessary to determine the length of the longest crack the plate can support without failure. A suitable geometric stress intensity factor for this geometry is given by Eq. (3.63); therefore, from Chapter 6, we have

$$K_{Ic} = 1.12\sigma_{max}\sqrt{\pi a_c}$$

$$40 = 1.12(20)\sqrt{\pi a_c}$$

$$a_c = 1.02 \text{ inches}$$

For this case, the Paris law has the form

$$\frac{da}{dN} = C(\Delta K)^n = C\left(1.12\Delta\sigma\sqrt{\pi a}\right)^n$$

Substituting this material's Paris-law parameters, we obtain

$$\frac{da}{dN} = 7 \times 10^{-10}(1.12)^4(20)^4\pi^2 a^2$$

Rearranging the foregoing equation yields

$$\frac{da}{a^2} = 7 \times 10^{-10}(1.12)^4(20)^4\pi^2 \, dN$$

which can be written in integral form as

$$\int_{a_i}^{a_c} \frac{da}{a^2} = 7 \times 10^{-10}(1.12)^4(20)^4\pi^2 \int_0^N dN$$

$$-\frac{1}{a}\Big|_{a_i}^{a_c} = 17.4 \times 10^{-4}\, N$$

$$N = 5200 \text{ cycles}$$

Note: The simple form of the geometric stress intensity factor in this example permitted a closed-form integration by the separation of variables method. In many cases, a procedure for analytic integration is not so obvious. Nonetheless it is always possible to recast Eq. (9.3) in finite form and integrate numerically. ∎

At this point it is worthwhile to make several observations about the Paris law. First, it is nonconservative. The form of the equation does not take into account the upturn in the fatigue crack growth rate curve (see Figure 9.9) as ΔK approaches the Stage III region and, as a result, overpredicts the number of cycles to failure. This observation might seem to be a serious objection to the use of the Paris law, but, in fact, relatively few cycles are spent in this final stage of crack growth compared with the long interval in the linear portion of the da/dN versus ΔK curve. As a result, this simple model provides a *rough* estimate of the number of cycles to failure and, if a generous safety factor is employed, a useful estimate, particularly for screening purposes. A second, and more serious objection, to the Paris law is the model's failure to account for nonzero R values. As shown in Figure 9.13, the effect of rising values of R is to shift the upper portion of the curve to the left. Since the Paris law does not compensate for changing R values, a separate calibration curve for each load ratio would need to be constructed (not an insignificant task) and the corresponding pair of fitting coefficients determined. There have been numerous alternative fatigue crack growth models proposed to correct for this failing of the Paris model. Cioclov [1977] lists more than 14 variations of the original Paris law—and that is only a partial list. In the following paragraphs we will discuss only a few of the better known models.

Walker [1970] proposed a crack propagation model of the form

$$\frac{da}{dN} = CK_{max}^p(\Delta K)^n \tag{9.4}$$

in order to incorporate the effect of R-ratio [i.e., $K_{max} = \Delta K/(1-R)$] into the Paris model. The essential feature of his model is the introduction of an additional degree of freedom (i.e., three constants to be determined by curve fitting of experimental results rather than Paris's two coefficients). Results taken from his paper, shown in Figure 9.14, demonstrate that this approach is rather effective for incorporating the

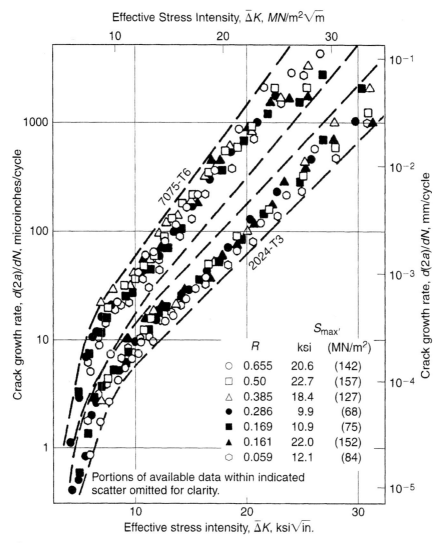

Figure 9.14 Crack growth data for 2024-T3 and 7075-T6 aluminum, corrected for R-ratio [Walker, 1970]. Copyright ASTM, reprinted with permission.

R-ratio into the model, but, like the Paris model, Walker's equation fails to predict the upturn in the crack growth rate at higher values of ΔK.

Forman, Kearney, and Engle [1967] took a different approach. They observed that, in the presence of nonzero R-values, the vertical asymptote in the plot of da/dN versus ΔK is of the form

$$\text{asymptote} = (1 - R)K_c \qquad (9.5)$$

and that the Paris model could be adapted to a model that incorporates Stage III features by dividing the classical Paris form [Eq. (9.3)], by the distance between the applied ΔK and the R-ratio modified asymptote, that is,

$$\frac{da}{dN} = \frac{C(\Delta K)^n}{(1 - R)K_c - \Delta K} \tag{9.6}$$

Despite having only two degrees of freedom, the Forman, Kearney, and Engle model does a good job of incorporating both the effect of R-ratio and the Stage III asymptote into the Paris law. Of the many variations of the Paris law proposed over the years, it has become the most popular of the low-order models and forms the kernel of the multiparameter model used in the NASA/NASGRO model, to be discussed in a later section. Finally, it is important to note that the constants C and n used in the above equations are to be interpreted in the generic sense and do not have the same value in each of the forms. For each model the constants are determined by a best-fit analysis of the data to the chosen form. Similarly, being empirically based models, it is inappropriate to use them outside the range for which experimental data exist.

Schwartz, Engsberg, and Wilson [1984], in a departure from the usual approach, observed that crack growth curves with a wide range, such as that of Figure 9.9, bear a strong resemblance to a hyperbolic sine function and proposed a four-parameter model based on that functional form:

$$\log(da/dN) = C_1 \sinh\left[C_2(\log(\Delta K) + C_3\right] + C_4 \tag{9.7}$$

Moreover, they assumed that for the case of aircraft turbine blade materials the "constants" were, in fact, functions of the frequency, temperature and stress ratio, except constant C_1, which appeared to be a material property. Using nonlinear regression analysis, they were able to determine a multiparameter fit to their data within frequencies from 0.00833 to 30 Hz, temperatures form 260° to 620°C, and stress ratios from -1 to 0.7. The model was tested on a set of 25 fatigue life tests covering a wide range of conditions within the data range and the average life prediction was within 1.5 percent of the actual data, but the range of error was as large as 40 percent. The Schwartz, Engsberg, and Wilson model refocuses attention to the fact that all fatigue crack growth models are nothing more than curve-fitting exercises to large databases. That being said, a fracture mechanics based model, using a suitable fitting function, has demonstrated itself to be a useful tool for predicting fatigue life that can, and should, be used as part of the design analysis when there is a possibility of fatigue failure. Fortunately, much of the model-building work has been done and is now automated. We will discuss these computer programs later, but first there are several other aspects of fatigue crack growth modeling that we need to discuss.

9.4 THE EFFECTS OF RESIDUAL STRESS ON CRACK GROWTH RATES

As we have described in earlier sections, fatigue crack growth is the result of dislocation mechanisms going on within the plastic zone. Simultaneously, and not unrelated, there are other mechanisms at work within the plastic zone that have a profound effect on the growth rate but are less well understood. The most important of these mechanisms is the consequences of the inelastic deformation just ahead of the crack tip. As the external forces are applied to a body containing a crack, the crack faces open, and the ligaments of material just ahead of the advancing crack are permanently elongated, because the local stress exceeds the proportional limit. The immediate result is crack blunting, but there are other consequences as well.

When the applied loads are reversed, the elastic stress field surrounding the confined plastic zone attempts to return to the natural state but is prohibited from doing so by the stretched ligaments. In order to restore equilibrium in the unloaded state, the elastic field compresses these ligaments and sets up a compressive residual stress state ahead of the crack tip. This residual stress field is an *elastic* field created by prior plastic deformation, and, since there are no externally applied forces, it must be in self-equilibrium at all times. In the region directly ahead of the crack, the component of stress normal to the crack face has the form shown in Figure 9.15, which illustrates that the compressive stress created by the collapsing stress field must be balanced by a tensile stress further away from the crack tip. Moreover, if the crack should somehow be allowed to advance into this field (e.g., by cutting),

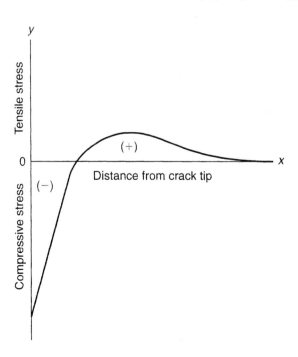

Figure 9.15 Residual stress distribution ahead of crack, due to a plastic zone.

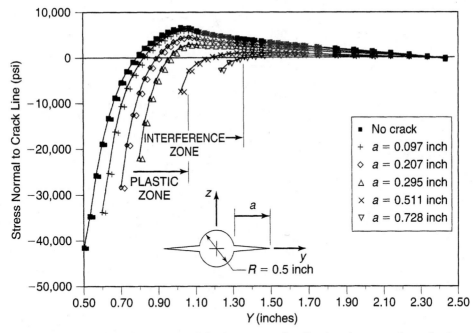

Figure 9.16 Finite-element model of stress redistribution in central cracked plate following an overload [Graham, 1988].

then the entire residual stress field will slowly diminish, maintaining equilibrium with each incremental advance of the crack. This effect is illustrated in Figure 9.16, based on finite element results for a plate containing a plastically deformed hole with the subsequent introduction of a crack [Graham, 1988]. The consequences of not compensating for this reduction in the residual stress field can have a significant impact on the calculation of fatigue life in the presence of residual stress fields but are often ignored.

Conceptually, the stress intensity range which determines the crack growth rate is the net result of the residual stress ahead of the crack and the stress due to the applied remote stress. For the case of constant amplitude fatigue crack growth the impact of the residual stress field on fatigue life prediction need not be treated independently, for two reasons. First, as the crack propagates ahead, the effects of the collapsing residual stress field behind the crack are compensated for by the reestablishment of a new residual stress state ahead of the crack, associated with the advancing plastic zone moving in front of the crack in a self-similar manner. Second, since a plastic zone of the same size and associated residual stress field existed in the test specimen used to determine experimentally the crack growth rate constants, the suppressing effect of the residual stresses is already incorporated into the experimental measurements. On the other hand, for variable amplitude loading, the self-similar nature of the plastic zone advance no longer exists, and the residual stress can play a significant role in the prediction analysis. This behavior, called the *load interaction effect*, greatly complicates the fatigue life analysis, and numerous mathematical schemes have been proposed to incorporate load interaction into the model. Regardless of the model chosen, accounting for load interaction greatly complicates the analysis, with the result that these effects are best analyzed using large-scale fatigue life prediction computer programs. We will discuss several such programs in a later section, but before that, we need to look at two specific influences: overload retardation and crack closure.

Overload Retardation

It was known as early as 1961 [Schijve] that isolated overloads have a temporary retardation effect on the growth of fatigue cracks. After the application of a sufficiently large overload, as illustrated in Figure 9.17, there is a period during which the crack growth rate is significantly lower; then the rate gradually increases until it is the same as the unretarded growth rate. The explanation for the retardation behavior is explained by the much larger compressive residual stress at the crack tip after the overload. Since the crack will not start to grow until the net stress at the crack tip is tensile, the effective stress intensity range is much smaller immediately after the overload. As the crack advances through the overload plastic zone, the redistribution of the residual field results in a decrease in the compressive stress at the tip, and the net ΔK slowly rises. Eventually, the crack grows out of the shadow of the overload plastic zone, and the crack growth rate returns to its unretarded value. If the process is repeated at just the correct intervals, there is a significant increase in the fatigue life. This effect is a reported benefit of periodic proof testing, but if it

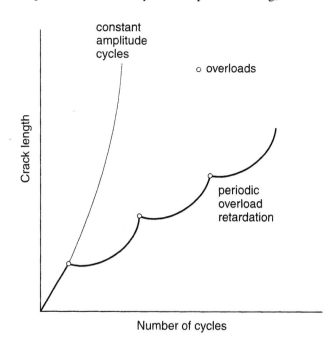

Figure 9.17 Fatigue-life enhancement due to periodic overloads.

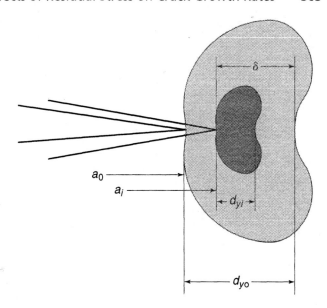

Figure 9.18 The Wheeler model of crack growth inside an overload plastic zone.

is repeated too frequently, the overload becomes the primary load, and most of the crack growth will occur during these isolated load excursions.

A mathematical model based on the progression of the plastic zone through the overload plastic zone was proposed by Wheeler [1972]. He suggested that

$$\left(\frac{da}{dN}\right)_{\text{retarded}} = \left(\frac{d_{yi}}{\delta}\right)^m \left(\frac{da}{dN}\right) \text{ for } d_{yi} < \delta \tag{9.8}$$

where d_{yi} is the instantaneous plastic zone diameter and δ is the distance from the current crack tip to the end of the overload plastic zone diameter d_{yo}, as illustrated in Figure 9.18. Based on experiments using a single-edge notched specimen on a high yield strength steel (0.45% C) and Ti–6Al–4V titanium alloy, Wheeler proposed values for the exponent m of 1.3 and 3.4, respectively.

For constant amplitude fatigue with occasional overloads, the Wheeler model works well. Good results also have been reported for block loading in which the amplitude remains fixed for a given period of time (many cycles per block) and then changes to another level, and the process is repeated over and over again. However, for truly variable amplitude loading, Wheeler proposed cycle-by-cycle crack growth summation using the retardation model of Eq. (9.8). However, the validity of this approach is problematic.

Crack Closure

Associated with the advance of a fatigue crack is plastic deformation at the crack tip. In addition to the residual stress ahead of the crack, these deformations produce effects behind the crack tip that have a profound effect on the crack propagation

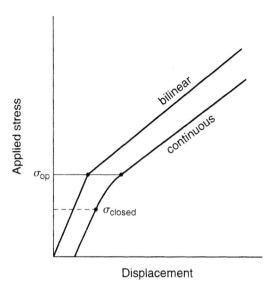

Figure 9.19 Load-displacement diagram of a cracked body exhibiting crack closure.

behavior. Elber [1970] demonstrated that the nonlinear load–displacement behavior of specimens with fatigue cracks was the result of crack closure prior to the full relaxation of the applied load. He argued that the premature recontacting of the crack faces was a direct consequence of the permanent deformation left in the wake of the advancing crack, because the narrow zone of stretched material left the crack faces incompatible.

In its simplest form, the load–deflection curve for a cracked body experiencing crack closure is bilinear, as shown in Figure 9.19, with a slope initially dictated by the closed crack, followed by a lower slope at higher loads, indicative of the actual crack length. In practice the crack does not open uniformly along its front, and there is a continuous transition from closed to open, making experimental determination of the opening stress difficult. In either event the closure load is defined as the onset of contact between the two crack faces—that is, the lower end of the upper linear slope, as shown in Figure 9.19. There is a difference between the crack opening load and the crack closing load, albeit small in many cases. The latter is easier to measure, but the former is probably more germane to the crack propagation rate. Experimental measurements, using high resolution microscopy techniques, of crack opening in aluminum have shown that the closed crack unfolds progressively toward the crack tip, but the event is nonproportional, with a large increase in load required to open the last 50 μm [Davidson, 1988].

In a later paper Elber [1971] observed that there can be no fatigue crack propagation until the applied stress, σ, exceeds the stress necessary to fully open the crack faces, σ_{op}. Accordingly, it was not the stress range, $\Delta\sigma$, but the effective stress range

$$\Delta\sigma_{\text{eff}} = \sigma_{\text{max}} - \sigma_{\text{op}} \qquad (9.9)$$

that governed the crack growth rate in the Paris law. That is,

$$\frac{da}{dN} = C(\Delta K_{\text{eff}})^n = C(U\Delta K)^n \tag{9.10a}$$

where

$$U = \frac{\sigma_{\text{max}} - \sigma_{\text{op}}}{\sigma_{\text{max}} - \sigma_{\text{min}}} = \frac{\Delta\sigma_{\text{eff}}}{\Delta\sigma} \tag{9.10b}$$

To determine U, Elber performed a series of constant amplitude tests on 2024-T3 aluminum, varying the stress intensity range, crack length (in center-cracked panels), and stress ratio. He found that all of the effects of the stress ratio could be incorporated into the closure-corrected Paris equation [Eq. (9.10a)] with a linear relation between U and R of the form

$$U = 0.5 + 0.4R \tag{9.11}$$

where $-0.1 < R < 0.7$. Elber also found that, for the linear portion of the crack propagation behavior, this form of R-ratio correction produced a smaller error than did the Forman equation, Eq. (9.6); however, the Elber form does not include Stage III behavior in the model. The success of the Elber's crack closure model in accounting for R-ratio effects can be seen in Figure 9.20, taken from Elber's 1971 paper, as applied to data from another source [Hudson, 1969].

Crack closure is affected by the properties of the material, geometry of the specimen, and the thickness. Since crack closure is due to the plastic zone, it should come as no surprise that material and loading variables that influence the size of the plastic zone should also have an influence on crack closure. For Ti–6Al–4V, Katcher and Kaplan [1974] found a smaller crack closure effect and suggested a relation of the form

$$U = 0.73 + 0.82R \tag{9.12}$$

for this titanium alloy. (Results for another titanium alloy were not significantly different.) For values of R greater than 0.35, they found no crack closure, that is, $U = 1$.

Over the years there have been numerous modifications to Elber's formula for various aluminum alloys, but the differences have been insignificant. Kumar [1992] made a systematic study of the crack closure variable U from the literature for the 20 years following Elber's work for aluminum and other alloy families. Of the more than 50 papers reviewed in this study, the majority determined that U was dependent on R only. Other papers indicated a dependence on K_{max} as well, but, while some of these papers found an increase in U with larger values of K_{max}, others found a decrease. To the first order, the assumption that $U = U(R)$ is not unreasonable. While the results for different alloy families were markedly different, the trend lines

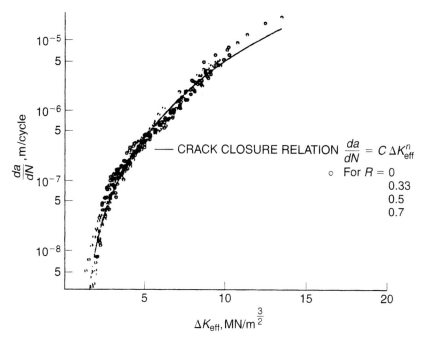

Figure 9.20 Crack growth rate versus effective ΔK for 2024-T3 aluminum (compare with Figure 9.13) [Elber, 1971].

were all nearly linear with R. Considering all of the other unknowns, the inclusion of second-order corrections to the closure model does not appear to be justified.

For variable amplitude loading the effects of crack closure become significant. In any given load cycle, it is reasonable to assume that there will be no crack growth unless

$$\sigma_{max} > \sigma_{op} \tag{9.13}$$

In principle, cycle-by-cycle calculations of the crack growth should be possible for variable amplitude fatigue with the aid of the crack closure concept; however, such calculations are tedious and subject to a variety of other influences. Nonetheless, any attempt to predict crack growth rates under variable amplitude conditions without including crack closure effects is doomed to failure.

In the years following Elber's discovery of plasticity-induced closure, other mechanisms of closure for special situations have been proposed. These mechanisms include crack-face surface roughness, oxidation, and grain boundary mismatch. These alternative mechanisms have been reviewed by McEvily [1988].

9.5 LIFE-PREDICTION COMPUTER PROGRAMS

A computer program that implements much of the fatigue crack growth analysis discussed in the previous sections, including the effects of crack closure and retar-

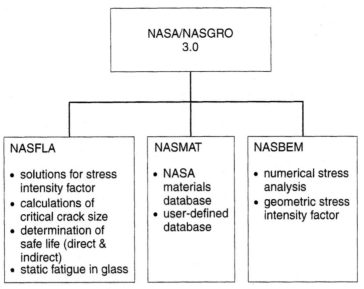

Figure 9.21   Modules in the NASA/NASGRO 3.0 fatigue-crack life-prediction program.

dation, has been implemented by the staff of the National Aeronautics and Space Administration (NASA) at the Lyndon B. Johnson Space Center in Houston, Texas. The program NASA/NASGRO 3.0 [2000], consists of three main modules, as shown in Figure 9.21.[4] The module that implements the fatigue analysis is NASFLA; the other two modules, NASBEM and NASMAT, perform stress-analysis calculations and maintain the extensive NASA fatigue database, respectively. The fatigue crack growth model used in the program is a variation of the Forman equation, Eq. (9.6), and is of the form [Forman and Mettu, 1992]

$$\frac{da}{dN} = \frac{C(1 - f)^n \Delta K^n \left(1 - \frac{\Delta K_{th}}{\Delta K}\right)^p}{(1 - R)^n \left(1 - \frac{\Delta K}{(1-R)K_c}\right)^q} \quad (9.14)$$

where the plasticity-induced crack closure function f, defined as

$$f = \frac{K_{op}}{K_{max}} \quad (9.15)$$

produces changes in the Forman crack growth equation comparable with those produced by the term U in the Elber closure model. In the program, f is expressed in power series form, using equations developed by Newman [1984].

---

[4]The program is available on the Internet at http:/mmptdpublic.jsc.nasa.gov/nasgro/NasgroMain.html.

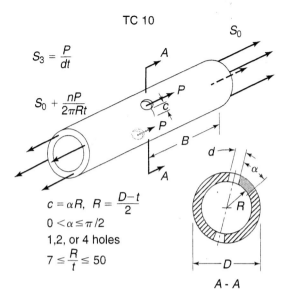

TC 10

$S_3 = \dfrac{P}{dt}$

$S_0 + \dfrac{nP}{2\pi Rt}$

$c = \alpha R,\ \ R = \dfrac{D-t}{2}$

$0 < \alpha \le \pi/2$

1,2, or 4 holes

$7 \le \dfrac{R}{t} \le 50$

S_0

$A - A$

Figure 9.22 Representative crack case (TC10) included in NASA/NASGRO 3.0.

The term raised to the power p accounts for the crack growth rate threshold behavior. This term tends toward zero as the applied ΔK approaches the threshold value, thereby providing the left asymptote in a full-range crack growth curve. For $p = 0$, Eq. (9.14) has the characteristics of the earlier Forman model, Eq. (9.6), which does not account for threshold behavior. Internal to the program are additional formulae for estimating the values of variables that may not be known from experiments (e.g., ΔK_{th} and K_c).

The crack growth rate model has considerable flexibility to model a wide variety of materials by virtue of its four empirical fitting constants C, n, p, and q, similar to the Schwartz model, Eq. (9.7). Appendix E provides the constants for several common metallic materials taken from the extensive NASMAT module database.

In addition to the large database of empirical curve-fitting constants, the program contains the geometric stress intensity factor relations for a large number of crack cases (see Table 9.1), which, in the program, are addressed by their identification code (the leftmost column of the table). A typical geometry, a through crack emanating from a hole in a pipe wall, crack case TC10, is shown in Figure 9.22. The manual that accompanies the program contains the details of the geometric stress intensity factor calculation, as well as the equation for determining the net section stress that the program uses to determine if yielding has occurred. The general form for the stress intensity relation used throughout the program is of the form

$$K = (S_0 F_0 + S_1 F_1 + S_2 F_2 + S_3 F_3)\sqrt{\pi a} \qquad (9.16)$$

Table 9.1 Crack Cases Included in NASGRO 3.0

	Through Cracks:
TC01:	Through crack at the center of a plate
TC02:	Through crack at the edge of a plate
TC03:	Through crack from an offset hole in a plate
TC04:	Through crack from a hole in a lug
TC05:	Through crack from a hole in a plate with a row of holes
TC06:	Through crack in a sphere
TC07:	Through crack in a cylinder (longitudinal direction)
TC08:	Through crack in a cylinder (circumferential direction)
TC09:	Through crack from a hole in a plate under combined loading
TC10:	Through crack from a hole in a cylinder (circumferential direction)
	Embedded Cracks:
EC01:	Embedded crack in a plate
	Corner Cracks:
CC01:	Corner crack in a rectangular plate
CC02:	Corner crack from a hole in a plate
CC03:	Corner crack from a hole in a lug
CC04:	Corner crack from a hole in a plate (one or two cracks)
	Surface Cracks:
SC01:	Surface crack in a rectangular plate—tension and/or bending
SC02:	Surface crack in a rectangular plate—nonlinear stress
SC03:	Surface crack in a spherical pressure vessel
SC04:	Longitudinal surface crack in a hollow cylinder—nonlinear stress
SC05:	Thumbnail crack in a hollow cylinder
SC06:	Circumferential crack in a hollow cylinder—nonlinear stress
SC07:	Thumbnail crack in a solid cylinder
SC08:	Thumbnail crack in a threaded, solid cylinder
SC09:	Circumferential crack at a thread root in a cylinder
SC10:	Circumferential crack in a threaded pipe—nonlinear stress
SC11:	Surface crack from a hole in a plate (one or two cracks)
SC12:	Surface crack from a hole in a lug (one or two cracks)
SC13:	Surface crack in a bolt-head fillet—shear bolt
SC14:	Surface crack in a bolt-head fillet—tension bolt
	Standard Specimens:
SS01:	Center-cracked tension specimen M(T)
SS02:	Compact-tension specimen C(T)
SS03:	Disc-shaped compact-tension specimen DC(T)
SS04:	Arc-shaped tension specimen A(T)
SS05:	Three-point bend specimen SE(B)
SS06:	Edge-cracked tension specimen SE(T)—constrained ends
SS07:	Notched round-bar specimen R-bar(T)—circumferential crack
SS08:	Notched plate with a surface crack
SS09:	Notched plate with a corner crack
SS10:	Notched plate with a through crack

where

$$S_0 = \text{applied tension (or compression)}$$
$$S_1 = \text{bending in the thickness direction (antiplane)}$$
$$S_2 = \text{bending in the width direction(in-plane)}$$
$$S_3 = \text{pin bearing pressure}$$

and F_0, F_1, F_2, and F_3 are the associated dimensionless shape functions for the combinations of geometry and loading [analogous to the shape functions $F(a/W)$ previously defined in Eq. (3.56)]. For the most part, the shape functions F_i were obtained from the literature, and for cases in which the functions are defined graphically, they have been fitted to appropriate functions for computational purposes.

The NASFLA module provides a systematic procedure (in a Windows^{TM} environment) for analyzing the fatigue crack propagation behavior of any of the geometry–loading configurations contained in the model, as well as for inputting user-defined crack cases, provided that the cases can be represented in a format consistent with the program logic. This module provides several calculation options, depending on the known values:

(a) *direct life prediction*: Given the initial crack size, the program computes the critical flaw size and determines the number of cycles to failure;

(b) *allowable load analysis*: Given the initial and final crack sizes, the program computes the maximum allowable load factors to achieve a preselected fatigue life;

(c) *inspection limit computation*: Given the critical crack size, the applied loads, and the preselected fatigue life, the program iteratively calculates the initial flaw that will result in failure.

In the event that the initial flaw size is not known, the program provides a table of standard minimum flaw sizes, based on the inspection method selected, which NASA considers acceptable for space flight hardware. These minimum inspectable flaw sizes, some of which are shown in Table 9.2, are those flaw sizes that have a high probability of detection when inspections are performed in accordance with proper specifications, (i.e., very conservative). For applications other than NASA hardware, alternative estimates of inspectable flaw sizes based on experience or nondestructive evaluation (NDE) staff guidance are probably more useful.

In addition to the degree of automation provided by the NASGRO program, which reduces otherwise complex integrations to straightforward interactive input tasks, the program becomes more compelling when load interaction effects in combination with variable amplitude loading are considered. These effects are integrated into the program through a series of five alternative models (described in the program manual).

Table 9.2 Default Initial NDE Flaw Sizes Used in NASGRO 3.0

Crack Case (see Table 9.1)	NDE Inspection Technique or Flaw-Size Criterion	Thickness Range (in) t	Crack Size (in)*** a	c
TC01, TC06, TC07, TC08 (open surface)	EC	$t \leq 0.050$	-	0.050
	P	$t \leq 0.050$	-	0.100
	P	$0.050 < t \leq 0.075$	-	$0.15 - t$
	MP	$t \leq 0.075$	-	0.125
TC02 (edge)	EC	$t \leq 0.075$	-	0.100
	P	$t \leq 0.100$	-	0.100
	MP	$t \leq 0.075$	-	0.250
TC03, TC04, TC05, TC09 (hole) TC10	EC	$t \leq 0.075$	-	0.100
	P	$t \leq 0.100$	-	0.100
	MP	$t \leq 0.075$	-	0.250
	HPD—driven rivet	any thickness	-	0.005
	HPD—other holes	$t \leq 0.050$	-	0.050
EC01	R	$0.025 \leq t \leq 0.107$	$0.35t$	0.075
	R	$t > 0.107$	$0.35t$	$0.7t$
	U	$t \geq 0.300$	0.065	0.065
CC01 (edge), CC05	EC	$t > 0.075$	0.075	0.075
	P	$t > 0.100$	0.100	0.100
	MP	$t > 0.075$	0.075	0.25
	U	$t > 0.100$	0.100	0.100
CC02, CC03 (hole), CC04	EC	$t > 0.075$	0.075	0.075
	P	$t > 0.100$	0.100	0.100
	MP	$t > 0.075$	0.075	0.25
	U	$t > 0.100$	0.100	0.100
	HPD—nondriven rivet	$t > 0.050$	0.050	0.050
SC01, SC02, SC03 (open surface) SC11, SC12	EC	$t > 0.050$	0.020	0.100*
			0.050	0.050 **
	P	$t > 0.075$	0.025	0.125*
			0.075	0.075 **
	MP	$t > 0.075$	0.038	0.188*
			0.075	0.125 **
	R	$0.025 \leq t \leq 0.107$	$0.7t$	0.075
		$t > 0.107$	$0.7t$	$0.7t$
	U	$t \geq 0.100$	0.030	0.150*
			0.065	0.065 **
SC04, SC05	EC (ext. and int.)	$t > 0.050$	0.020	0.100*
			0.050	0.050 **
	P (ext.)	$t > 0.075$	0.025	0.125*
			0.075	0.075 **
	MP (ext.)	$t > 0.075$	0.038	0.188*
			0.075	0.125 **
	R (ext. and int.)	$0.025 \leq t \leq 0.107$	$0.7t$	0.075
		$t > 0.107$	$0.7t$	$0.7t$
	U (ext. and int.)	$t \geq 0.100$	0.030	0.150*
			0.065	0.065 **
SC06	EC (ext. and int.)	$t > 0.050$	0.020	-
	P (ext.)	$t > 0.075$	0.025	-
	MP (ext.)	$t > 0.075$	0.038	-
	R (ext. and int.)	$0.025 \leq t \leq 0.107$	$0.7t$	-
	U (ext. and int.)	$t \geq 0.100$	0.030	-

Notes:

EC = eddy current	R = radiographic	MP = magnetic particle
P = dye penetrant	U = ultrasonic	HPD = hole penetration defect (max.)
*minimum crack depth	**maximum crack depth	***1 in = 25.4 mm

There is a similar crack growth prediction program developed by the U.S. Air Force that has a somewhat more user-friendly interface, but less comprehensive capabilities. AFGROW 4.0 [Harter, 1999] uses the same NASMAT database as NAS-GRO, but offers different options for modeling fatigue.[5] Both programs share some of the same classical crack geometry solutions, but AFGROW has some crack cases that are more germane to aircraft that are not found in NASGRO. An interesting feature of AFGROW is its use of Microsoft's Component Object Model (COM) programming capability, which allows the program to be called from any Windows™ program. When AFGROW is used in conjunction with a spreadsheet program, complex macros can be written to perform parametric life prediction analyses to study the effects of systematic changes in any of the variables.

As with any computer program intended to implement the engineering design process, the program is only as good as the numerical models implemented within and the accuracy of the coding. Although these programs are well thought out and have been in use in some form for a number of years, the reader is cautioned to execute several trial examples for which solutions are known or can be computed independently and that closely model the problem of interest. As both programs are continually undergoing revision, it is important for the user to consult their respective Web sites for updates before using them to begin a solution to a new problem.

9.6 MEASURING FATIGUE PROPERTIES: ASTM E 647

Fatigue crack growth rates, and hence the fitting parameters in any of the growth rate models we have discussed in this chapter, are among the most sensitive of all fracture mechanics related properties. As a result, the use of generic properties (including those listed in Appendix E) for anything other than short-term crack growth predictions or *rough* estimates of long-term life is problematic.

Some of the factors known to influence fatigue crack growth rates include environment, thickness of the material, and temperature. There are extensive data in the literature on the influence of environment, especially humidity and salt water. Water vapor, by virtue of its corrosive attack on the chemically active crack tip, has a moderate to strong effect on aluminum, ceramics, glass, and plastics. Regardless of the thickness of the material, fatigue cracks always start normal to the maximum tension. In thin sheets (i.e., plane stress), a transition to slant (shear) fracture occurs before the final failure. All other factors being equal, the crack growth rate for a given material will be lower in plane stress than in plane strain, because of the influence of the larger plastic zone in the former case. In very thick materials, surface flaws are more likely to occur than through cracks, and the growth rates in

---

[5] The program is available for download at http://fibec.flight.wpafb.af.mil/fibec/afgrow.html.

the thickness direction will be different from those in the surface direction, for the same reason. The effects of moderate temperature changes are not dramatic, but colder temperatures tend to improve fatigue performance by decreasing the kinetic mobility of atoms in the lattice; however, colder temperatures may simultaneously adversely affect K_c.

The fatigue crack propagation properties of metal alloys depend on the processing variables in addition to the specifics of the composition. Compared with rolled plate, castings tend to have lower strength and poorer fatigue performance. Forgings and rolled products can have fatigue properties that are strongly orthotropic. Heat treatment affects microstructure, which, in turn, influences the slip band motion and hence fatigue crack growth. As a general rule, increasing the yield stress decreases the ductility and degrades the fatigue properties. The effects of work hardening are not systematic and can be beneficial or detrimental, depending on the material and the degree of working. For critical applications, batch-to-batch variations of nominally the same product can be significant. Because of all of these factors, the probability of finding data relevant to the service conditions you might expect is small. Consequently, it is often necessary to measure the crack growth rate of your particular material under conditions simulating the conditions of intended use.

Ideally, fatigue crack growth rates should be measured under constant applied stress intensity conditions, that is, constant ΔK. Several test specimens which meet that requirement have been described in earlier chapters; however, for a variety of reasons, none have met the rigorous requirements of the ASTM Standards community. Instead, the currently accepted test methods are, for the most part, extensions of the previously certified geometries modified for fatigue crack growth measurements. The details are described in the ASTM Standard *Standard Test Method for Measurement of Fatigue Crack Growth Rate*, E 647–99.

Unlike the fracture toughness standard, in which the crack length is measured after the completion of the test, this standard requires that the crack length be measured at regular intervals which are small enough that ΔK can be replaced by its average value over the crack-length interval while the test is in progress. Three methods have been proposed: visual, compliance, and electrical potential. The recommended procedure for visual measurement of the crack length is with a travelling microscope with $20\times$ to $50\times$ power. Measurements are made on both sides of the polished specimen at small intervals of crack growth, typically two to four percent of W. The required intervals depend on the crack length and become smaller as the crack length increases, to compensate for the more rapid change in the value of K for the longer cracks. The test method suggests that these optical measurements be made without stopping the fatigue cycling. The other two methods are indirect and infer the instantaneous crack length from either the compliance measured during an unloading cycle or the change in electrical potential across the crack face. In principle, these measurements can be made on every load cycle for a continuous record of the crack length. Appropriate equations for converting the measurements to crack length are given in the Standard E 647. These latter two methods simplify the test procedure by reducing the manual labor involved, at the expense of increased instrumentation and the need for computerization.

Three specimen types are permitted for the fatigue crack growth rate measurements: the compact tension, C(T); the center-cracked panel, M(T); and the eccentrically loaded single edge-crack specimen, ESE(T). The first two types have the same planar geometry as their fracture toughness measurement equivalents specified in ASTM Standards E 399 and E 338, respectively, the only difference being the thickness, which is thinner for fatigue tests than for toughness measurement, to minimize crack-front curvature. For the C(T) geometry, the preferred thickness B is in the range $W/20 \leq B \leq W/4$, and for the M(T) geometry, $B < W/8$. In order to provide a long range over which measurements can be taken, the initial crack length is only 20 percent of the specimen width. Measurements can continue as long as the remaining ligament satisfies the following equations:

$$W - a \geq \left(\frac{4}{\pi}\right)\left(\frac{K_{max}}{\sigma_{ys}}\right) \qquad \text{for the C(T) and ESE(T) specimens} \qquad (9.17a)$$

$$W - 2a \geq \frac{1.25 P_{max}}{B\sigma_{ys}} \qquad \text{for the M(T) specimen} \qquad (9.17b)$$

These limits are necessary to ensure that the behavior is still elastic. For materials with high work-hardening characteristics, the yield stress is replaced by the effective flow stress $\sigma_{FS} = (\sigma_{ys} + \sigma_{ULT})/2$.

The third geometry, the ESE(T) specimen, shown in Figure 9.23, is a new geometry for fracture mechanics studies and has several advantages over the other two geometries. The elongated design of the ESE(T) geometry provides additional working space compared with the C(T) geometry, to which displacement or strain gages can be applied for automated crack-length determination. In comparison to the M(T) geometry, the specimen with its off-center loading, requires lower applied force, which results in a lower net section stress. Additional differences are noted in ASTM Standard E 647–99. For the geometry shown in Figure 9.23, the geometric stress intensity factor is given by

$$K = \frac{P}{B\sqrt{W}}F\left(\frac{a}{W}\right) \qquad (9.18)$$

where

$$F\left(\frac{a}{W}\right) = \alpha^{1/2}(1.4 + \alpha)[1 - \alpha]^{-3/2}G$$

where $\alpha = \frac{a}{W}$, and

$$G = 3.97 - 10.88\alpha + 26.25\alpha^2 - 38.9\alpha^3 + 30.15\alpha^4 - 9.27\alpha^5$$

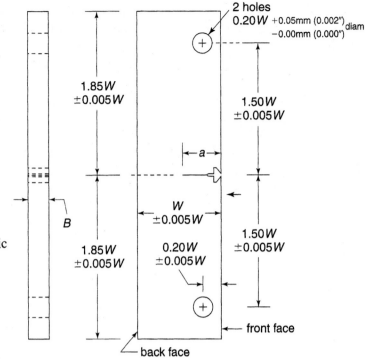

Figure 9.23 The eccentric single-edge notched specimen [ESE(T)] used for determining the fatigue-crack growth rate in ASTM standard E 647. Copyright ASTM, reprinted with permission.

The corresponding expressions for the C(T) and M(T) specimens have been presented earlier as Eqs. (8.3) and (3.62), respectively.

The ESE(T) geometry is well suited to studying crack closure, and a procedure has been described for determining K_{op}, by using a strain gage mounted on the back face at the crack line. The method compares the experimentally measured compliance with the theoretical compliance for a fully open crack, which, in terms of the measured strain ϵ, is given by

$$a/W = N_0 + N_1 X + N_2 X^2 + N_3 X^3 + N_4 X^4 \qquad (9.19a)$$

where $X = \log(|\epsilon|/P)BWE$ and the constants are

$$N_0 = 0.09889$$

$$N_1 = 0.41967$$

$$N_2 = 0.06751 \qquad (9.19b)$$

$$N_3 = -0.07018$$

$$N_4 = 0.01082$$

For any of the approved geometries the standard procedure is to perform the test at constant force amplitude, ΔP, with the stress ratio and frequency held constant. Since all of the geometries exhibit rising values of K with increasing crack length, a wide range of ΔK values will result. This procedure tends to break down for desired crack growth rates below 10^{-8} m/cycle. For very low crack growth rates, approaching threshold values, a load shedding procedure is preferred; however, care must be taken to ensure that prior plastic zones do not result in retardation. Recommendations are given in the standard to avoid pitfalls with the procedure.

In using ASTM Standard E 647–99 to measure the crack growth rate you must take several precautions. Although any specimen can be used to determine the growth rate, the thickness should match that of the application, since effects due to thickness can occur. Because of crack growth retardation caused by residual stresses, test specimens removed from stock that is not completely stress relieved can result in biased measurements. Finally, the crack growth rate for small fatigue cracks can be markedly different from that of long cracks, and the use of data from this standard can lead to nonconservative life predictions for small cracks. In this context, a crack is considered to be small if its length is small compared with microstructural features, if it is small compared with the size of the plastic zone, or if its length is less than 1 mm.

REFERENCES

ASTM Standard E 647–99, 1999, "Standard Test Method for Measurement of Fatigue Crack Growth Rates," *Annual Book of ASTM Standards*, Vol. 03.01, American Soc. for Testing and Materials, West Conshohocken, PA

ASTM Standard E 1823–96, 1996, "Standard Terminology Relating to Fatigue and Fracture Testing," *Annual Book of ASTM Standards*, Vol. 03.01, American Soc. for Testing and Materials, West Conshohocken, PA.

Cioclov, D., 1977, *Mechanica Ruperii Materialelor (Fracture Mechanics)*, Editura Academiei Republicii Socialiste România, Bucharest, Rumania.

Cottrell, A. H., 1962, "Introductory Review of the Basic Mechanisms of Crack Propagation," *Proceedings of the Crack Propagation Symposium*, Vol. 1, Cranfield College of Aeronautics, United Kingdom, pp. 1–7.

Davidson, D. L., 1988, "Plastically Induced Fatigue Crack Closure," *Mechanics of Fatigue Crack Closure*, ASTM STP 982, American Soc. for Testing and Materials, Philadelphia, pp. 44–61.

Dieter, G. E., 1986, *Mechanical Metallurgy*, 3rd ed., McGraw-Hill Book Co., New York, p. 384.

Elber, W., 1970, "Fatigue Crack Closure Under Cyclic Tension," *Engineering Fracture Mechanics*, Vol. 2(1), pp. 37–45.

Elber, W., 1971, "The Significance of Crack Closure," *Damage Tolerance in Aircraft Structures*, ASTM STP 486, American Soc. for Testing and Materials, Philadelphia, PA, pp. 230–242.

Forman, R. G., Kearney, V. E., and Engle, R. M., 1967, "Numerical Analysis of Crack Propagation in Cyclic-Loaded Structures," *J. Basic Engineering*, Vol. 89, pp. 459–464.

Forman, R. G. and Mettu, S. R., 1992, "Behavior of Surface and Corner Cracks Subjected to Tensile and Bending Loads in Ti–6l–4V Alloy," *Fracture Mechanics, Twenty-Second Symposium*, Vol. 1, STP 1131, American Soc. for Testing and Materials, Philadelphia, pp. 519–546.

Forsyth, P. J. E., 1956, "The Basic Mechanism of Fatigue and Its Dependence on the Initial State of a Material," *Proceedings of the International Conference on Fatigue in Metals*, Institution of Mechanical Engineers, London, pp. 535–537.

Forsyth, P. J. E., 1962, "A Two Stage Process of Fatigue Crack Growth," *Proceedings of the Crack Propagation Symposium*, Vol. 1, Cranfield College of Aeronautics, United Kingdom, pp. 76–94.

Graham, S. M., 1988, *Stress Intensity Factors for Bodies Containing Initial Stress*, Ph.D. Dissertation, University of Maryland, College Park, MD.

Harter, J. A., 1999, *AFGROW Users Guide and Technical Manual*, AFRL-VA-WP-TR–1999–3016, Air Force Research Laboratory, Air Vehicles Directorate, Wright-Patterson AFB, OH.

Hudson, C. M., 1969, "Effect of Stress Ratio on Fatigue-Crack Growth in 7075-T6 and 2024-T3 Aluminum-Alloy Specimens," NASA TN D–5390, National Aeronautics and Space Administration.

Katcher, M. R., and Kaplan, M., 1974, "Effect of R-Factor and Crack Closure on Fatigue Crack Growth for Aluminum and Titanium Alloys," *ASTM STP 559*, American Soc. for Testing and Materials, Philadelphia, pp. 264–282.

Kumar, R., 1992, "Review on Crack Closure for Constant Amplitude Loading in Fatigue," *Engineering Fracture Mechanics*, Vol. 42(2), pp. 389–400.

Laird, C., and Smith, G. C., 1962, "Crack Propagation in High Stress Fatigue," *Phil. Mag.*, Vol. 7, pp. 847–857.

Laird, C., 1967, "The Influence of Metallurgical Structure on the Mechanisms of Fatigue Crack Propagation," *Fatigue Crack Propagation*, ASTM STP 415, American Soc. for Testing and Materials, Philadelphia, pp. 131–180.

Laird, C., 1979, "Mechanisms and Theories of Fatigue," *Fatigue and Microstructure*, ASM International, Materials Park, OH, pp. 149–203.

Manson, S. S., 1960, *Thermal Stress and Low-Cycle Fatigue*, McGraw-Hill Book Co., New York.

McEvily, A. J., 1988, "On Crack Closure in Fatigue Crack Growth," *Mechanics of Fatigue Crack Closure*, ASTM STP 982, American Soc. for Testing and Materials, Philadelphia, pp. 35–43.

NASA/NASGRO 3.0, 2000, *Fatigue Crack Growth Computer Program, "NASGRO" Version 3.0*, JSC–22267B, March 2000, Johnson Space Center, Houston, TX.

Newman, J. C., Jr., 1984, "A Crack Opening Stress Equation for Fatigue Crack Growth," *Int. J. Fracture*, Vol. 24, No. 3, pp. R131–R135.

Paris, P. C., Gomez, R. E., and Anderson, W. E., 1961, "A Rational Analytic Theory of Fatigue," *The Trend in Engineering*, Vol. 13, No. 1, University of Washington, Seattle, pp. 9–14.

Paris, P., and Erdogan, F., 1963, "A Critical Analysis of Crack Propagation Laws," *J. Basic Engineering*, Vol. 85, pp. 528–534.

Schijve, J., 1961, "Fatigue Crack Propagation in Light Alloy Steel Material and Structures," *Advances in Aeronautical Sciences*, Pergamon Press, New York, pp. 387–408.

Schwartz, B. J., Engsberg, N. G., and Wilson, D. A., 1984, "Development of a Fatigue Crack Propagation Model of Incoloy 901," *Fracture Mechanics, Fifteenth Symposium*, ASTM STP 833, American Soc. for Testing and Materials, Philadelphia, pp. 218–241.

Walker, K., 1970, "The Effect of Stress Ratio During Crack Propagation and Fatigue for 2024-T3 and 7075-T6 Aluminum," *Effect of Environment and Complex Load History on Fatigue Life*, ASTM STP 462, American Soc. for Testing and Materials, Philadelphia, pp. 1–14.

Wheeler, O.E., 1972, "Spectrum Loading and Crack Growth," *J. Basic Engineering*, Vol. 94, pp. 181–186.

Wood, W. A., 1956, "Failure of Metals Under Cyclic Strain," *Proceedings of the International Conference on Fatigue in Metals*, Institution of Mechanical Engineers, London, pp. 531–554.

Wood, W. A., 1958, "Recent Observations on Fatigue Failure in Metals," *Symposium on Basic Mechanisms of Fatigue*, ASTM STP 237, American Soc. for Testing and Materials, Philadelphia, pp. 110–121.

EXERCISES

9.1 Determine the constants in the Paris law [Eq. (9.3)] for the linear portion of the $R = 0.75$ data in Figure E9.1. (Note: The tabulated data corresponding to Figure E9.1 are available at http://www.wam.umd.edu/~sanford.)

9.2 Determine the constants in the Walker model [Eq. (9.4)] for the $R = 0.33$ data in Figure E9.1.

9.3 Determine the constants in the Walker model [Eq. (9.4)] for the $R = 0.75$ data in Figure E9.1.

9.4 Determine the constants in the Forman–Kearney–Engle model [Eq. (9.6)]

(a) for the $R = 0.75$ data in Figure E9.1.

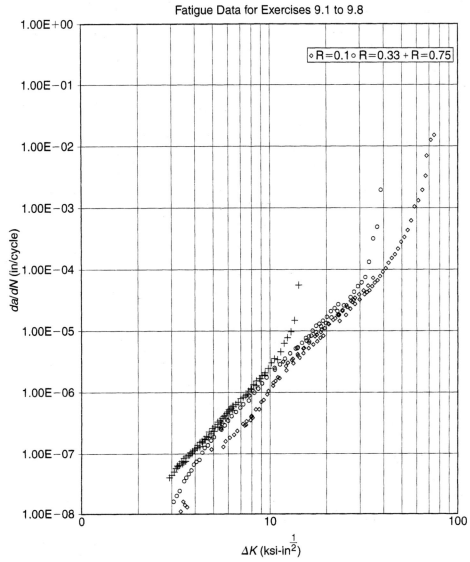

Figure E9.1 Fatigue-crack growth data for a metallic alloy
($K_{Ic} = 60$ ksi-in$^{1/2}$ and $\Delta K_{th} = 3.5$ ksi-in$^{1/2}$) obtained from
compact-tension specimens under plane-strain conditions.

(b) for the $R = 0.33$ data in Figure E9.1.

(c) for the $R = 0.10$ data in Figure E9.1.

Comparing the results of parts (a), (b), and (c), comment on the ability of the
Forman–Kearney–Engle model to account for the effect of R-ratio.

9.5 Determine the constants in the Forman–Kearney–Engle model [Eq. (9.6)], incorporating all of the data from $R = 0.10, 0.33$, and 0.75 in Figure E9.1.

9.6 Determine the constants in the hyperbolic sine model [Eq. (9.7)] for the $R = 0.33$ data in Figure E9.1.

9.7 Determine the constants in the hyperbolic sine model [Eq. (9.7)], for the $R = 0.10$ data in Figure E9.1.

9.8 Assuming that the crack closure function f is equal to 0.30 for $R = 0.10, 0.40$ for $R = 0.33$, and 0.75 for $R = 0.75$, fit the data of Figure E9.1 to the NASA NASGRO 3.0 model [Eq. (9.14)].

9.9 Given the Paris law constants $C = 8.8 \times 10^{-8}$ and $n = 2.3$, for ΔK measured in ksi$\sqrt{\text{in}}$ and da/dN in inches, determine the corresponding constants for ΔK measured in MPa$\sqrt{\text{m}}$ and da/dN in millimeters.

9.10 A wide aluminum plate ($K_{Ic} = 27$ ksi$\sqrt{\text{in}}$) contains a hole of diameter D, with a crack of length a *on one side only*. Assuming that the geometric stress intensity factor can be approximated by $K = \sigma\sqrt{\pi/2}\sqrt{D+a}$, where σ is the remotely applied stress, determine the number of cycles to failure for remote stress varying between 10 ksi and 20 ksi in a plate with a 0.5 inch diameter hole. Assume Paris-law constants of 2.26×10^{-7} and $n = 4$ for ΔK measured in ksi$\sqrt{\text{in}}$.

9.11 A standard (1T) compact-tension specimen to be tested in accordance with ASTM Standard E 399 has a machined notch to a depth of $a/W - 0.40$. For a material of similar composition, $K_{Ic} = 40$ ksi$\sqrt{\text{in}}$, $\sigma_{ys} = 65$ ksi, and Paris law coefficients of $C = 12 \times 10^{-11}$ and $n = 3$ in English units,

    (a) determine the load range $\Delta\sigma$ required to fatigue crack the specimen at an initial ΔK range of 30 ksi$\sqrt{\text{in}}$. Assume that $R = 0$, and include the effects of the plastic zone.

    (b) for a load range of 5 kips, *estimate* the number of cycles required to extend the crack to a depth of $a/W = 0.45$. Assume that the Paris law applies, and neglect the influence of the plastic zone.

CHAPTER 10

Designing Against Fracture

10.1 INTRODUCTION

Now that linear elastic fracture mechanics has grown into a mature science, it would seem obvious to incorporate its principles into the routine design process. The reality, however, is that, with the exception of the aerospace and nuclear power industries, fracture mechanics based design is the exception rather than the rule. More likely, fracture mechanics arguments are considered as part of a post mortem analysis of a failed component returned from the field. This point is brought home by the fact that there are more books devoted to case studies of failed components than there are textbooks that mention the use of fracture mechanics for design. A notable exception is the book by Barsom and Rolfe [1987].

Design engineering has risks associated with it. By definition, each new design poses new challenges and the potential for new problems. For this reason much of design engineering employs the incremental method of design—that is, the new design is a small variant of a proven earlier design. While this approach may seem to be the safest course, there are hidden dangers. As a design goes through many stages of revision, it is possible to lose sight of the origin of some of the initial design features. For example, it might be overlooked that a section of seemingly excess thickness might have been purposefully introduced as a crack arresting section. Also, design revisions might introduce a problem that was nonexistent in earlier versions. Chapter 1 presented a historical example wherein the design of larger boilers around the turn of the 20th century to fuel the needs of the growing industrial revolution was obtained by simple scaling up of successful boilers already in use. While the laws of elasticity do scale, the increased size failed to account for the plane-stress to plane-strain transition in the material, with its attendant dramatic decrease in fracture toughness. Of course, without the aid of fracture mechanics, which had not yet been developed, such a behavior could not have been anticipated.

The dominant design criterion to be met is the lowest cost per unit. From a management perspective, this criterion is interpreted to mean the lowest *manufacturing* cost per unit, but from an engineering perspective, the lowest *life cycle* cost per unit is more appropriate. Included in this latter criterion are such factors as: maintenance, warranty repair or replacement, inspection costs, and potential liability. The latter factor, liability costs, is often termed *risk assessment* and in many firms is a management decision, not an engineering one. However, the added cost of fracture mechanics analysis can most often be justified on the basis of reducing risk.

10.2 FRACTURE MECHANICS IN CONVENTIONAL DESIGN

In conventional design engineering practice the focus is on the stress analysis of the problem. Material selection is based on such factors as cost, availability, and adequate yield strength to support the computed stresses. To ensure that the manufactured component or structure is able to perform its intended function, nondestructive inspection (NDI), as part of the quality control process, is sometimes specified. The role of inspection is to ensure that, to some definition, the item is "defect-free." In contrast, the importance of NDI is enhanced in fracture mechanics based design, and the component or structure is designed to function *despite* the presence of flaws. The process is further complicated by the fact that a new material property, the material's resistance to fracture, needs to be included in the decision process. Often, desirable yield properties are in conflict with desirable fracture properties. It is a fact of life that in many cases, materials have increased yield stress at the expense of fracture resistance and/or ductility. In addition, materials tailored to have desirable fracture properties tend to cost more than their run-of-the-mill counterparts. Finally, fracture mechanics based design places a burden on the designer to select a meaningful flaw location, type, and size into the stress analysis. We will discuss the implications of nondestructive evaluation in a subsequent section.

For the case of statically loaded structures the incorporation of fracture mechanics into the design stream is not particularly difficult. The process is generally done in two steps. The component is designed based on strength arguments, ignoring the fracture mechanics implications, and suitable design parameters (thickness, for example) and material are selected. A separate analysis using fracture mechanics principles is performed, and the design variables and material are redetermined based on fracture criteria. After both analyses are performed, the results of one approach are compared with the results of the alternative analysis to ensure that the design is safe. A successful design is one that satisfies both analyses with the same set of variables.

In the abstract the process seems complicated, but, in practice, it is quite straightforward. We will use an example [adapted from Reid and Baikie, 1986] to illustrate the point. Let us consider the design of a long, large-bore, pressurized pipe. Because of the large internal diameter needed (2.5 m) the choice of material is limited, and

Table 10.1 Engineering Properties of Candidate Materials

Property	Steel A	Steel B
Yield strength, MPa	695	325
Ultimate strength, MPa	800	500
Fracture toughness, MPa-m$^{1/2}$	100	120
Cost per ton, $	790	430

we are forced to select between two materials with germane properties, as listed in Table 10.1

The maximum operating pressure in the pipe is 8 MPa, and the desired factor of safety (FS) against yield is 1.5. The pipe will be fabricated from rolled and welded flat plate. After stress relief treatment, the residual stress normal to the longitudinal weld, $\sigma_{residual}$, is expected to be on the order of 70 MPa.

For a thin-walled cylinder, the radial stress is zero, and the Tresca yield condition reduces to

$$\frac{PR}{t} + \sigma_{residual} = \frac{\sigma_y}{FS} \tag{10.1}$$

Substituting the appropriate values for each of the two steels into Eq. (10.1), we find that the minimum thicknesses required for design against yield failure are

$$t_A = 26 \text{ mm} \quad \text{and} \quad t_B = 72 \text{ mm} \tag{10.2}$$

Assuming that the fabrication costs are the same for both steels, the choice of which steel to use becomes an economic one. Although the required thickness of Steel B is nearly three times that of Steel A, Steel B is significantly less expensive on a bulk basis. The comparative costs can be written as follows

$$\frac{\text{Cost using steel B}}{\text{Cost using steel A}} = \frac{\text{volume}_B}{\text{volume}_A} \cdot \frac{\$_B}{\$_A} = \frac{t_B}{t_A} \cdot \frac{\$_B}{\$_A} = 1.5 \tag{10.3}$$

Since the final cost of the pipe using Steel B is 50 percent higher than that of pipe using Steel A with the added bonus that the Steel A pipe weighs only one third of the weight of the Steel B pipe, the thinner, high yield strength steel (Steel A) is the clear choice for this application, based on a yield failure criterion.

In order to develop the corresponding design based on fracture mechanics criteria, we need to assemble additional information. For the purposes of our analysis, we will assume that the critical crack geometry is an infinitely long longitudinal edge crack of depth a. Of the various types of cracks that may occur in this application, the long edge-crack has the largest geometric stress intensity factor and therefore is the most demanding on the design. A suitable geometric stress intensity factor for this type of crack problem is [Gross and Srawley, 1964]

$$K = Y\left(\frac{a}{t}\right)\sigma\sqrt{\pi a} \qquad (10.4a)$$

where

$$Y\left(\frac{a}{t}\right) = 1.12 - 0.231\left(\frac{a}{t}\right) + 10.55\left(\frac{a}{t}\right)^2 - 21.71\left(\frac{a}{t}\right)^3 + 30.38\left(\frac{a}{t}\right)^4$$

If we neglect the effect of the residual stress, which is primarily on the surface, the critical crack length can be determined from

$$\left(\frac{a_c}{t}\right)Y\left(\frac{a_c}{t}\right)^2 = \frac{(FS)^2 K_{Ic}^2}{\pi\sigma_y^2 t} \qquad (10.4b)$$

Using the thicknesses determined from the yield criteria analyses, we can solve Eq. (10.5) graphically, as shown in Figure 10.1, and the results are

$$a_c = 6.5\text{ mm for steel A}\quad\text{and}\quad a_c = 26\text{ mm for steel B} \qquad (10.5)$$

We should note that the crack lengths calculated by this method are the notational crack lengths (see Example 6.4) and that the physical crack lengths are somewhat smaller.

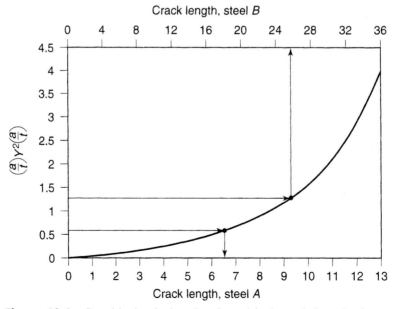

Figure 10.1 Graphical solution for the critical crack lengths for steels A and B.

Are either of these crack lengths acceptable? Ideally, we would like the critical crack length to be greater than the thickness, called the *leak-before-break criterion*, but that is not the case for either steel in this example. For this application ultrasonic inspection is determined to be the method of choice. NDE experts have indicated that the lower limit of inspectability for this application is 6 mm at the 90-percent confidence level. Since the critical crack length for Steel *A* is at the limits of inspectability, Steel *A* is unacceptable in a thickness of 26 mm. Repeating the calculation for larger thicknesses for Steel *A* reveals that increasing the thickness to as large as 39 mm increases the critical crack length only slightly. Beyond 39 mm, Steel *A* loses its economic advantage, and, all things considered, Steel *B* is the better choice, based on considerations of unstable fracture.

Summarizing the results of the various analyses, we find that, in the minimum thickness needed to satisfy the yield criterion, Steel *A* has an unacceptably small critical crack length and therefore must be rejected, despite its significant cost advantage. Comparing the steels on an equal cost basis provides only marginal improvement in the resistance to unstable fracture and places a heavy burden on NDE to ensure safe operation. In contrast, Steel *B* has a critical crack length well in excess of the inspectability limit, and some relaxation of the inspection requirement may be justified if the cost savings justify such a decision. Therefore, despite the weight penalty, Steel *B* in minimum thickness of 72 mm is the only alternative that satisfies both the yield and unstable fracture criteria. This example, taken from an actual case history, demonstrates that conventional design engineering can result in a design that lacks important safety considerations. If static pressure loading were the only consideration, the design analysis would be complete at this stage, and use of Steel *B* at a thickness of 72 mm (or more) would be the logical choice.

If there are alternating loads, even at low frequencies, fatigue crack growth needs to be considered. In this example the pipe in question is subjected to alternating pressures of 5.6 MPa an estimated 20 times per day. Considering the uncertainties, the use of a sophisticated fatigue crack growth model is hardly justified for low *R*-ratio conditions such as those just described, and the Paris Law is adequate for an initial screening. Using this simple law for fatigue crack growth and assuming an initial crack length equal to the inspection limit (6 mm), we find that a numerical integration results in an estimate of the time to failure of about seven years. This period is ample time around which to develop a periodic inspection plan. Because of the need for these inspections, the installation should allow access to all of the areas needing inspection, particularly the weld seams.

10.3 THE ROLE OF NDE IN DESIGN

In conventional design engineering practice, nondestructive evaluation is considered a part of the quality control process and is the responsibility of the manufacturing department. In contrast, designing with fracture mechanics concepts is of little value if the defects cannot be measured, or the costs to complete the required inspections are prohibitively expensive. For this reason it is important that an NDE specialist be involved in the design from the outset. This person (or group) can assist the designer in establishing the minimum defect size, depending on the application. A 2-mm semi-elliptical surface crack is a reasonable requirement for a turbine disk where eddy current methods are the norm, but, as illustrated in the design example discussed earlier in this chapter, would be unreasonable for the inspection limit of a large pipe under field conditions where ultrasonics is the method of choice. In addition the choice of inspection method and initial defect size depends strongly on the service loads. For statically loaded structures the detection limit is the critical crack size reduced by the required factor of safety, but for periodically loaded structures, the requirement is for an initial defect small enough to support the applied loads until the next inspection—a much more stringent requirement.

There are two possible approaches to determining inspection intervals: active or passive. In the passive approach the choice of inspection interval is decided as part of the design analysis. The usual assumption is that the interval should be selected so that a flaw just smaller than the inspection limit will not grow to critical size before the next inspection, with some safety margin included in the calculation. If the part passes the inspection, it is returned to service until the next regularly scheduled inspection. The other approach is to base the decision as to the next inspection interval on the results of the most recent nondestructive examination. Using the size of any observed flaws, one recalculates the expected life and sets the next inspection for some fraction thereof. For safety the inspection interval should be n times the residual life, so that even if an inspection is missed or the defect is undetected in an inspection, the structure will perform safely until the next inspection—or beyond. The more critical the application, the larger the value of n to be employed. For pressure vessels a value of five is not uncommon.

As an alternative to NDE at regular intervals, it is common practice to substitute proof tests at regular intervals. The proof load is selected sufficiently high so as to cause failure in the presence of subcritical cracks that could grow to critical size before the next proof test. The use of proof testing is controversial. On one hand, proof testing before the part is placed in service could weed out potentially dangerous parts that managed to slip past the quality controls at the manufacturer's plant. The counterargument is that the proof test permits parts with defects just below the critical size (at the proof load) to be released into the field. Moreover, the proof test itself may encourage the growth of a crack that otherwise may have remained small.

Some people believe that periodic proof testing is beneficial. The argument is that if the part survives the overload, any existing cracks will have been subjected to plastic strains, and the residual stress field surrounding the crack will provide some shielding of the crack. (Recall the Wheeler model of crack growth retardation previously discussed.)

In a study of the benefits of proof testing with which I was involved in the late 1970s, we examined the question of the advisability of continuing periodic proof testing of highly specialized weight-handling equipment used by the Navy for handling shipborne nuclear missiles. The practice prior to the study was to subject these rigging fixtures (slings, spreader bars, etc.) to an initial inspection and proof test before placing them in service. The parts were tracked and then returned from the fleet at regular intervals for cleaning, reinspection, proof testing, and repainting before returning them to the fleet. Aside from the cost of maintaining the testing and rework facility (several million dollars per year), there was the added cost of transporting the parts back and forth and of maintaining a documentation trail. The assignment given to us at the Naval Research Laboratory was to determine if the proof testing was inducing damage into the weight-handling equipment. After studying the use patterns of the equipment, we determined that some of the parts were used so rarely that the proof tests were the primary loads. As a result, the overloads were the service loads, and the beneficial effects of retardation were lost. Therefore, we determined that it was more cost effective and safer to inspect and proof test each part at the manufacturer's plant under the supervision of a Navy inspector and then release the part, with a service life printed directly on it, to the fleet. At the end of its service life each part was to be rendered inoperable and discarded. The point of this discussion is that the designer should let the analysis determine the conclusion rather than predefine a desired outcome and constrain the analysis to that task.

10.4 U.S. AIR FORCE DAMAGE-TOLERANT DESIGN METHODOLOGY

Arguably, the most carefully thought out and well-documented fracture mechanics based design procedure is the methodology spelled out in MIL-A–83444, *Airplane Damage Tolerance Design Requirements* [1974].[1] Developed by the Air Force in the mid–1970s and used throughout the U.S. Department of Defense, this methodology makes the assumption that cracks exist in *all* primary aircraft structures and includes initial flaw sizes based on the limits of inspectability. The design methodology presupposes that the cracks will propagate either by load-induced fatigue crack growth or environmentally-assisted crack growth and requires that these cracks not grow to critical size within a specified period, based on inspectability. A companion document, MIL-HDBK–1530(USAF) [1996], provides implementation guidelines and documentation requirements for an overall damage-tolerant control plan within the limitations imposed by principles of damage-tolerant design.

---

[1] This document is now obsolete as a stand-alone design requirement, but its essential elements have been incorporated into a comprehensive damage tolerant design handbook being prepared by the Air Force.

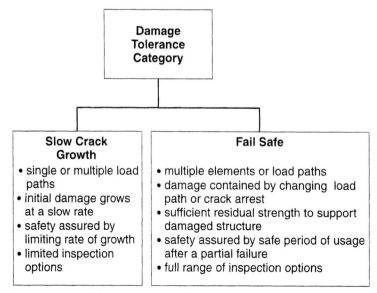

Figure 10.2 Key features of slow-growth versus fail-safe design categories.

The initial step in applying the specification is the selection of the appropriate design concept from the two choices permitted. Structures are divided into two categories (see Figure 10.2), based on the number of load paths or crack arrest capability. Structures that contain only one load path must be qualified under the category of *slow crack growth* and must rely solely on slow crack growth as the mechanism to prevent failure. Consequently, the emphasis must be on reducing design stress levels and careful material selection. On the other hand, structures with multiple load paths or built-in crack arrest features, defined as *fail safe*, have only to demonstrate that they are and will remain structurally sound until the next inspection—this is, that they have sufficient residual strength in the presence of a failure of one load path. A structure can be considered fail safe only if it is able to take advantage of its redundancy. For example, consider the structure shown in Figure 10.3 of a fitting containing multiple lugs. In the event of failure of one of the lugs, the load would be redistributed among the remaining lugs, and the structure would appear to qualify as fail safe; however, should a crack develop in the shaft, there is no redundancy. Consequently, this structure is classified as slow growth. Even for structures that have clearly defined multiple load paths, the analysis of all of the possible combinations of partial failure (a requirement for fail safe certification) may be prohibitive, in which case a slow growth analysis may be substituted.

The requirement for inspection and the coupling of allowable crack growth to the inspection interval is integral to the design philosophy implemented in MIL-A–83444. Six degrees of inspection and their associated intervals are defined:

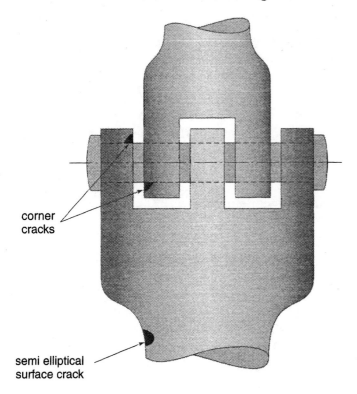

corner
cracks

semi elliptical
surface crack

Figure 10.3 Crack
locations in a lug
with multiple arms.

1. *In-Flight Evident Inspectable.* The nature and extent of damage during a flight is such that the flight crew is immediately and unmistakably aware that a significant event has occurred and that the mission should be discontinued. In this case, the event itself is the inspection.

2. *Ground-Evident Inspectable.* A structure is said to be ground-evident inspectable if the nature and extent of the damage is such that it is unmistakably obvious to ground personnel upon casual observation of the aircraft. It is implied that this inspection occurs once per flight.

3. *Walkaround Inspection.* This degree of inspection is similar to the previous one, except that ground personnel purposefully conduct a visual inspection of the exterior from the ground without removing any access panels or using any inspection aids. Presumably, defects smaller than in the previous case could be detected by trained personnel looking for damage. This level of inspection is typically performed once every 10 flights.

4. *Special Visual Inspection.* At this level, personnel conduct a detailed close visual inspection, looking specifically for defects. Magnification and special lighting, as well as removal of access panels, are part of this inspection, but the removal of paints, sealants, or other coatings and the use of established NDI procedures are not employed. This inspection typically occurs annually.

5. *Depot or Base Level Inspection.* This level is a thorough nondestructive inspection in the conventional sense. Established inspection aids including dye penetrant, X-ray, and ultrasonics are employed, and the entire structure is accessed and examined. At this level, some disassembly may be required to perform the needed inspections. Inspections to this degree are performed only at one fourth the design service lifetime intervals.

6. *In-Service Noninspectable Structures.* If either the critical damage size or the accessibility is such that the structure cannot not be inspected by any of the previous methods, the structure is classified as noninspectable. Inspection is performed at the point of manufacture, and the structure is expected to perform without failure for the entire design lifetime.

Of these six levels of inspection, only the last two (depot- or base-level and noninspectable) apply to structures designed to the slow crack growth standard. Moreover, to ensure that a defect missed in inspection does not grow to critical dimensions before the next inspection, the structure is to be designed such that the minimum lifetime is at least two inspection intervals. For multiple load path structures, the design requirement is that the residual strength after a load path failure plus the accumulated damage in the remaining load paths be sufficient to sustain the struc ture without failure for a minimum period of unrepaired service, as specified in Table 10.2.

The establishment of the applied loads on the structure or component is the most uncertain part of aircraft design. All aircraft, and particularly fighter aircraft, undergo large variations in the applied forces, depending on the current flight conditions. The magnitude of the forces is determined by extensive load spectra data gathered for each of the classes of aircraft and for each mission segment. Because of the stochastic nature of a typical flight mission, the magnitude of the forces will vary widely between flights. Accumulated data tabulated in MIL-A–008866 [1975] represent this fact in the form of the number of load occurrences in excess of a specified force level, expressed per 1000 flight hours. Since the probability that a particular maximum mean load will be seen by the structure at some point be-

Table 10.2 Minimum Service Requirements, Based on Inspection Level

INSPECTION LEVEL	MINIMUM PERIOD OF UNREPAIRED SERVICE
In-flight evident	Return to base
Ground evident	One flight
Walkaround	Five inspection intervals
Special visual	Two inspection intervals
Depot or base level	Two inspection intervals
Noninspectable	Does not apply to fail safe designs

Table 10.3 Design Load for Each Degree of Inspectability

INSPECTION CLASS	TYPICAL INTERVAL	MAXIMUM LOAD DETERMINED FROM:
In-flight evident	1 flight	100 flights
Ground evident	1 flight	100 flights
Walkaround	10 flights	1000 flights
Special visual	one year	50 years
Depot or base level	$\frac{1}{4}$ lifetime	5 lifetimes
Noninspectable	one lifetime	20 lifetimes

tween inspections increases as the inspection interval increases (and thus the damage incurred increases), the Air Force's damage tolerant design requirement for the design limit load is based on multiples of the inspection interval as tabulated in Table 10.3

Given these crack-length allowables and the maximum design load, the fracture mechanics analysis proceeds along customary lines for fatigue crack propagation. The analysis requires an accurate knowledge of the geometric stress intensity factor for the candidate design, baseline fatigue crack growth properties for the proposed material in the relevant environment, and a projection of the anticipated cyclic stress history of the structure or component. All of this information is incorporated into a suitable fatigue crack growth model, and the resultant expression is integrated to determine the computed time to failure. Finally, the computed time to failure is compared with the inspection interval multiple from, Table 10.3, and an accept/reject decision for the candidate design is made. Although this process sounds straightforward, the complexities of the analysis make the task tedious. To assist in implementing the damage tolerant design methodology, the Air Force has prepared a comprehensive manual, *USAF Damage Tolerant Design Handbook: Guidelines for the Analysis of Damage Tolerant Aircraft* [Gallager et al., 1984],[2] in which all of the intricacies of the fracture mechanics analysis are described. This handbook, in essence a short course in fracture mechanics targeted toward the airframe designer, is intended to provide the analyst with the latest thinking on the various aspects of fracture mechanics analysis. A summary of the topics treated in the handbook, shown in Table 10.4, demonstrates the broad scope of the manual. However, a word of caution to the inexperienced practitioner of the fracture mechanics art, this voluminous handbook is, at best, overwhelming and, at worst, dangerous in that it can provide a false sense of security. On the other hand, the *Damage Tolerant Design Handbook* is a carefully thought-out guide, developed and refined over many years, and, when interpreted correctly, is a valuable tool for applied fracture mechanics that can be incorporated into the design analysis of a variety of structures and components, not just airframes.

---

[2]The handbook is available in .pdf format from www.udri.udayton.edu/DTDHReview/.

Table 10.4 Damage Tolerance Design Handbook: Summary of Chapters

Chapter	Title	Description
1	Introduction and Methodology Fundamentals	Overview of handbook, introduction to damage-control fundamentals of fracture mechanics, schemes for determining the stress intensity factor, selected stress intensity factors, and alternative analysis methods
2	Summary of Requirements	Contains a review of MIL-A–83444, including examples for clarity, data to support specific requirements, and assumptions and rationale where limited data exist
3	Damage Size Considerations	Discusses appropriate NDI practice, state-of-the-art procedures, demonstration programs to qualify NDI, in-service NDI practice, and specific examples illustrating how damage is assumed to exist in structures
4	Determination of Residual Strength	Summarizes theory, methods, assumptions, material data, and test verification and gives examples for estimating the final fracture strength or crack arrest potential of cracked structures
5	Analysis of Damage Growth	Describes current practice for estimating the rate of crack growth as a function of time and cyclic and sustained load occurrence; gives examples indicating limitations of methods, use of material data and suggested testing to support predictions and establish confidence
6	Damage Tolerance Analysis— Sample Problems	Provides detailed analysis of typical structural examples, illustrating methodology and assumptions required
7	Damage Tolerance Testing	Describes methods and recommended tests to verify methods and full-scale testing to verify residual strength and slow crack growth rates
8	Individual Airplane Tracking	Describes current methods available to account for usage variations for individual aircraft, based on a crack growth model
9	Guidelines for Damage Tolerant Design and Fracture Control Planning	Describes methods and procedures for development and implementation of a damage-tolerance control plan as required in MIL-HDBK–1530 [1996]
10	Repair Guidelines	Describes the factors that should be considered when designing a repair, in order to ensure that the basic damage tolerance present in the original structure is not degraded by the repair

10.5 DESIGNING BY HINDSIGHT: CASE STUDIES

Design engineering, more than any other branch of engineering, is subject to that particular variant of Murphy's law known as the *law of unintended consequences*. As new concepts develop, the lack of historical precedent can lead to problems that the designer failed to anticipate. One way to prepare for this eventuality is to see how others have handled problems when they occurred. All too often, failures are never reported, but, when they are, in the form of published case studies, they can provide valuable lessons. Even though the problems solved in case studies may not directly apply to a designer's particular dilemma, the case studies provide mental training to condition the designer to develop meaningful hypothetical problems that can then be examined before the fact. The use of case studies as a teaching tool was standard practice at one time in engineering schools to help future engineers develop the logic skills vital to their success, but this practice has been lost in modern education. The next few examples provide some insights into the design process that have been gained from case studies.

In a complex structure the stress intensity factor depends on the load path. In multiple load path structures, the worst combination of loads needs to be considered, including the possibility of unintended load paths. For example, consider the aluminum barge hatch cover analyzed by Kaplan, Willis, and Barnett [1986]. These covers contain six hold-down dogs used to seal the compartment so that the powdered contents (cement in this case) could be pumped out under pressure. It is common practice to release one or more of these dogs to assist in venting the compartment at the end of the unloading cycle. This action breaks the neoprene seal so that the pressure in the compartment can dissipate more quickly. On one such occasion in August 1975, a failure occurred immediately upon release of the second dog. A fracture mechanics analysis performed after the fact (see Figure 10.4) demonstrated that safe conditions would exist unless one of the hinge legs was broken off. This situation is not uncommon in service, and the design engineer should have considered such a scenario. Aside from the obvious operational issue, there is a separate lesson to be learned from this example: Court opinions in product liability cases routinely have held that it is the engineer's responsibility not only to design structures that are safe when used as intended, but also to design against abuses that might reasonably be expected to occur. In this example, cast aluminum lugs that formed the hinge attachments on the hatch cover could reasonably be expected to break off due to rough handling by dock workers.

The study by Pearson and Dooman [1986] demonstrates the importance of performing a thorough post mortem failure analysis. They studied the failure of a truck-mounted propane tank. The tank failed along one of the girth welds connecting the domes to the rolled and welded cylindrical center section. Fortunately, the truck was parked and unattended when the failure occurred. Initial examination of the fracture surface showed evidence of existing cracks, and there was some temptation to attribute the failure to such cracks and terminate the analysis. However, a

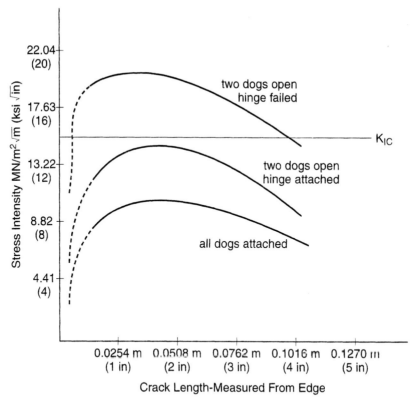

Figure 10.4 Failure analysis of an aluminum hatch cover for various configurations [Kaplan and Barnett, 1986]. Copyright ASTM, reprinted with permission.

preliminary calculation assuming a semi-elliptical surface flaw of the size known to exist showed that the pressure required to cause failure would have been in excess of three times the operating pressure. Since the tank contained a pressure relief valve, excess pressure seemed unlikely, and the possibility of substandard material was considered. However, ASTM E 399 test specimens made from a sample of the tank wall demonstrated that the material toughness was more than adequate for normal stress conditions in service. Once again, attention was directed at an overpressure event, but even if the pressure relief valve failed—which it did, due to excessive corrosion—the increase in pressure would have simply liquefied the propane, unless the tank was overfilled. Unfortunately, there was evidence that the tank had been overfilled. Even then, failure should not have occurred unless the tank had been heated. Although the tank was in the sun when it exploded, thermodynamic analysis indicated that the temperature rise would have been be insufficient to raise the pressure in the tank enough to cause failure—unless a supplemental heat source were added. As it turned out, the truck had been left idling while the operator went inside for lunch. The heat given off by the muffler supplied the additional heat needed to cause failure! This example demonstrates the importance of considering all possible scenarios when conducting a failure analysis. Even improbable mechanisms should not be ruled out without supporting analysis.

Although the design engineer can specify the material and quality levels required, he/she cannot guarantee that the material used meets specification requirements. In critical applications, ASTM or other accepted standards, along with appropriate NDI procedures, can reduce the incidence of substandard materials, but in less critical applications, the costs associated with these preventative measures may not be justified. As a consequence, many of the failures observed in the field are the result of imperfections in the material, and the focus of much of the published case study literature is directed at the metallurgy, not the mechanics aspects of the failure. A proper study of this aspect of failure analysis is best performed with the assistance of a metallurgist. A more thorough discussion of this topic can be found in any of the several excellent volumes of case studies prepared by the American Society for Metals [Esaklul, 1992] that address the metallurgical aspects of fracture. Additional information can be found in the *ASM Handbook*, Vol. 11 [1986] and Vol. 19 [1996].

REFERENCES

ASM Handbook, 1986, Volume 11: "Failure Analysis and Prevention," ASM International, Materials Park, OH.

ASM Handbook, 1996, Volume 19: "Fatigue and Fracture," ASM International, Materials Park, OH.

Barsom, J. R., and Rolfe, S. T., 1987, *Fracture and Fatigue Control in Structures*, 2nd ed., Prentice Hall, Englewood Cliffs, NJ.

Esaklul, K. A., Ed., 1992, *Handbook of Case Histories in Failure Analysis*, ASM International, Materials Park, OH.

Gallager, J. P., Giesler, F. J., Berens, A. P., and Engle, R. M., 1984, "USAF Damage Tolerant Design Handbook: Guidelines for the Analysis of Damage Tolerant Aircraft Structures," AFWAL-TR–82–3073, Wright-Patterson Air Force Base, OH.

Gross, B., and Srawley, J. E., 1964, "Stress Intensity Factors for a Single-Edge-Notch Tension Specimen by Boundary Collocation," NASA TN D–2395.

Kaplan, M. P., Willis, T., and Barnett, R. L., 1986, "A Pressure Vessel Hatch Cover Failure: A Design Analysis," *Case Histories Involving Fatigue and Fracture Mechanics*, ASTM STP 918, American Soc. for Testing and Materials, Philadelphia, pp. 46–64.

MIL-A–008866B (USAF), 1975, *Airplane Strength and Rigidity Reliability Requirements: Repeated Loads and Fatigue*, 27 August 1975.

MIL-A–83444 (USAF), 1974, *Airplane Damage Tolerance Requirements*, 2 July 1974.

MIL-HDBK–1530 (USAF), 1996, *Aircraft Structural Integrity Program: General Guidelines for*, 31 October 1996.

Pearson, H. S., and Dooman, R. G., 1986, "Fracture Analysis of Propane Tank Explosion," *Case Histories Involving Fatigue and Fracture Mechanics*, ASTM STP 918, American Soc. for Testing and Materials, Philadelphia, pp. 65–77.

Reid, C. N., and Baikie, B. L., 1986, "Choosing a Steel for Hydroelectric Penstocks," *Case Histories Involving Fatigue and Fracture Mechanics*, ASTM STP 918, American Soc. for Testing and Materials, Philadelphia, pp. 102–121.

EXERCISES

10.1 Reanalyze the example of the 2.5 m pressurized pipe example described in the chapter, assuming a desired factor of safety of 2.0. What are the consequences of this change on the choice of steel for this application?

10.2 For the example of the 2.5 m pipe example described in the chapter, determine the number of cycles to failure for Steel B given the Paris Law constants $C = 2 \times 10^{-11}$ and $n = 3$, where stress is measured in MPa and crack growth is measured in m/cycle. Make this determinations for inspection limits of

  (a) 6 mm.

  (b) 10 mm.

10.3 Two steels are being considered for the construction of a 30-in diameter pressure tank to hold compressed gas at 1000 psi. The properties of the two steels are: Steel A, $\sigma_{ys} = 50$ ksi and $K_{Ic} = 90$ ksi $\sqrt{\text{in}}$, and Steel B, $\sigma_{ys} = 130$ ksi and $K_{Ic} = 55$ ksi $\sqrt{\text{in}}$. For a desired factor of safety against yield of 2.0 and against fracture of 6.0, select the optimal steel for this application, and determine the minimum thickness required. Assume that the costs of the two steels are comparable and that magnetic particle inspection will be used.

10.4 A double-edge cracked rectangular plate that is 2 in wide by 1 in thick has been designed for an applied stress of half the yield stress (i.e., FS $= 2$). Determine the mode of failure if

  (a) $\sigma_{ys} = 90$ ksi and $K_{Ic} = 45$ ksi $\sqrt{\text{in}}$.

  (b) $\sigma_{ys} = 60$ ksi and $K_{Ic} = 60$ ksi $\sqrt{\text{in}}$.

10.5 A tension link, shown in Figure E10.1, is used in earth-moving equipment and is subjected to cyclic stresses which vary from 10 ksi to 35 ksi. The link contains an internal crack that is 4 in long. If the link is made from A517 steel for which $\sigma_{ys} = 110$ ksi and $K_{Ic} = 170$ ksi $\sqrt{\text{in}}$, do the following:

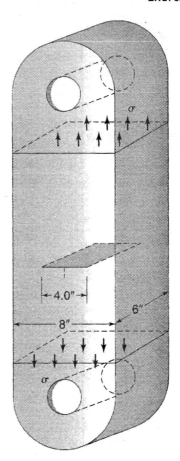

Figure E10.1 Tension
link.

(a) Determine the factor of safety against net section yield in the presence of the crack.

(b) Determine the factor of safety against brittle fracture for the same conditions.

(c) If the maximum stress were increased (due to an accidental overload), which would occur first, net section yield or brittle fracture? Explain.

(d) For Paris law coefficients of $C = 0.66 \times 10^{-8}$ and $n = 2.25$, where stress is measured in ksi and crack growth in inches, *estimate* the number of cycles needed to grow the crack by the amount 0.1 in.

(e) If you were the engineer responsible for analyzing the life of this part, what would be your recommendation for continued service of this part? Defend your decision, based on sound engineering principles.

CHAPTER 11

Elastoplastic Fracture

11.1 INTRODUCTION

By now, the astute reader should have observed that, except for some discussion of the behavior of materials, the subject of linear elastic fracture mechanics is an applied elasticity study of bodies containing cracks. Limited plasticity at the crack tip was incorporated into our model through the use of the small scale yielding argument. We argued that the region is small enough that once we compensate for the effects of the plastic zone on the equilibrium of the overall body by introducing the concept of a notational crack tip, the rest of the behavior is completely elastic. This approach is more than adequate for a wide variety of problems, notably fatigue problems for which the applied loads are generally less than 30 percent of the yield stress. The LEFM approach also applies to materials that are naturally elastic: glass, ceramics, plastics (except those that are highly plasticized), and high strength metals with limited amounts of ductility prior to fracture.

The linear model does not apply when the combination of geometry and loading is such that the plastic zone is not confined—for example, a ductile plate containing a central crack. For this geometry the plastic zones radiate from the crack tips at roughly 45° and extend to the free boundaries. Slip-line theory is more appropriate for analysis in this case. Also, we should not expect the linear theory to extend to the case of large-scale plastic behavior with extensive deformation prior to failure wherein the net section containing the defect approaches the yield stress of the material. Included in this class of problems are some very critical applications that demand thorough analysis. For example, in the nuclear pressure vessel industry, material toughness is of paramount importance, due to the long term degradation of the material caused by neutron bombardment over the lifetime of the pressure vessel. Despite the high initial toughness, subcritical flaws develop, mainly in weld metal,

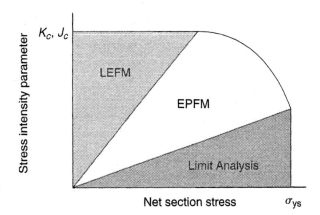

Figure 11.1 Failure Analysis Diagram (FAD) for a ductile material. (The region inside the locus is considered safe.)

due to fatigue and stress corrosion cracking. As a result, fracture failure must be considered in the event of an accident, but the extent of plastic deformation at the crack tip precludes the use of the LEFM approach. As illustrated in Figure 11.1, there is a region between linear elastic fracture mechanics and plastic collapse that needs to be categorized. It is this region to which the development of a theory of elastoplastic fracture mechanics (EPFM) is directed. Unfortunately, a direct approach incorporating a rigorous plasticity theory has proven elusive, and alternative approaches that expand and generalize the already firmly grounded linear elastic fracture approach have been adopted.

11.2 NONLINEAR ELASTIC BEHAVIOR

We can extend the concepts of energy of fracture developed in Chapter 7 to the class of nonlinear, but still elastic, materials by again considering the energy available for crack extension in terms of areas in the load–load point displacement diagram. In this case the load–deflection diagram has the appearance shown in Figure 11.2. The area under the curve is the strain energy U, and the area above the curve, called the complementary strain energy, U^*, is related to the potential energy, Π, of the system by the relation

$$U^* = P\delta - U = -\Pi \tag{11.1}$$

For a two-dimensional body of area, A, with surface tractions, T_i, prescribed over a portion of the bounding surface, Γ, the potential energy of the body is given by

$$\Pi = \int_A W \, dA - \int_\Gamma T_i u_i \, ds \tag{11.2}$$

where W is the strain-energy density defined in Eqs. (2.18).

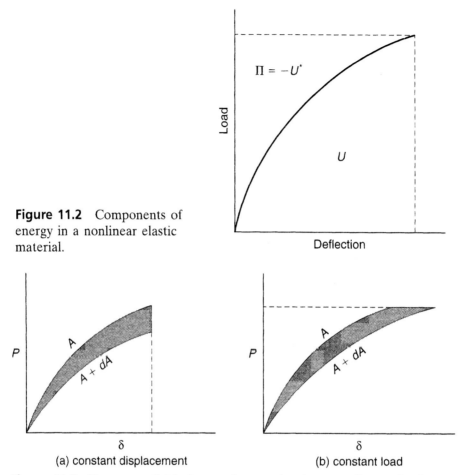

Figure 11.2 Components of energy in a nonlinear elastic material.

Figure 11.3 Energy available for crack extension in a non-linear elastic body under (a) constant displacement and (b) constant load.

By analogy with the linear elastic case, we can define an energy release rate for nonlinear elastic bodies, J, as the area on the load–displacement diagram between crack areas A and $A + dA$, as shown in Figure 11.3 for (a) constant displacement or (b) constant load divided by the change in crack area, dA,

$$J = \left| \frac{\partial \Pi}{\partial A} \right| \qquad (11.3)$$

In Eq. (11.3), the absolute value has been used to ensure that $J > 0$. For the special case of a linear elastic body, $P\delta = 2U$ and $U = -\Pi$. Then, from Eqs. (7.16) or (7.18),

$$J = \left| \frac{\partial U}{\partial A} \right| = \mathcal{G} = \frac{K^2}{E'} \qquad (11.4)$$

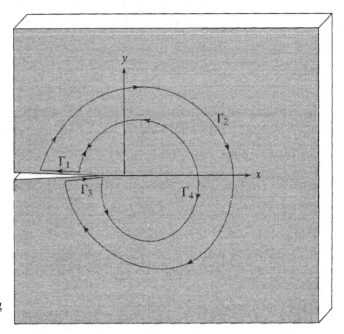

Figure 11.4 A J-integral integration path not encircling the singularity.

Alternatively, we can obtain a formal definition of J by differentiating Eq. (11.2):

$$J = -\frac{d\Pi}{dA} = \frac{1}{B}\int_A \frac{dW}{da}\, dA - \frac{1}{B}\int_\Gamma T_i \frac{du_i}{da}\, ds \tag{11.5}$$

After some manipulations [e.g., Kanninen and Popelar, 1985], this equation can be written in the form

$$J = \int_\Gamma \left(W\, dy - T_i \frac{\partial u_i}{\partial x}\, ds \right) \tag{11.6}$$

The integrand in Eq. (11.6) is the energy momentum tensor developed by Eshelby [1970], and, as such, this integral has the property that $J = 0$ on any closed contour not encircling a singularity within an elastic solid (a conservative system). Rice [1968a, 1968b] was the first to adapt the principle of conservation of energy momentum to a two-dimensional body containing a singularity. He considered the contour Γ shown in Figure 11.4, where the segments Γ_1 and Γ_3 are chosen to be marginally on the material side of the traction-free crack interface. We can write the condition for conservation of energy momentum as

$$J_\Gamma = J_{\Gamma_1} + J_{\Gamma_2} + J_{\Gamma_3} + J_{\Gamma_4} = 0 \tag{11.7}$$

where $\Gamma = \Gamma_1 + \Gamma_2 + \Gamma_3 + \Gamma_4$; however, $J_{\Gamma_1} = J_{\Gamma_3} = 0$, since $dy = 0$ and $T_i = 0$ on these segments. Therefore,

$$J_{\Gamma_2} + J_{\Gamma_4} = 0 \quad \text{or} \quad J_{\Gamma_2} = -J_{\Gamma_4} = J_{-\Gamma_4} \tag{11.8}$$

where the notation, $-\Gamma_4$, denotes a line integral taken in the direction opposite to Γ_4. Since Γ_2 and $-\Gamma_4$ are any two arbitrary line integrals surrounding the crack tip, both taken in the same direction, Eqs. (11.8) demonstrates that J is path independent. In recognition of his contribution, the quantity J is often described as *Rice's J-Integral*.

Path independence of the J-integral offers a variety of opportunities for determining its value and, with the aid of Eq. (11.4), the corresponding geometric stress intensity factor for the linear elastic case. The J-integral method for determining K is often used in finite element analysis. Since the integral can be taken at a distance somewhat removed from the crack tip, this approach lessens the need for a highly refined mesh at the crack tip or the use of singular elements. However, care must still be taken to ensure that the increased stiffness of the numerical model does not lead to an underestimate of the fracture parameters. Also, judicious choices for the integration path can often simplify its calculation. The J-integral method has applications in analytical computations as well. As the next example illustrates, the stress intensity factor for some bodies with finite boundaries can be determined conveniently.

Example 11.1   A narrow strip of material with elastic modulus E and height h is rigidly attached to parallel platens, as shown in Figure 11.5. If the upper platen is displaced by an amount δ, determine the geometric stress intensity factor for an edge crack midway between the platen faces.

We have

$$J = J_{\Gamma_1} + J_{\Gamma_2} + J_{\Gamma_3} + J_{\Gamma_4} + J_{\Gamma_5} \tag{11.9}$$

but along Γ_2 and Γ_4 $dy = 0$, and the displacements are constant along the length. As a result, $\partial u_{1,2}/dx = 0$ and

$$J_{\Gamma_2} = J_{\Gamma_4} = 0 \tag{11.10}$$

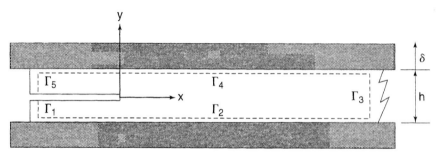

Figure 11.5 A thin strip between two rigid platens with a fixed displacement.

Also, on the free edges Γ_1 and Γ_5, far removed from the crack tip, the surface tractions and $\partial u_{1,2}/dx$ are zero; therefore,

$$J_{\Gamma_1} = J_{\Gamma_5} = 0 \qquad (11.11)$$

If path Γ_3 is taken sufficiently far from the crack tip, then the strain state ϵ_y becomes uniform in accordance with St. Venant's principle and $\partial u_{1,2}/dx = 0$. Then J_{Γ_3} reduces to

$$J_{\Gamma_3} = \int_0^h W \, dy = Wh \qquad (11.12)$$

For a linear elastic solid in plane stress,

$$J_{\Gamma_3} = \frac{1}{2}\epsilon_y\sigma_y h = \frac{1}{2}\left(\frac{\delta}{h}\right) \cdot \frac{E}{1 - \nu^2}\frac{\delta}{h} \cdot h = \frac{E}{2(1 - \nu^2)}\frac{\delta^2}{h} = \frac{K^2}{E} \qquad (11.13)$$

Solving for the geometric stress intensity factor, we find that

$$K = \frac{E\delta}{\sqrt{2(1 - \nu^2)h}} \qquad (11.14)$$

■

With the introduction of the J-integral, the linear theory of fracture mechanics has been extended to nonlinear elastic bodies. The theory is completely rigorous. Unfortunately, there are few engineering applications for the generalization, except as an alternative method to determine the stress intensity factor. The primary interest in the J-integral lies in its application to an approximate theory for elastoplastic fracture, to be discussed next.

11.3 CHARACTERIZING ELASTOPLASTIC BEHAVIOR

The extension of the theory of nonlinear fracture mechanics to include elastoplastic materials in which the size of the plastic zone exceeds the small scale yielding approximation is based on the deformation theory of plasticity. The argument for the deformation plasticity model is that, provided no unloading is permitted to occur, the behavior of an elastoplastic material and a nonlinear elastic material is indistinguishable to an outside observer. By this, we mean that any external measure of the deformation—for example, a load–deflection diagram, such as Figure 11.2—would look the same for either material; the mechanisms going on inside the two materials are markedly different, but outwardly there is no difference. For this argument to be valid, the stress must be nondecreasing everywhere, and the stress components must remain in fixed proportion as the deformation proceeds. This latter condition is not strictly satisfied, due to plasticity-induced stress redistribution, but outside the plastic region the stresses are assumed to be nearly proportional for simple loadings. Finite element studies using the incremental theory of plasticity [McMeeking, 1977] support this assumption. With these restrictions, the J-integral has been proposed as a fracture mechanics parameter for elastoplastic materials. Eshelby [1970] argued that so long as the J-integral is computed along a path outside the region in which plastic deformation had taken place (i.e. in "good" material, as he termed it), it is necessary only that the region have recoverable energy content. He further argued that "the energy does not even necessarily need to be recoverable if we do not call the material's bluff by unloading to see if it is."

Despite the similarities between nonlinear and elastoplastic behavior, there are important differences, even if we do not permit any unloading to occur. Let us reexamine Figure 11.3 for the elastoplastic case. Since we can not permit any unloading, the change in the load–deflection diagram caused by a crack length change cannot, in principle, be measured with a single specimen. Therefore, we will consider two specimens identical in every respect except for their crack lengths A and $A + dA$, respectively, each loaded to the same total deflection. From the deformation theory of plasticity argument, the change in potential energy is given by the shaded area in Figure 11.3(a) just it as was for the case of a nonlinear elastic material. However, unlike the elastic case, not all of this energy is available to form new surfaces. In the elastoplastic case, much of the energy is nonrecoverable, and we cannot interpret J as an energy measure of crack extension.

Working independently, Hutchinson [1968] and Rice and Rosengren [1968] proposed an alternative interpretation of J as a measure of the stress state under some conditions. They all assumed that the elastoplastic material could be modeled as a material fitting the Ramberg–Osgood flow rule,

$$\frac{\epsilon}{\epsilon_o} = \frac{\sigma}{\sigma_o} + \alpha \left(\frac{\sigma}{\sigma_o} \right)^n \tag{11.15}$$

where σ_o and ϵ_o are the flow stress and the corresponding flow strain $(= \sigma_o/E)$, respectively, measured in true stress–strain units, and n is the strain hardening exponent. Typical Ramberg–Osgood stress–strain curves for several values of exponent n are shown in Figure 11.6.

Since J is path independent, we are free to evaluate it along a circular path surrounding the crack tip—that is, $r = $ constant—as shown in Figure 11.7. For this path, $dy = r \cos \theta \, d\theta$ and $ds = r \, d\theta$, and Eq. (11.6) can be written as

$$J = r \int_{-\pi}^{\pi} \left(W \cos \theta - T_i \frac{\partial u_i}{\partial x_i} \right) d\theta \tag{11.16a}$$

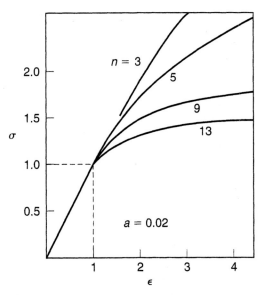

Figure 11.6 Ramberg–Osgood
constitutive model of a nonlinear or
elastoplastic material.

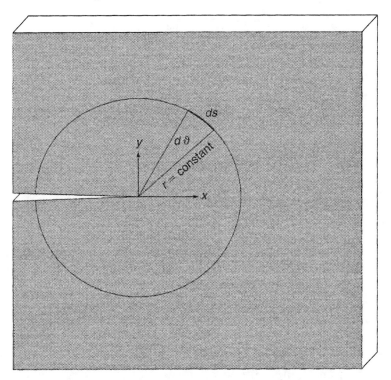

Figure 11.7 J-integral circular path centered at the crack tip.

or

$$\frac{J}{r} = \int_{-\pi}^{\pi} \left(W \cos\theta - T_i \frac{\partial u_i}{\partial x_i} \right) d\theta \tag{11.16b}$$

Since the terms in the integrand of Eqs. (11.16) are the product of stress and strain, and J is independent of the choice of r, Eq. (11.16b) can be satisfied if

$$\sigma_{ij}\epsilon_{ij} \propto \left(\frac{J}{r} \right) h_{ij}(n, \theta) \tag{11.17}$$

where $h_{ij}(n, \theta)$ are suitable functions describing the angular variations in the stresses and strains. In the very near-field of the crack tip, the linear term in Eq. (11.15) is generally small compared with the power hardening term and can be neglected. With this approximation, Eqs. (11.15) and (11.17) can be combined to yield

$$\sigma_{ij} = \sigma_o \left(\frac{J}{\alpha \sigma_o \epsilon_o I_n r} \right)^{\frac{1}{1+n}} f_{ij}(n, \theta) \tag{11.18a}$$

$$\epsilon_{ij} = \alpha \epsilon_o \left(\frac{J}{\alpha \sigma_o \epsilon_o I_n r} \right)^{\frac{n}{1+n}} g_{ij}(n, \theta) \tag{11.18b}$$

where f_{ij} and g_{ij} describe the angular variation of the stress and strain, respectively, in the near field of the crack tip and I_n is an integration constant that depends on n and the state of stress. Extensive tables of the angular functions f_{ij} and g_{ij} have been generated by Shih [1993]. I_n has been approximated in series form [Saxena, 1998] as

$$I_n = 6.568 - 0.4744\,n + 0.040\,n^2 - 0.001262\,n^3 \qquad \text{for plane strain} \tag{11.19a}$$

$$I_n = 4.546 - 0.2827\,n + 0.0175\,n^2 - 0.4516 \cdot 10^{-4}n^3 \quad \text{for plane stress} \tag{11.19b}$$

From Eqs. (11.18), we can observe that J plays the same role in elastoplastic fracture mechanics that K plays in linear elastic fracture mechanics—namely, it describes the strength of the singularity, that is, the stress–strain fields scale with J. The order of the singularity is determined by the strain hardening exponent. Note that when $n = 1$, these equations reduce to their corresponding forms for linear elastic materials. The stress–strain fields described by Eqs. (11.18) are called *HRR fields* and play an important role in the extension of nonlinear fracture mechanics into the elastoplastic domain.

In 1972, Begley and Landes proposed the use of the J-integral as a plane-strain, elastoplastic fracture criterion. The combination of the HRR crack-tip stress field interpretation of the J-integral, along with the ability to measure its value experimentally from potential energy arguments, led them to suggest that, by analogy, there must be a parameter, J_{Ic}, for elastoplastic materials that is comparable with K_{Ic} for linear elastic materials. The suggested procedure for determining the J-integral is similar to that previously discussed for nonlinear elastic materials, with one important exception. Since the deformation theory of plasticity does not permit unloading, the area between the two load–deflection curves, as shown in Figure 11.3, for two adjacent crack lengths cannot be determined from a single specimen. Instead, Begley and Landes proposed that a series of specimens of increasing crack length be used to construct a set of master curves for each combination of material and specimen. In their experiments the area under the load-vs.-load point deflection curve up to a fixed total displacement was determined with a polar planimeter (with modern digital instrumentation the area would be computed on the fly, using a suitable integration rule) for each specimen. If we were to follow the definition of J, from Eq. (11.3) literally, we would integrate the area above the load–load point displacement curves; however, since Begley and Landes were concerned only with the difference between curves, the area under the curves was more convenient to measure. The computed potential energy was then plotted as a function of the crack area and numerically differentiated to determine $J = -\Delta\Pi/\Delta A$. The critical value, J_{Ic}, was the value of J corresponding to the total displacement at the onset of crack extension.

To be a valid material property the measured value of J_{Ic} should be independent of the specimen used to measure it. In a companion paper Landes and Begley [1972; see also Begley and Landes, 1976] determined J_{Ic} for a rotor steel from two specimen geometries having significantly different plastic slip-line fields. Their results showed that, to within experimental error, J_{Ic} was independent of specimen geometry over the full range from elastic to fully plastic conditions. More recent studies have shown that measurements of the value of the J-integral under less than plane-strain conditions may depend on the specimen's geometry.

Over the years the recommended procedure for measuring J_c has undergone a series of evolutionary changes. The currently adopted procedure is described in ASTM Standard E 1820–99, *Standard Test Method for Measurement of Fracture Toughness* [1999].[1] In many respects, this test method is similar to ASTM Standard E 399, described in Chapter 8, except that, at the expense of additional instrumentation, it permits the determination of a J_c value when the stringent requirements for a valid K_{Ic} are not met. This test method requires continuous measurement of both the load–line displacement (needed for J_c) and the crack mouth opening displacement (used to determine K_{Ic}). The procedure for determining the germane fracture parameter is the same for all three of the certified specimen types: the single-edge

---

[1] This Standard replaces ASTM Standards E 813, *Standard Test Method for J_{Ic}, A Measure of Fracture Toughness* (discontinued 1997); E 1152, *Standard Test Method for Determining J–R Curves* (discontinued 1996); and E 1737, *Standard Test Method for J-Integral Characterization of Fracture Toughness* (discontinued 1998).

bend specimen [SE(B)], the compact-tension specimen [C(T)], and the disk-shaped compact-tension specimen [DC(T)]. Since this test method has as one of its options the determination of a valid K_{Ic}, the precracking and fixturing requirements are the same as described in Chapter 8. After the test is completed, K_Q is computed, and the relevant tests are performed to determine if $K_Q = K_{Ic}$. If this test fails, J_c is computed from

$$J = J_{el} + J_{pl} = \frac{K^2(1 - \nu^2)}{E} + J_{pl} \tag{11.20}$$

where K is calculated using $a = a_o$ (i.e., the fatigue precracked crack length) and J_{pl} is given by

$$J_{pl} = \frac{\eta A_{pl}}{B_N(W - a_o)} \tag{11.21}$$

where η is a factor that depends on the specimen's geometry and A_{pl} is the "plastic component" of the area under the load–load point displacement curve, shown in Figure 11.8. The J calculated at the final point of instability is denoted J_Q. If J_Q satisfies certain requirements spelled out in the ASTM Standard E 1820–99, $J_Q = J_c$; otherwise, the notation J_u is applied to signify that the determined value of J does not meet the necessary criteria to be considered in-plane size independent. Note that J_c may be dependent on thickness.

ASTM Standard E 1820 goes beyond the measurement of unstable crack extension (i.e., J_c) to characterize stable crack extension, referred to as *stable tearing*. In contract to the single point value, associated with the onset of instability, the stable-tearing behavior is characterized by a continuous fracture toughness (J)-versus-crack extension (Δa)-curve, such as the one illustrated in Figure 11.9. The resulting J–R curve is analogous to the K–R curve concept discussed in Chapter 6. The procedure

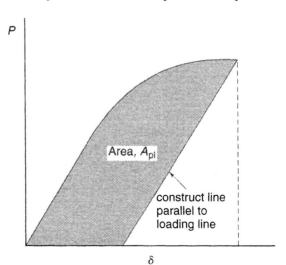

Figure 11.8   Plastic area A_{pl} used to compute J_{pl}.

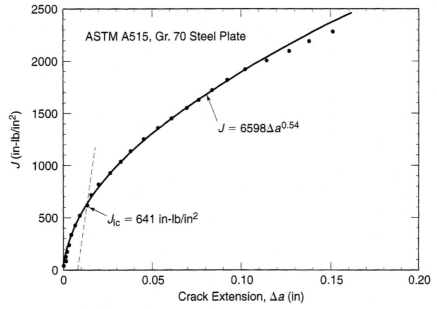

Figure 11.9   Typical J–R curve determined using the procedures described in ASTM Standard E 1820 (courtesy R. E. Link, U.S. Naval Academy).

for determining the $J–R$ curve is based on the elastic compliance method, where multiple points are determined from a single specimen. This procedure requires a higher degree of signal resolution than that required for determination of J_c or K_{Ic} within the standard. The difficulties notwithstanding, $J–R$ curves serve a useful role in characterizing material behavior in high-toughness materials.

In the United States the use of the J-integral to characterize elastoplastic fracture is almost universal; however, for historical reasons, an alternative measure, the crack tip opening displacement (CTOD), is more widely used in the United Kingdom and on the European continent. In principle, the CTOD can be measured directly at the crack tip. In practice, its value is inferred from crack mouth measurements, and ASTM Standard E 1820 includes procedures for reporting it. However, the CTOD and J are not independent quantities. Although various definitions of the crack tip opening displacement, denoted δ_t, have been proposed, we will, for the purposes of our discussion, use the one proposed by Tracy [1976], illustrated in Figure 11.10, as the intercept of two $45°$ lines originating at the deformed crack tip and intersecting the crack profile. Integrating Eq. (11.18b) and evaluating it for $\theta = \pi$ to determine the opening profile of the crack yields an expression of the form

$$\begin{Bmatrix} u_1 \\ u_2 \end{Bmatrix} = \alpha\epsilon_o \left(\frac{J}{\alpha\epsilon_o\sigma_o I_n} \right)^{\frac{n}{n+1}} r^{\frac{1}{n+1}} \begin{Bmatrix} k_{11}(n,\pi) \\ k_{22}(n,\pi) \end{Bmatrix} \qquad (11.22)$$

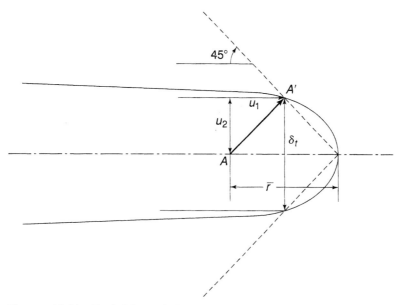

Figure 11.10 Definition of the crack-tip opening displacement (CTOD) [after Tracy, 1976].

where k_{ii} represents the angular variation in u_i. Note that points on the crack profile in Figure 11.10 are displaced in both the x- and y-directions relative to their undeformed positions (point A). Letting

$$\delta_t = 2u_2(\bar{r}, \theta) = 2[\bar{r} - u_1(\bar{r}, \theta)] \tag{11.23}$$

where $\bar{r}$ is not known *a priori*, and solving Eqs. (11.22) and (11.23) together to eliminate $\bar{r}$ results in an expression for δ_t of the form

$$\delta_t = d_n \frac{J}{\sigma_o} \tag{11.24}$$

Shih [1981] evaluated d_n using finite element analysis for a wide range of values of n and σ_o/E. His results for plane stress are shown in Figure 11.11. For plane strain, the values are about 20 percent lower. As a result, the CTOD approach for characterizing elastoplastic fracture and the J-integral are alternative representations of the same measure. For perfectly plastic behavior, $n \to \infty$, and Eq. (11.24) for plane stress has the particularly simple form

$$\delta_t = \frac{J}{\sigma_o} \tag{11.25}$$

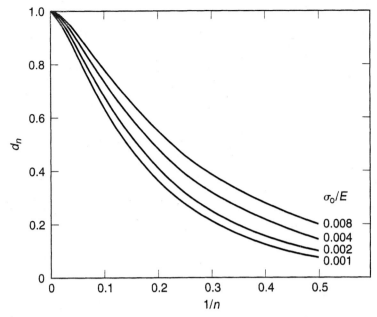

Figure 11.11   Dependence of d_n on the strain-hardening exponent n, and σ_0/E for plane-stress conditions [adapted from Shih, 1981].

11.4 COMMENTS ON THE *J*-INTEGRAL IN ELASTOPLASTIC FRACTURE MECHANICS

The extension of the J-integral approach for characterizing fracture from a nonlinear elastic parameter to an elastoplastic one required the introduction of several very significant assumptions. It is worthwhile to review them here:

(a) In an elastoplastic material, energy momentum is not necessarily conserved on an arbitrary closed contour not containing a singularity. As a result, we cannot use $J = 0$ to prove path independence, and the magnitude of J may depend on the integration path.

(b) The change in potential energy between two crack states, represented by the shaded areas in Figure 11.3, is not equal to the energy required to form new surfaces. Some, if not most, of the loss in potential energy goes to irreversible losses, such as heat, and there is no known algorithm for allocating this area among the various energy-absorbing mechanisms.

(c) The use of the deformation theory of plasticity to model the behavior of a structure undergoing elastoplastic deformation is appropriate only as long as there is no unloading anywhere in the structure. Crack extension, by its nature, violates this assumption—that is, the formation of a crack involves the separation of two planes into new, stress-free surfaces. Hence, the stress at the crack tip must be redistributed to the remaining portions of the body.

(d) The interpretation of J as a measure of the crack tip stress state depends on the applicability of the Ramberg–Osgood constitutive model to the material. It is tacitly assumed that ductile metals can be modeled by this flow rule without testing the rule's validity in each case. Not all materials can be modeled by a smoothly varying transition from the linear to the nonlinear region. One obvious example is certain mild steels with clearly defined upper and lower yield points.

(e) The development of the HRR crack tip stress field concept ignores the linear portion of the flow model under the assumption that in the very near-tip region, the strains are large enough that the contribution of the linear portion is small compared with the total deformation, but small enough that the small-strain theory of elasticity is still applicable. In principle, but not in practice, this assumption would rule out the use of the J-integral as an alternative characterizing parameter for materials that just barely miss satisfying the requirements for valid K_{Ic} as described in ASTM Standard E 1820.

(f) The HRR stress field is not unique. There are an infinite number of ways to satisfy Eq. (11.17), corresponding to an infinite number of constitutive models for elastoplastic materials. By this, we mean that constitutive relations are not intrinsic to nature—that is, they are not "laws" of nature in the same sense as, for example, stress or strain transformation laws, but rather they are empirical relations chosen to model the observed behavior of the material they purport to represent.

Despite the lack of as firm a mathematical foundation as enjoyed by linear elastic fracture mechanics, elastoplastic fracture mechanics plays an important role in the analysis and prevention of failure within its domain of influence.

Just as there are restrictions on the stress intensity factor K_c in the linear case to ensure that the measured critical value is specimen independent, similar restrictions exist for J_c. Since the shape of the load–load point deflection curve for a structure undergoing elastoplastic deformation depends on both the structure (and its slip-line field) and the constitutive behavior, there is no reason to presuppose that the change in potential energy (i.e., the shaded area in Figure 11.3) will be structure independent. In the elastic case, we insisted that the plastic zone be fully contained within the singularity-dominated zone as a necessary (but not sufficient) condition to employ a one parameter (i.e., K) representation to the stress state. The corresponding requirement for an elastoplastic crack-tip stress field is called J-*dominance*, and the confined field is not the plastic zone, but rather the region of finite strain. Hutchinson [1983] has studied the size of the J-dominance zone in relation to the crack-tip opening displacement and, through Eq. (11.25), to the magnitude of J. Using the finite element results of McMeeking [1977], he determined that one condition for a valid determination of J was that

$$R > 3\delta_t \tag{11.26}$$

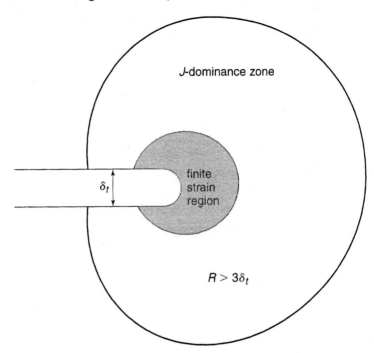

Figure 11.12 The *J*-dominance zone in elastoplastic fracture.

where R is the "radius" of the J-dominated zone, depicted schematically in Figure 11.12.

A second requirement is that the fracture process zone be fully contained within the J-dominated zone. Since the primary mechanism of ductile crack growth is that of void nucleation and coalescence, which are themselves finite strain events, this requirement is also satisfied by Eq. (11.26). Based on finite element studies, Hutchinson estimated R to be on the order of 20 to 25 percent of the corresponding plastic zone diameter $2r_y$ computed from Eq. (6.29). For the specific cases of the standard bend specimen and the compact geometry, the radius of the J-dominated zone is approximately 7 percent of the uncracked ligament.

At the present time the primary application of J_c in elastoplastic fracture mechanics is as a comparative measure of fracture resistance when plastic deformation renders the K_c concept moot. The development of a J-controlled fracture criterion, similar to $K_{\text{applied}} = K_c$ developed in Chapter 6 for the linear elastic case, requires that J_{applied} be determined for each combination of geometry and material being considered. Even if nonlinear elasticity concepts are employed rather than elastoplastic theory, the task is formidable, and few solutions exist. As a result, finite element methods are usually employed to determine estimates of J where the benefit of path independence is used to simplify the computation of the line integral.

Not to be forgotten in all this process is the fact that crack instability in the elastoplastic domain is only one of several competing mechanisms of failure. As illustrated in Figure 11.1, when the net section stress approaches approximately 80 percent of the yield stress, the difference between the EPFM estimate of failure load and that predicted by classical limit theory is not significantly different. In this event, computing the maximum load-carrying capacity of the structure by using the much simpler limit theory and applying a generous factor of safety may be a reasonable alternative to the more costly and time-consuming fracture mechanics calculations based on J.

REFERENCES

ASTM Standard E 1820–99a, 1999, "Standard Test Method for Measurement of Fracture Toughness,"*Annual Book of ASTM Standards*, Vol. 03.01, American Soc. for Testing and Materials, West Conshohocken, PA.

Begley, J. A., and Landes, J. D., 1972, "The J Integral as a Fracture Criterion," *Fracture Toughness, Proceedings of the 1971 National Symposium on Fracture Mechanics, Part II*, ASTM STP 514, American Soc. for Testing and Materials, Philadelphia, pp. 1–20.

Begley, J. A., and Landes, J. D., 1976, "Serendipity and the J Integral," *Int. J. of Fracture*, Vol. 12, pp. 764–766.

Eshelby, J. D., 1970, "Energy Relations and the Energy-Momentum Tensor in Continuum Mechanics," *Inelastic Behavior of Solids*, McGraw-Hill, New York, pp. 77–115.

Hutchinson, J. W., 1968, "Singular Behaviour at the End of a Tensile Crack in a Hardening Material," *J. Mech. Phys. Solids*, Vol. 16, pp. 13–31.

Hutchinson, J. W., 1983, "Fundamentals of the Phenomenological Theory of Nonlinear Fracture Mechanics," *J. Applied Mechanics*, Vol. 50, pp. 1042–1051.

Kanninen, M. F., and Popelar, C. H., 1985, *Advanced Fracture Mechanics*, Oxford University Press, New York.

Landes, J. D., and Begley, J. A., 1972, "The Effect of Specimen Geometry on J_{Ic}," *Fracture Toughness, Proceedings of the 1971 National Symposium on Fracture Mechanics, Part II*, ASTM STP 514, American Soc. for Testing and Materials, Philadelphia, pp. 24–39.

McMeeking, R.M., 1977, "Finite Deformation Analysis of Crack-Tip Opening in Elasto-Plastic Materials and Implications for Fracture," *J. Mech. Physics Solids*, Vol. 25, pp. 357–381.

Rice, J. R., 1968a, "A Path Independent Integral and the Approximate Analysis of Strain Concentration by Notches and Cracks," *J. Applied Mechanics*, Vol. 35, pp. 379–386.

Rice, J. R., 1968b, "Mathematical Analysis in the Mechanics of Fracture," *Fracture, An Advanced Treatise: Volume II, Mathematical Fundamentals*, Academic Press, New York, pp. 191–311.

Rice, J. R., and Rosengren, G. F., 1968, "Plane Strain Deformation Near a Crack Tip in a Power-Law Hardening Material," *J. Mech. Phys. Solids*, Vol. 16, pp. 1–12.

Saxena, A., 1998, *Nonlinear Fracture Mechanics for Engineers*, CRC Press, Boca Raton, FL.

Shih, C. F., 1981, "Relationship between the J-integral and the Crack Opening Displacement for Stationary and Extending Cracks," *J. Mech. Physics Solids*, Vol. 29, pp. 305–326.

Shih, C. F., 1993, *Tables of Hutchinson, Rice and Rosengren Singular Field Quantities*, MRL E–147 Report, Brown University, Providence, RI.

Tracy, D. M., 1976, "Finite Element Solutions for Crack Tip Behavior in Small Scale Yielding," *J. Eng. Materials and Technology*, Vol. 98, pp. 146–151.

EXERCISES

11.1 Derive an expression for the J-integral for the Dugdale model (i.e., an internal crack with a strip yield zone) described in Chapter 6.

11.2 Determine the J-integral for a semi-infinite strip of height h, as shown in Figure E11.1, subjected to in-plane bending moments M acting on the free end in a linear elastic material.

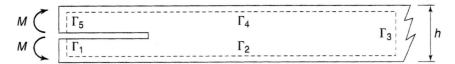

Figure E11.1

11.3 A standard 1T aluminum compact specimen with $W = 2.01$ in is fatigue pre-cracked to $a/W = 0.48$ following ASTM E 399 procedures. After completion of a fracture-toughness test, the load–load line displacement (V_M) was fitted to the function $V_M = (1.3P + 0.005P^2) \times 10^{-6}$ inches, and failure occurred at 7000 lb. Calculate J_C.

11.4 Perform a finite element analysis of a strip of finite width in tension containing a central crack of length $2a/W = 0.6$. Using a contour of your choice, determine the J-integral, and compare your result with established values.

Comprehensive Exercises

A.1 GENERAL COMMENTS

The comprehensive problems presented in this appendix provide an opportunity for the reader to apply the principles he or she has learned to open-ended design situations in fracture mechanics. Like in real-world design decisions, not all of the information needed to solve the problem is predefined, and numerous assumptions will need to be made. It is the responsibility of the problem solver to seek additional references and make educated guesses when no data can be found. While the goal is to solve the problems using the best available method, it is also worthwhile to estimate the result by using simple analytical models. Quite often, these estimates (before the days of handheld computers, the estimates were called "back-of-the-envelope calculations") can provide order-of-magnitude results that might highlight errors hidden in more complex calculation schemes. Moreover, it is simply good engineering practice to use more than one approach to solve problems. The more critical the application, the greater the emphasis on multiple solution schemes to flush out sources of error. Finally, do not ignore the possibility of alternative modes of failure. Fracture failure is only one of many competing mechanisms leading to failure, and all possible mechanisms must be considered or dismissed with justification.

EXERCISES

A.1 A crane hook, illustrated in Figure A.1, with a rated capacity of 55 tons (110,000 lb) is used to lift cargo containers in a dockside operation. It is estimated that the hook undergoes 1000 loading cycles per month at the rated capacity. As a result of contact with the mating swivel collar of the hook assembly, the shaft

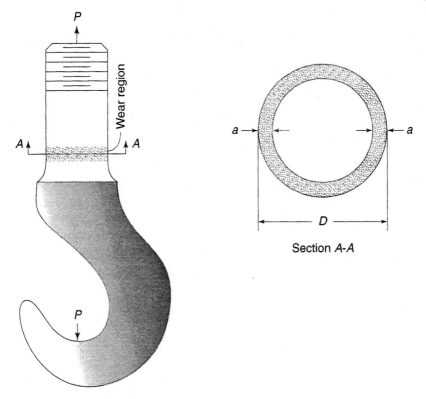

P

Wear region

A   A

a   a

D

Section A-A

Figure EA.1 Crane hook.

of the hook experiences scoring at the location shown in the figure. This area is likely to become the site of a circumferential crack that will ultimately result in failure of the hook. The following properties are available:

diameter D = 2.00 in;

yield stress σ_{ys}, = 65 ksi;

fracture toughness K_{Ic} = 60 ksi$\sqrt{\text{in}}$;

Forman constants C = 8.8 × 10^{-08} for K in ksi$\sqrt{\text{in}}$, and n = 2.3.

(a) Assuming that the smallest circumferential defect that can be detected in the region of wear is 0.050 in in depth, determine the number of cycles to failure of the crane hook in this application.

(b) As a preventative measure, it has been proposed that the crane hook be proof-tested to 150 percent of the rated capacity before being placed in service, and annually thereafter. Determine the number of cycles to failure for the same initial defect as in part (a), including the effects of overload retardation in your analysis. Assume a Wheeler exponent of 1.3 for this material.

(c) Based on the results of parts (a) and (b), what would you recommend as a procedure to ensure safe life of the crane hook? Assume that periodic inspection is not possible.

A.2 The closed-ended cylinder shown in Figure A.2 is to be used as a pressure accumulator. The cylinder is made of 2-mm thick 7075-T651 aluminum plate rolled to a diameter of 500 mm and held together by a riveted lap joint, as shown the figure. The rivet holes are 14 mm in diameter, and spaced 50 mm apart along the entire length (1.5 meters) of the cylinder. The rivet head is 17 mm in diameter, and inspection is possible only on visible portions of the riveted joint. The results of a separate stress analysis indicate that the region of overlap is sufficient for free-end effects to be ignored. Details of the end closures are not given, and you may assume that they are sufficiently strong and attached such that failure of the end closures will not occur. The cylinder contains a rubber liner that is filled with hydraulic fluid. In operation the pressure on the fluid varies between 300 kPa and 600 kPa for each use cycle.

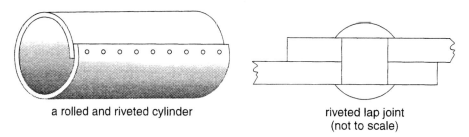

a rolled and riveted cylinder   riveted lap joint
                                                 (not to scale)

Figure EA.2   Riveted pressure cylinder.

(a) Derive an expression for the geometric stress intensity factor for cracks emanating from the rivet holes by the method of superposition applied to Westergaard stress functions. Incorporate suitable corrections to account for the stress raiser effect of the hole if it is deemed significant. Compare your result with solutions to similar problems presented in the literature.

(b) Assuming that the cracks grow at an equal rate from all holes, determine the critical crack length and the number of cycles to failure at the operational use cycle.

(c) Is visual inspection a viable option for prevention of failure? If so, what inspection interval do you recommend?

A.3 The attachment lug shown in Figure A.3 is used in a construction elevator with a rated load capacity of 10 tons. The "dead" load supported by the lug is 5000 lb. The normal duty cycle of the elevator is 20,000 cycles per year, with reinspection

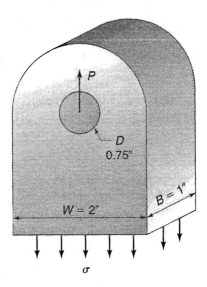

Figure EA.3 Attachment lug.

every five years. In order to provide sufficient hardness for the bearing surface for the pin, a high-yield-strength material must be used. The material chosen is A579 (Grade 75) steel. The following properties are available;

yield stress σ_{ys} = 180 ksi;

ultimate tensile strength σ_{UTS} = 190 ksi;

fracture toughness K_{Ic} = 90 ksi$\sqrt{in}$;

Paris-law constants C = 0.15 × 10^{-07} for ΔK in ksi $\sqrt{in}$ and n = 2.0.

As the engineer responsible for quality control issues, your task is to determine the inspection criteria for this part.

(a) Recognizing that fatigue life assessments require accurate representations of the geometric stress intensity factor, develop a suitable expression for the stress intensity factor for the given geometry with a crack initiating from one side of the hole's diameter. Plot the shape function [see Eq. (6.4)] $Y(a/W)$ versus a/W.

(b) Using a suitable fatigue law, construct a plot of the crack length versus the number of cycles from initiation to failure. To be conservative, neglect the effects of crack closure and any threshold behavior.

(c) Construct a plot of maximum initial flaw size versus safety factor (based on fracture toughness) over the range of SF = 1 to 4. Specify for the

NDE engineer the inspection location, maximum acceptable flaw size, and "recommended" inspection procedure for this part at a safety factor of 2.

(d) Does your recommendation include the possibility that a critical defect will be missed in one inspection cycle?

A.4 Repeat Exercise A.3 for the case of cracks initiating from both sides of the hole. Which of the two cases is the defining one for quality control inspection?

A.5 A long thin-walled steel pressure vessel is used as an accumulator in an industrial plant. The vessel was rolled and welded into shape and fitted with spherical end caps. The tank has an outer diameter of 21.45 in and is 1.00 in thick. The vessel contains hydraulic fluid, and a minimum pressure of 400 psi is maintained at all times. Pressure surges in the tank increase the pressure to 4000 psi on a regular basis. During routine inspection, a longitudinal surface crack is discovered on the outside wall. The crack has a length along the surface of 1.00 in and ultrasonic inspection reveals that it is 0.25 in deep. As a consulting engineer, you have been asked to determine how long the vessel can continue to be used. The manufacturer of the pressure vessel has provided the following properties:

yield stress $\sigma_{ys} = 90$ ksi;

fracture toughness $K_{Ic} = 75$ ksi$\sqrt{\text{in}}$;

threshold fracture toughness $\Delta K_{th} = 20$ ksi$\sqrt{\text{in}}$;

Paris-law constants $C = 4.0 \times 10^{-11}$ for K in ksi$\sqrt{\text{in}}$, and $n = 3.5$.

(a) Using the Newman–Raju empirical equation for the stress intensity factors for a semielliptical surface flaw [Eq. (3.91)], plot the changes in the crack profile as the crack continues to grow into the cylinder wall (i.e., plot a/c versus a/t).

(b) How many additional cycles can the vessel withstand before failure? Will the failure be a leak-before-break failure or a sudden fracture?

(c) What is your recommendation to increase the life of the pressure vessel at reasonable cost? Qualitatively justify your decision, and describe any potential problems your solution might create.

Complex-Variable Method in Elasticity

B.1 COMPLEX NUMBERS

Complex numbers play an important role in the description of points in the xy-plane. Consider the point P shown in Figure B.1. We can describe the point with the ordered pair (x, y) or the radius, r, and angle, θ, measured from the horizontal axis. Both measures are widely used in real-variable theory, and the choice of which system to use depends on the nature of the problem.

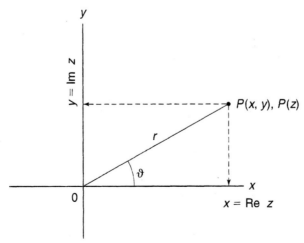

Figure B.1 Coordinate descriptions of a point in the xy-plane.

Alternatively, we can define a variable, z, such that

$$z = x + iy \tag{B.1}$$

where $i \equiv \pm\sqrt{-1}$ is called the *complex operator*. Note that the operation $\sqrt{-1}$ is undefined in the real-number system. With this convention, the Cartesian coordinates become

$$x = Re\ z \quad \text{and} \quad y = Im\ z \tag{B.2}$$

where the notation Re and Im denote "the real part of" and "the imaginary part of," respectively.

Associated with the complex number system is a well-defined set of rules for algebraic operations. Given two complex numbers $z_1 = x_1 + iy_1$ and $z_2 = x_2 + iy_2$,

1. Equality:

$$z_1 = z_2 \quad \Longleftrightarrow \quad x_1 = x_2 \quad \text{and} \quad y_1 = y_2 \tag{B.3}$$

   That is, two complex numbers are equal if and only if *both* their real and imaginary parts are equal.

2. Addition/subtraction:

$$z_1 + z_2 = (x_1 + x_2) + i(y_1 + y_2) \tag{B.4a}$$

$$z_1 - z_2 = (x_1 - x_2) + i(y_1 - y_2) \tag{B.4b}$$

3. Multiplication:

$$z_1 z_2 = (x_1 + iy_1) \cdot (x_2 + iy_2)$$

$$= x_1 x_2 - y_1 y_2 + i(x_1 y_2 + x_2 y_1) \tag{B.5}$$

   That is, multiplication of complex numbers follows the usual rules for multiplication of binomials, provided that i^2 is interpreted as the real number -1.

4. Polar coordinates:
   Since

$$x = r \cos\theta \quad \text{and} \quad y = r \sin\theta \tag{B.6a}$$

$$z = x + iy = r(\cos\theta + i \sin\theta) \tag{B.6b}$$

However,

$$\cos \theta \ = \ 1 \ - \ \frac{\theta^2}{2!} \ + \ \frac{\theta^4}{4!} \ - \ \frac{\theta^6}{6!} \cdots \tag{B.7a}$$

$$\sin \theta \ = \ \theta \ - \ \frac{\theta^3}{3!} \ + \ \frac{\theta^5}{5!} \ - \ \frac{\theta^7}{7!} \cdots \tag{B.7b}$$

$$e^\alpha \ = \ 1 \ + \ \alpha \ + \ \frac{\alpha^2}{2!} \ + \ \frac{\alpha^3}{3!} \cdots \tag{B.7c}$$

So, if we let $\alpha = i\theta$ in Eq. (B.7c) and regroup terms, then we have

$$\cos \theta \ + \ i \sin \theta \ = \ e^{i\theta} \tag{B.7d}$$

Equation (B.7d) is called the *Euler Identity*. Substituting Eq. (B.7d) into Eq. (B.6b), we find that a complex number can be expressed in its polar form as

$$x \ + \ iy \ = \ re^{i\theta} \tag{B.8}$$

The polar form is particularly useful for certain algebraic operations—for example,

$$z_1 z_2 \ = \ (r_1 e^{i\theta_1})(r_2 e^{i\theta_2}) \ = \ r_1 r_2 e^{i(\theta_1 + \theta_2)} \tag{B.9a}$$

$$\frac{z_1}{z_2} \ = \ \frac{r_1 e^{i\theta_1}}{r_2 e^{i\theta_2}} \ = \ \frac{r_1}{r_2} e^{i(\theta_1 - \theta_2)} \tag{B.9b}$$

$$z^n \ = \ \left[re^{i\theta} \right]^n \ = \ r^n e^{in\theta} \tag{B.9c}$$

There is an additional property of the complex-number system that we need to describe, namely the *complex conjugate*: For every complex number $z = x + iy$, there is associated another complex number $\bar{z} = x - iy$. This latter number is called the *complex conjugate* of the former. The complex conjugate is formed by replacing i by $-i$ wherever it appears. From Figure B.2, we observe that the complex conjugate $\bar{z}$ is a mirror reflection of the point z about the x-axis. Because of this property, complex functions (to be discussed next) that are symmetric about the x-axis have the same form using either z or $\bar{z}$. Figure B.2 also shows that for symmetry about the y-axis, we have similar forms if we replace z by $-\bar{z}$. Finally, note that

$$z\bar{z} \ = \ x^2 + y^2 \ = \ r^2 \ = \ |z|^2 \tag{B.10}$$

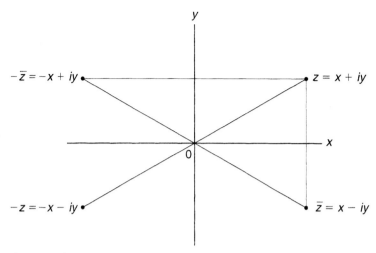

Figure B.2 Role of z-conjugate in symmetry in the complex plane.

Equation (B.10) is the complex-variable equivalent of the dot product in vector algebra.

From Eq. (B.10), we observe that complex numbers have vector-like properties in the xy-plane. This similarity is also exhibited with respect to addition. The point P in Figure B.3 can be described by either z or $z_0 + z_1$, since $z = z_0 + z_1$. If we let z_0 be the location (x', y') of a crack tip in an arbitrary xy-coordinate system, then z_1 can be interpreted as the crack-tip coordinate description. The use of local coordinates is particularly useful in implementing the Westergaard method of stress analysis developed in Chapter 3.

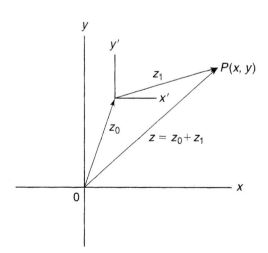

Figure B.3 Local coordinate systems in the complex plane.

B.2 COMPLEX FUNCTIONS

In the theory of real functions, we interpreted the quantity $f(x, y)$ as a real function F of two real variables x and y over some domain. Likewise we can define a function $Z(z)$ as a complex function—that is, a function that has both real and imaginary parts of one complex variable, z. However, complex functions are more than simply the complex analog of real functions. They possess unique properties that make them particularly valuable in elasticity theory.

If $Z(z)$ is a single-valued function, then, when $Z(z)$ is evaluated at an arbitrary point z anywhere in the domain of Z, the result will be, in general, a complex number. That is,

$$Z(z) = Re\, Z(z) + i\, Im\, Z(z) \tag{B.11}$$

which can be written as

$$Z(x + iy) = \phi(x, y) + i\, \psi(x, y) \tag{B.12}$$

Therefore, the real and imaginary parts of a complex function $Z(z)$ are *real* functions. However, $\phi(x, y)$ and $\psi(x, y)$ are not necessarily independent.

Example B.1   Determine the real and imaginary parts of

$$Z(z) = \sin z \tag{B.13}$$

$$\sin z = \sin(x + iy)$$

$$= \sin x \cos iy + \cos x \sin iy \tag{B.14}$$

$$= \sin x \cosh y + i \cos x \sinh y$$

Therefore,

$$Re\, \sin z = \sin x \cosh y$$

$$Im\, \sin z = \cos x \sinh y \tag{B.15}$$

∎

There is a special class of functions of the complex variable z that deserve attention. These functions are called *analytic functions*. The function $Z(z)$ is analytic if, over some domain, the value of Z possesses a unique derivative at every point in the domain. This property is analogous to the concept of continuity in real-variable theory. Isolated points at which a function is not analytic are called *singular points*. As long as we do not attempt to evaluate an otherwise analytic function at its isolated singular points, the theory of analytic functions still applies. We can, however, consider the limit of an analytic function as we *approach* a singular point. In fact, this latter task is among the most interesting from a fracture mechanics perspective.

We will now develop some rules for the differential calculus of analytic functions that will prove useful in applying complex-variable theory to elasticity. Since $z = x + iy$, it follows that

$$\frac{\partial z}{\partial x} = 1 \quad \text{and} \quad \frac{\partial z}{\partial x} = i \tag{B.16}$$

but, since Z is a function of the complex variable z, and not x and y, we need to invoke the chain rule for differentiation to determine the deriviatives of Z with respect to either of the real variables x or y. Therefore,

$$\frac{\partial Z(z)}{\partial x} = \frac{\partial Z(z)}{\partial z} \cdot \frac{\partial z}{\partial x} = Z' \tag{B.17a}$$

$$\frac{\partial Z(z)}{\partial y} = \frac{\partial Z(z)}{\partial z} \cdot \frac{\partial z}{\partial y} = i\,Z' \tag{B.17b}$$

where the prime denotes differentiation of the complex function with respect to its variable z. Alternatively, from Eq. (B.11), we have

$$\frac{\partial Z(z)}{\partial x} = \frac{\partial Re\,Z(z)}{\partial x} + i\frac{\partial Im\,Z(z)}{\partial x} \tag{B.18a}$$

$$\frac{\partial Z(z)}{\partial y} = \frac{\partial Re\,Z(z)}{\partial y} + i\frac{\partial Im\,Z(z)}{\partial y} \tag{B.18b}$$

Since we require that the derivative of an analytic function be unique, Eqs. (B.17) impose a restriction on the derivatives of Z with respect to the real variables, namely

$$i\frac{\partial Z(z)}{\partial x} = \frac{\partial Z(z)}{\partial y} \tag{B.19}$$

Applying this constraint to Eqs. (B.18) and equating the real and imaginary parts of the result, we find that

$$\frac{\partial Re\,Z(z)}{\partial y} = -\frac{\partial Im\,Z(z)}{\partial x} \tag{B.20a}$$

$$\frac{\partial Im\,Z(z)}{\partial y} = \frac{\partial Re\,Z(z)}{\partial x} \tag{B.20b}$$

These equations, called the *Cauchy–Riemann conditions*, provide the necesary and sufficient conditions for a complex function $Z(z)$ to be analytic.

Example B.2   Prove that $\sin z$ is an analytic function.

From the results of Example B.1 [i.e., Eqs. (B.15)], we have

$$\frac{\partial\,Re\,Z(z)}{\partial y} = \sin x \sinh y = -\frac{\partial\,Im\,Z(z)}{\partial x} \tag{B.21a}$$

$$\frac{\partial\,Im\,Z(z)}{\partial y} = \cos x \cosh y = \frac{\partial\,Re\,Z(z)}{\partial x} \tag{B.21b}$$

Since the Cauchy–Riemann conditons are satisfied, $\sin z$ is an analytic function.

∎

If we combine Eqs. (B.17) and Eqs. (B.18), we obtain a set of relations between the real and imaginary parts of Z differentiated with respect to the real variables x and y and the same functions differentiated with respect to the complex variable z:

$$\frac{\partial\,Re\,Z(z)}{\partial y} = -\frac{\partial\,Im\,Z(z)}{\partial x} = -Im\,Z' \tag{B.22a}$$

$$\frac{\partial\,Im\,Z(z)}{\partial y} = \frac{\partial\,Re\,Z(z)}{\partial x} = Re\,Z' \tag{B.22b}$$

These relations provide a convenient way to transform differential operations in elasticity theory described in real variables to their counterparts in terms of the complex variable z.

Analytic functions possess another property that we will find most useful in our development of the Airy stress function method of stress analysis. If we differentiate Eq. (B.20a) with respect to y and Eq. (B.20b) with respect to x and add the resulting equations, we find that

$$\left(\frac{\partial^2}{\partial x^2} + \frac{\partial^2}{\partial y^2}\right) Re\,Z(z) = \nabla^2 Re\,Z = 0 \tag{B.23}$$

where we have taken advantage of the fact that the unique derivative property of analytic functions allows us to interchange the order of differentiation with respect to x and y. Interchanging the order of differentiation from the previous development leads to

$$\left(\frac{\partial^2}{\partial x^2} + \frac{\partial^2}{\partial y^2}\right) Im\,Z(z) = \nabla^2 Im\,Z = 0 \tag{B.24}$$

Consequently, the real and imaginary parts of any analytic function are harmonic functions, and by inference, all harmonic functions are solutions of the biharmonic equation

$$\nabla^2 \, \nabla^2 \, \phi \; = \; \nabla^4 \phi \; = \; 0 \tag{B.25}$$

The biharmonic equation is fundamental to the theory of elasticity in two dimensions. The Airy stress function, discussed in Chapter 3, satisfies the biharmonic equation. Consequently, all analytic functions are potential Airy stress functions. Unlike the case for potential stress functions in real-variable theory, it is not necessary to prove that the chosen stress function satisfies the biharmonic equation; it is necessary only to prove that the function is analytic—a much easier task. Equally important is the fact that, given any analytic functions whose real and imaginary parts independently are solutions to a particular problem of interest, the sum of such functions is also a solution, by the principle of superposition. Thus, we can build a solution to a particular problem by adding together independent solutions, each of which adds (or subtracts) a particular feature to the problem.

Finally, the Cauchy–Riemann conditions provide a basis for integration of complex functions with respect to real variables. So, rewriting Eqs. (B.22)

$$Im \, W \; = \; -\frac{\partial}{\partial y} Re \, \widetilde{W} \; = \; \frac{\partial}{\partial x} Im \, \widetilde{W} \tag{B.26a}$$

$$Re \, W \; = \; \frac{\partial}{\partial y} Im \, \widetilde{W} \; = \; \frac{\partial}{\partial x} Re \, \widetilde{W} \tag{B.26b}$$

where the tilde over the function W denotes the integral of W. Note that a different level of integration has been used in the last pair of equations to emphasize the role of the Cauchy–Riemann relations in the integration process. As an application of this property, consider an integral of the form

$$\int y \, \cdot \, Im \, W' \, dy \tag{B.27a}$$

Recall the integration-by-parts relation

$$\int u \, dv \; = \; uv \; - \; \int v \, du \tag{B.27b}$$

Let $u = y$ then $du = dy$ and $dv = Im \, W' \, dy$. Then, from Eq. (B.22b) $v = -\, Re \, W$. With these substitutions,

$$\int y \, \cdot \, Im \, W' \, dy \; = \; -y \, Re \, W \; + \; \int Re \, W \, dy$$

$$= \; -y \, Re \, W \; + \; Im \, \widetilde{W} \tag{B.27c}$$

The complex-variable method has a rich history in the two-dimensional theory of elasticity, much of it following the publication in English of Muskhelishvili's treatise [1963] on the subject. Unfortunately, understanding and implementing his methods requires a comprehensive knowledge of complex-variable theory, including such topics as contour integration and conformal mapping. In contrast, only the elementary operations involving complex variables and their derivatives described in this appendix are needed to implement the generalized Westergaard method, despite the fact that the basic formulations of the two complex variable approaches are equivalent [Sanford, 1979]. For the reader interested in a more in-depth presentation of the complex-variable method, the text by Churchill [1960] is recommended.

REFERENCES

Churchill, R. V., 1960, *Complex Variables and Applications*, 2nd ed., McGraw-Hill Book Co., New York.

Muskhelishvili, N. I., 1963, *Some Basic Problems of the Mathematical Theory of Elasticity*, P. Noordhoff, Ltd., Groningen, The Netherlands.

Sanford, R. J., 1979, "A Critical Re-Examination of the Westergaard Method for Solving Opening-Mode Crack Problems," *Mechanics Research Communications*, 6(5), pp. 289–294.

EXERCISES

B.1 Given the complex function $Z(z) = z^2 \cos z$,
   (a) determine the real and imaginary parts of $Z(z)$;
   (b) determine if $Z(z)$ is analytic over some domain (proof required).

B.2 Given the complex function $Z(z) = z \ln z^2$,
   (a) determine the real and imaginary parts of $Z(z)$;
   (b) determine if $Z(z)$ is analytic over some domain (proof required).

B.3 Given the complex function $Z(z) = e^y(\cos x + i \sin x)$,
   (a) determine the real and imaginary parts of $Z(z)$;
   (b) determine if $Z(z)$ is analytic over some domain (proof required).

B.4 Given the complex function $Z(z) = e^{-y}(\cos x + i \sin x)$,
   (a) determine the real and imaginary parts of $Z(z)$;
   (b) determine if $Z(z)$ is analytic over some domain (proof required).

An Abbreviated Compendium of Westergaard Stress Functions

In this appendix, the Westergaard stress functions for some elementary crack problems are tabulated. For each entry, the Westergaard function for the opening mode, $Z_I(z)$; the forward shear mode, $Z_{II}(z)$; and their integrals are given, along with the corresponding geometric stress intensity factor(s) K_I and K_{II}. Note that concentrated forces are given per unit thickness.

The case presented in this appendix have been adapted with permission from *The Stress Analysis of Cracks Handbook*, Third Edition, by Tada, Paris, and Irwin, published by the American Society of Mechanical Engineers (2000). The case numbers correspond to their counterparts in *The Stress Analysis of Cracks Handbook*. For a more comprehensive listing of Westergaard stress functions, as well as extensive solutions for the geometric stress intensity factor K, the reader is referred to the cited handbook.

REFERENCES

Tada, H., Paris, P. C., and Irwin, G. R., 2000, *The Stress Analysis of Cracks Handbook*, 3rd ed., ASME Press, American Soc. Mechanical Engineers, New York.

Case 3.6

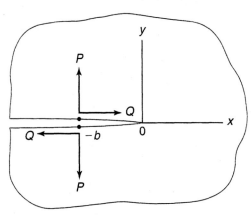

$$\begin{Bmatrix} Z_I(z) \\ Z_{II}(z) \end{Bmatrix} = \frac{1}{\pi} \begin{Bmatrix} P \\ Q \end{Bmatrix} \frac{1}{z+b} \sqrt{\frac{b}{z}}$$

$$\begin{Bmatrix} \tilde{Z}_I(z) \\ \tilde{Z}_{II}(z) \end{Bmatrix} = -\frac{2}{\pi} \begin{Bmatrix} P \\ Q \end{Bmatrix} \tan^{-1} \sqrt{\frac{b}{z}}$$

$$\begin{Bmatrix} K_I \\ K_{II} \end{Bmatrix} = \frac{\sqrt{2}}{\sqrt{\pi b}} \begin{Bmatrix} P \\ Q \end{Bmatrix}$$

Case 3.7

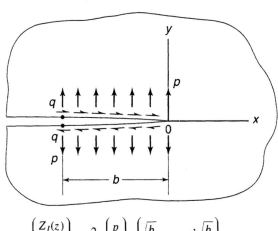

$$\begin{Bmatrix} Z_I(z) \\ Z_{II}(z) \end{Bmatrix} = \frac{2}{\pi} \begin{Bmatrix} p \\ q \end{Bmatrix} \left\{ \sqrt{\frac{b}{z}} - \tan^{-1} \sqrt{\frac{b}{z}} \right\}$$

$$\begin{Bmatrix} \tilde{Z}_I(z) \\ \tilde{Z}_{II}(z) \end{Bmatrix} = \frac{2}{\pi} \begin{Bmatrix} p \\ q \end{Bmatrix} b \left\{ \sqrt{\frac{z}{b}} - \left(1 + \frac{z}{b}\right) \tan^{-1} \sqrt{\frac{b}{z}} \right\}$$

$$\begin{Bmatrix} K_I \\ K_{II} \end{Bmatrix} = \frac{2}{\pi} \begin{Bmatrix} p \\ q \end{Bmatrix} \sqrt{2\pi b}$$

Case 4.5

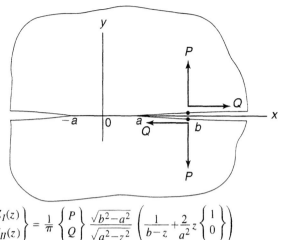

$$\left\{\begin{matrix} Z_I(z) \\ Z_{II}(z) \end{matrix}\right\} = \frac{1}{\pi} \left\{\begin{matrix} P \\ Q \end{matrix}\right\} \frac{\sqrt{b^2-a^2}}{\sqrt{a^2-z^2}} \left(\frac{1}{b-z} + \frac{2}{a^2} z \left\{\begin{matrix} 1 \\ 0 \end{matrix}\right\} \right)$$

$$\left\{\begin{matrix} \widetilde{Z}_I(z) \\ \widetilde{Z}_{II}(z) \end{matrix}\right\} = \frac{1}{\pi} \left\{\begin{matrix} P \\ Q \end{matrix}\right\} \left[\sin^{-1} \frac{bz-a^2}{a(b-z)} - \frac{2}{a^2} \sqrt{b^2-a^2}\sqrt{a^2-z^2} \left\{\begin{matrix} 1 \\ 0 \end{matrix}\right\} \right]$$

$$\left\{\begin{matrix} K_I \\ K_{II} \end{matrix}\right\}_{\pm a} = \frac{1}{\sqrt{\pi a}} \left\{\begin{matrix} P \\ Q \end{matrix}\right\} \sqrt{b^2-a^2} \left(\frac{1}{b \mp a} \pm \frac{2}{a} \left\{\begin{matrix} 1 \\ 0 \end{matrix}\right\} \right)$$

Case 4.6

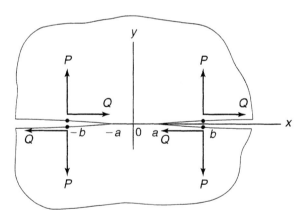

$$\left\{\begin{matrix} Z_I(z) \\ Z_{II}(z) \end{matrix}\right\} = \frac{2}{\pi} \left\{\begin{matrix} P \\ Q \end{matrix}\right\} \frac{b\sqrt{b^2-a^2}}{(b^2-z^2)\sqrt{a^2-z^2}}$$

$$\left\{\begin{matrix} \widetilde{Z}_I(z) \\ \widetilde{Z}_{II}(z) \end{matrix}\right\} = \frac{2}{\pi} \left\{\begin{matrix} P \\ Q \end{matrix}\right\} \tan^{-1} \sqrt{\frac{1-(a/b)^2}{(a/z)^2-1}}$$

$$\left\{\begin{matrix} K_I \\ K_{II} \end{matrix}\right\}_{\pm a} = \frac{2}{\sqrt{\pi a}} \left\{\begin{matrix} P \\ Q \end{matrix}\right\} \frac{b}{\sqrt{b^2-a^2}}$$

Case 4.7

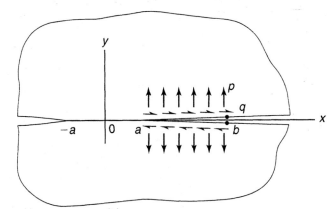

$$\begin{Bmatrix} Z_I(z) \\ Z_{II}(z) \end{Bmatrix} = \frac{1}{\pi} \begin{Bmatrix} p \\ q \end{Bmatrix} \left\{ \frac{\sqrt{b^2-a^2} + f\left(\frac{b}{a}\right)z}{\sqrt{a^2-z^2}} - \cos^{-1}\frac{bz-a^2}{a(b-z)} \right\}$$

$$\begin{Bmatrix} \widetilde{Z}_I(z) \\ \widetilde{Z}_{II}(z) \end{Bmatrix} = \frac{1}{\pi} \begin{Bmatrix} p \\ q \end{Bmatrix} \left\{ -f\left(\frac{b}{a}\right)\sqrt{a^2-z^2} + (b-z)\sin^{-1}\frac{bz-a^2}{a(b-z)} \right\} \quad \text{where } f\left(\frac{b}{a}\right) = \left\{ \begin{array}{c} \frac{b}{a}\sqrt{\left(\frac{b}{a}\right)^2-1} \\[2mm] \cosh^{-1}\frac{b}{a} \end{array} \right\}$$

$$\begin{Bmatrix} K_I \\ K_{II} \end{Bmatrix}_{\pm a} = \frac{1}{\sqrt{\pi a}} \begin{Bmatrix} p \\ q \end{Bmatrix} \left\{ \sqrt{b^2-a^2} \pm af\left(\frac{b}{a}\right) \right\}$$

Case 4.8

$$\begin{Bmatrix} Z_I(z) \\ Z_{II}(z) \end{Bmatrix} = \frac{2}{\pi} \begin{Bmatrix} p \\ q \end{Bmatrix} \left\{ \sqrt{\frac{b^2-a^2}{a^2-z^2}} - \tan^{-1}\sqrt{\frac{b^2-a^2}{a^2-z^2}} \right\}$$

$$\begin{Bmatrix} \widetilde{Z}_I(z) \\ \widetilde{Z}_{II}(z) \end{Bmatrix} = \frac{2}{\pi} \begin{Bmatrix} p \\ q \end{Bmatrix} \left\{ b\tan^{-1}\sqrt{\frac{1-(a/b)^2}{(a/z)^2-1}} - z\tan^{-1}\sqrt{\frac{(b/a)^2-1}{1-(z/a)^2}} \right\}$$

$$\begin{Bmatrix} K_I \\ K_{II} \end{Bmatrix}_{\pm a} = \frac{2}{\sqrt{\pi a}} \begin{Bmatrix} p \\ q \end{Bmatrix} \sqrt{b^2-a^2}$$

Case 4.9

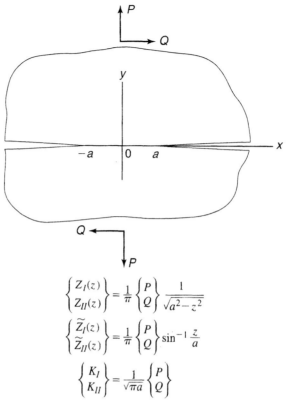

$$\left\{\begin{array}{c} Z_I(z) \\ Z_{II}(z) \end{array}\right\} = \frac{1}{\pi} \left\{\begin{array}{c} P \\ Q \end{array}\right\} \frac{1}{\sqrt{a^2 - z^2}}$$

$$\left\{\begin{array}{c} \widetilde{Z}_I(z) \\ \widetilde{Z}_{II}(z) \end{array}\right\} = \frac{1}{\pi} \left\{\begin{array}{c} P \\ Q \end{array}\right\} \sin^{-1} \frac{z}{a}$$

$$\left\{\begin{array}{c} K_I \\ K_{II} \end{array}\right\} = \frac{1}{\sqrt{\pi a}} \left\{\begin{array}{c} P \\ Q \end{array}\right\}$$

Case 4.10

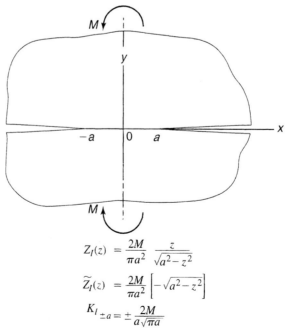

$$Z_I(z) = \frac{2M}{\pi a^2} \frac{z}{\sqrt{a^2 - z^2}}$$

$$\widetilde{Z}_I(z) = \frac{2M}{\pi a^2} \left[-\sqrt{a^2 - z^2} \right]$$

$$K_{I\,\pm a} = \pm \frac{2M}{a\sqrt{\pi a}}$$

Case 5.1

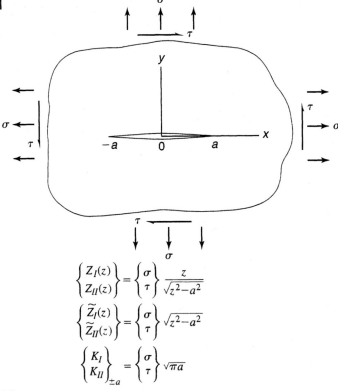

$$\begin{Bmatrix} Z_I(z) \\ Z_{II}(z) \end{Bmatrix} = \begin{Bmatrix} \sigma \\ \tau \end{Bmatrix} \frac{z}{\sqrt{z^2 - a^2}}$$

$$\begin{Bmatrix} \widetilde{Z}_I(z) \\ \widetilde{Z}_{II}(z) \end{Bmatrix} = \begin{Bmatrix} \sigma \\ \tau \end{Bmatrix} \sqrt{z^2 - a^2}$$

$$\begin{Bmatrix} K_I \\ K_{II} \end{Bmatrix}_{\pm a} = \begin{Bmatrix} \sigma \\ \tau \end{Bmatrix} \sqrt{\pi a}$$

Case 5.10

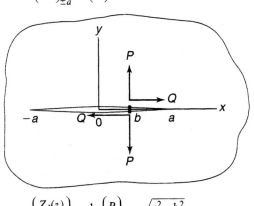

$$\begin{Bmatrix} Z_I(z) \\ Z_{II}(z) \end{Bmatrix} = \frac{1}{\pi} \begin{Bmatrix} P \\ Q \end{Bmatrix} \frac{\sqrt{a^2 - b^2}}{(z - b)\sqrt{z^2 - a^2}}$$

$$\begin{Bmatrix} \widetilde{Z}_I(z) \\ \widetilde{Z}_{II}(z) \end{Bmatrix} = \frac{1}{\pi} \begin{Bmatrix} P \\ Q \end{Bmatrix} \sin^{-1} \frac{bz - a^2}{a(z - b)}$$

$$\begin{Bmatrix} K_I \\ K_{II} \end{Bmatrix}_{\pm a} = \frac{1}{\sqrt{\pi a}} \begin{Bmatrix} P \\ Q \end{Bmatrix} \sqrt{\frac{a \pm b}{a \mp b}}$$

Case 5.11

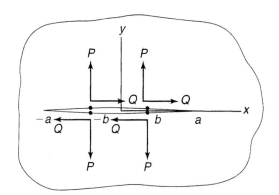

$$\begin{Bmatrix} Z_I(z) \\ Z_{II}(z) \end{Bmatrix} = \frac{2}{\pi} \begin{Bmatrix} P \\ Q \end{Bmatrix} \frac{\sqrt{a^2 - b^2}}{(z^2 - b^2)\sqrt{1 - (a/z)^2}}$$

$$\begin{Bmatrix} \widetilde{Z}_I(z) \\ \widetilde{Z}_{II}(z) \end{Bmatrix} = \frac{2}{\pi} \begin{Bmatrix} P \\ Q \end{Bmatrix} \tan^{-1} \sqrt{\frac{z^2 - a^2}{a^2 - b^2}}$$

$$\begin{Bmatrix} K_I \\ K_{II} \end{Bmatrix}_{\pm a} = \frac{2}{\sqrt{\pi a}} \begin{Bmatrix} P \\ Q \end{Bmatrix} \frac{a}{\sqrt{a^2 - b^2}}$$

Case 5.12

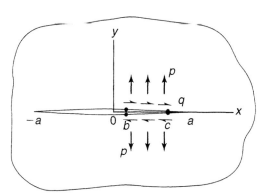

$$\begin{Bmatrix} Z_I(z) \\ Z_{II}(z) \end{Bmatrix} = \frac{1}{\pi} \begin{Bmatrix} p \\ q \end{Bmatrix} \left[\sin^{-1} \frac{a^2 - cz}{a(z - c)} - \sin^{-1} \frac{a^2 - bz}{a(z - b)} + \frac{\sin^{-1}\frac{c}{a} - \sin^{-1}\frac{b}{a}}{\sqrt{1 - (a/z)^2}} - \frac{\sqrt{a^2 - c^2} - \sqrt{a^2 - b^2}}{\sqrt{z^2 - a^2}} \right]$$

$$\begin{Bmatrix} \widetilde{Z}_I(z) \\ \widetilde{Z}_{II}(z) \end{Bmatrix} = \frac{1}{\pi} \begin{Bmatrix} p \\ q \end{Bmatrix} \left[(z - c)\sin^{-1} \frac{a^2 - cz}{a(z - c)} - (z - b)\sin^{-1} \frac{a^2 - bz}{a(z - b)} + \left(\sin^{-1} \frac{c}{a} - \sin^{-1} \frac{b}{a} \right)\sqrt{z^2 - a^2} \right]$$

$$\begin{Bmatrix} K_I \\ K_{II} \end{Bmatrix}_{\pm a} = \frac{1}{\pi} \begin{Bmatrix} p \\ q \end{Bmatrix} \sqrt{\pi a} \left[\sin^{-1} \frac{c}{a} - \sin^{-1} \frac{b}{a} \mp \left(\sqrt{1 - (c/a)^2} - \sqrt{1 - (b/a)^2} \right) \right]$$

Case 5.13

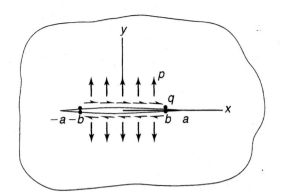

$$\begin{Bmatrix} Z_I(z) \\ Z_{II}(z) \end{Bmatrix} = \frac{2}{\pi} \begin{Bmatrix} p \\ q \end{Bmatrix} \left[\frac{\sin^{-1}\dfrac{b}{a}}{\sqrt{1-(a/z)^2}} - \tan^{-1}\sqrt{\frac{1-(a/z)^2}{(a/b)^2-1}} \right]$$

$$\begin{Bmatrix} \widetilde{Z}_I(z) \\ \widetilde{Z}_{II}(z) \end{Bmatrix} = \frac{2}{\pi} \begin{Bmatrix} p \\ q \end{Bmatrix} \left[\left(\sin^{-1}\frac{b}{a}\right)\sqrt{z^2-a^2} - z\tan^{-1}\sqrt{\frac{1-(a/z)^2}{(a/b)^2-1}} + b\tan^{-1}\sqrt{\frac{(z/a)^2-1}{1-(b/a)^2}} \right]$$

$$\begin{Bmatrix} K_I \\ K_{II} \end{Bmatrix} = \frac{2}{\pi} \begin{Bmatrix} p \\ q \end{Bmatrix} \sqrt{\pi a}\left(\sin^{-1}\frac{b}{a}\right)$$

Case 7.1

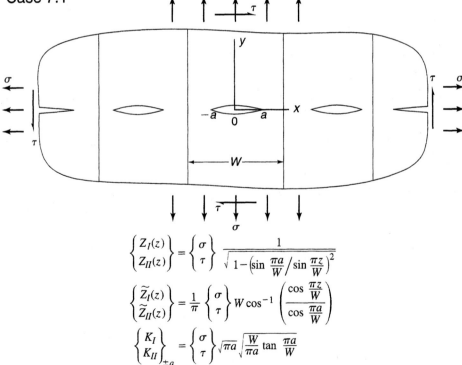

$$\begin{Bmatrix} Z_I(z) \\ Z_{II}(z) \end{Bmatrix} = \begin{Bmatrix} \sigma \\ \tau \end{Bmatrix} \frac{1}{\sqrt{1-\left(\sin\dfrac{\pi a}{W}\Big/\sin\dfrac{\pi z}{W}\right)^2}}$$

$$\begin{Bmatrix} \widetilde{Z}_I(z) \\ \widetilde{Z}_{II}(z) \end{Bmatrix} = \frac{1}{\pi} \begin{Bmatrix} \sigma \\ \tau \end{Bmatrix} W\cos^{-1}\left(\frac{\cos\dfrac{\pi z}{W}}{\cos\dfrac{\pi a}{W}}\right)$$

$$\begin{Bmatrix} K_I \\ K_{II} \end{Bmatrix}_{\pm a} = \begin{Bmatrix} \sigma \\ \tau \end{Bmatrix} \sqrt{\pi a}\sqrt{\frac{W}{\pi a}\tan\frac{\pi a}{W}}$$

Case 7.7

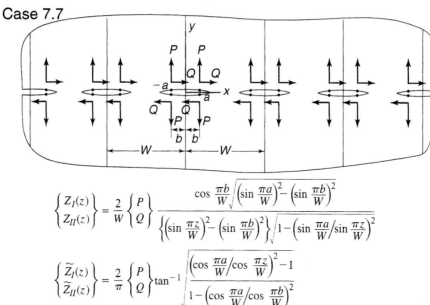

$$\begin{Bmatrix} Z_I(z) \\ Z_{II}(z) \end{Bmatrix} = \frac{2}{W} \begin{Bmatrix} P \\ Q \end{Bmatrix} \frac{\cos\frac{\pi b}{W}\sqrt{\left(\sin\frac{\pi a}{W}\right)^2 - \left(\sin\frac{\pi b}{W}\right)^2}}{\left\{\left(\sin\frac{\pi z}{W}\right)^2 - \left(\sin\frac{\pi b}{W}\right)^2\right\}\sqrt{1 - \left(\sin\frac{\pi a}{W}\Big/\sin\frac{\pi z}{W}\right)^2}}$$

$$\begin{Bmatrix} \widetilde{Z}_I(z) \\ \widetilde{Z}_{II}(z) \end{Bmatrix} = \frac{2}{\pi} \begin{Bmatrix} P \\ Q \end{Bmatrix} \tan^{-1}\sqrt{\frac{\left(\cos\frac{\pi a}{W}\Big/\cos\frac{\pi z}{W}\right)^2 - 1}{1 - \left(\cos\frac{\pi a}{W}\Big/\cos\frac{\pi b}{W}\right)^2}}$$

$$\begin{Bmatrix} K_I \\ K_{II} \end{Bmatrix} = \frac{2}{W} \begin{Bmatrix} P \\ Q \end{Bmatrix}\sqrt{\pi a}\sqrt{\frac{W}{\pi a}\tan\frac{\pi a}{W}}\,\frac{\cos\frac{\pi b}{W}}{\sqrt{\left(\sin\frac{\pi a}{W}\right)^2 - \left(\sin\frac{\pi b}{W}\right)^2}}$$

Case 7.9

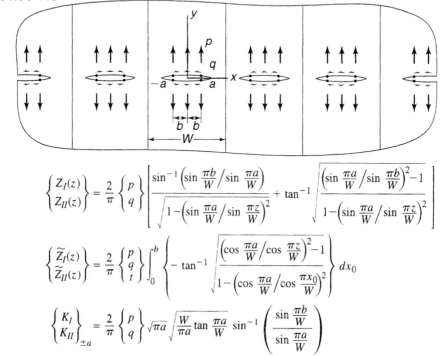

$$\begin{Bmatrix} Z_I(z) \\ Z_{II}(z) \end{Bmatrix} = \frac{2}{\pi} \begin{Bmatrix} p \\ q \end{Bmatrix}\left[\frac{\sin^{-1}\left(\sin\frac{\pi b}{W}\Big/\sin\frac{\pi a}{W}\right)}{\sqrt{1 - \left(\sin\frac{\pi a}{W}\Big/\sin\frac{\pi z}{W}\right)^2}} + \tan^{-1}\sqrt{\frac{\left(\sin\frac{\pi a}{W}\Big/\sin\frac{\pi b}{W}\right)^2 - 1}{1 - \left(\sin\frac{\pi a}{W}\Big/\sin\frac{\pi z}{W}\right)^2}} \right]$$

$$\begin{Bmatrix} \widetilde{Z}_I(z) \\ \widetilde{Z}_{II}(z) \end{Bmatrix} = \frac{2}{\pi} \begin{Bmatrix} p \\ q \\ t \end{Bmatrix}\int_0^b \left\{ -\tan^{-1}\sqrt{\frac{\left(\cos\frac{\pi a}{W}\Big/\cos\frac{\pi z}{W}\right)^2 - 1}{1 - \left(\cos\frac{\pi a}{W}\Big/\cos\frac{\pi x_0}{W}\right)^2}} \right\} dx_0$$

$$\begin{Bmatrix} K_I \\ K_{II} \end{Bmatrix}_{\pm a} = \frac{2}{\pi} \begin{Bmatrix} p \\ q \end{Bmatrix}\sqrt{\pi a}\sqrt{\frac{W}{\pi a}\tan\frac{\pi a}{W}}\,\sin^{-1}\left(\frac{\sin\frac{\pi b}{W}}{\sin\frac{\pi a}{W}}\right)$$

Fracture Properties of Engineering Materials

Adapted from NASA Report JSC–22267B, March 2000

Disclaimer

The data in the following tables have been adapted from information compiled by the National Aeronautics and Space Administration (NASA). The data are believed to be representative of the material described, but, due to the wide variations possible, the values should not be relied upon to accurately represent the properties of a specific batch of material. Critical design decisions should be based on results taken from the actual lot of material to be used for fabrication, tested under conditions that as accurately as possible mirror service conditions.

The data are provided "as is" without warranty of any kind, either expressed or implied. Neither the author nor the publisher shall be liable for any damages arising from the use of these illustrative data.

Notation

Code: the identifier code used in NASMAT to identify the complete data set from which the fracture properties were determined.

YS: yield stress σ_{ys}.

UTS: ultimate tensile strength.

K_{Ic}: critical plane-strain stress intensity factor for through cracks.

Note: In SI units the plane-strain fracture toughness is given in MPa-mm$^{1/2}$. To convert to MPa-m$^{1/2}$ divide the values given by 31.623.

Material; Condition; Environment*	Code†	U. S. Customary Units (ksi, ksi√in)			SI Units (MPa, MPa√mm)		
		YS	UTS	K_{Ic}	YS	UTS	K_{Ic}
[A] Iron, Alloy or Cast							
ASTM Specification							
A536 Grd 80–55–06							
As cast	A1AC50AB1	58	80	32	400	552	1112
[B] ASTM spec. grd. Steel							
A10 Series							
A36							
Plt (Dyn K_{Ic}, < 500 Hz); LA, HHA, 3% NaCl	B0CB10AB1	44	78	70	303	538	2432
ES Weld & HAZ (Dyn K_{Ic}, < 500 Hz); LA, HHA, 3% NaCl	B0CZK1AB1	44	78	70	303	538	2432
A200 Series							
A203 Grd E (3.5% Ni)							
Plt	B2CE12AB1	71	80	170	490	552	5907
Plt; −100°F	B2CE12LA7	82	96	200	565	662	6949
A216 Grd WCC							
Casting	B2GC51AB1	48	80	150	331	552	5212
A300 Series							
A302 Grd B							
Plt	B3AB12AB1	55	90	100	379	621	3475
A372 Type IV							
Forg	B3GD21AB1	65	105	100	448	724	3475
A387 Grd 22, Cl 2							
Plt	B3IQ10AB1	50	75	100	345	517	3475
A400 Series							
A469 Cl 4							
Forg	B4JD26AB1	85	105	170	586	724	5907
A469 Cl 5							
Forg	B4JE20AB1	95	115	170	655	793	5907
A500 Series							
A508 Cl2 & Cl3							
Forg	B5AC21AB1	65	100	100	448	689	3475
A514 Typ F							
Plt	B5BF10AB1	105	120	85	724	827	2953
GMA SR Weld	B5BFC2AB1	100	115	85	689	793	2953
A517 Grd F (T1 Steel)							
Plt	B5DF12AB1	100	125	100	689	862	3475
A533-B, Cl1 & Cl2							
Plt	B5HD10AB1	70	100	150	483	689	5212
SMA Weld	B5HDF1AB1	60	90	100	414	621	3475
A553 Typ I							
Plt	B5QA12AB1	95	110	180	655	758	6254
Plt; −320°F	B5QA12LA4	150	175	100	1034	1207	3475
[B] ASTM Spec.-Grade Steel							
A500 Series							
A579 Grd 75 (12% Ni)							
Forg	B5VW20AB1	180	190	90	1241	1310	3127
A588 Grd A & Grd B							
Plt	B5XA11AB1	55	80	100	379	552	3475
Plt; 3% NaCl, > 0.2 Hz	B5XA11WB1	55	80	100	379	552	3475
A645 (5% Ni)							
Plt	B6GA12AB1	75	105	180	517	724	6254
Plt; −320°F	B6GA12LA4	110	165	80	758	1138	2780

*Unless noted, assume Lab Air (LA) environment and any orientation except S-T, S-L, C-R, C-L, and R-L.

† NASA NASGRO 3.0 material identifier code.

Material; Condition; Environment*	Code†	U. S. Customary Units (ksi, ksi $\sqrt{\text{in}}$)			SI Units (MPa, MPa$\sqrt{\text{mm}}$)		
		YS	UTS	K_{Ic}	YS	UTS	K_{Ic}
[C] AISI—SAE Steel							
AISI 10xx–12xx Steel							
Low Carbon 1005–1012							
Hot-rolled plt	C1AB11AB1	25	45	70	172	310	2432
Low-Carbon 1015–1025							
Hot-rolled plt	C1BB11AB1	30	58	70	207	400	2432
AISI 43xx–48xx Steel							
4330V MOD							
180–200 UTS; Plt & Forg	C4BS10AB1	175	190	110	1207	1310	3822
200–220 UTS; Plt & Forg	C4BT10AB1	195	210	80	1344	1448	2780
220–240 UTS; Plt & Forg	C4BU10AB1	215	230	65	1482	1586	2259
4340							
160–180 UTS; Plt & Forg	C4DC21AB1	155	170	135	1069	1172	4691
180–200 UTS; Plt & Forg; HHA	C4DD11AD1	175	190	110	1207	1310	3822
180–200 UTS; Plt & Forg	C4DD21AB1	175	190	110	1207	1310	3822
200–220 UTS; Plt & Forg	C4DE11AB1	195	210	80	1344	1448	2780
200–220 UTS; Plt & Forg; −50°F	C4DE11LB7	200	220	55	1379	1517	1911
220–240 UTS; Plt & Forg	C4DF11AB1	215	230	65	1482	1586	2259
220–240 UTS; Plt & Forg; −50°F	C4DF11LB7	225	240	45	1551	1655	1564
240–280 UTS; Plt & Forg	C4DG11AB1	240	260	55	1655	1793	1911
240–280 UTS; Plt & Forg; −50°F	C4DG11LB7	250	270	40	1724	1862	1390
[D] Misc. U.S. Spec.-Grade Steel							
SAE Spec. Steel							
0030 Cast	D5AC50AB1	44	72	70	303	496	2432
[E] Trade/Common-Name Steel							
Ultra High Strength Steel							
18 Ni Maraging							
250 Grd; Plt & Forg	E1AD10AB1	240	260	75	1655	1793	2606
300 Grd; Plt & Forg	E1AE10AB1	280	290	70	1931	1999	2432
300M							
270–300 UTS; Plt & Forg	E1BF21AB1	240	285	55	1655	1965	1911
AF1410							
220–240 UTS; Plt & Forg; −65°F	E1CC12AA7	240	250	110	1655	1724	3822
220–240 UTS; Plt & Forg; LA, HHA / DW > 1Hz	E1CC12AB1	220	230	135	1517	1586	4691
D6AC							
220–240 UTS; Plt & Forg; Nom. $K_{Ic}(70)$; −40°F	E1DD10AA8	225	240	50	1551	1655	1737
220–240 UTS; Plt & Forg; Nom. $K_{Ic}(70)$	E1DD10AB1	215	230	70	1482	1586	2432
220–240 UTS; Plt & Forg; Nom. $K_{Ic}(70)$; HHA/DW > 0.1 Hz	E1DD10AD1	220	230	70	1517	1586	2432
220–240 UTS; Plt & Forg; High $K_{Ic}(90)$	E1DJ10AB1	215	230	90	1482	1586	3127
HP–9–4–20							
190–210 UTS; Plt & Forg; L-T, T-L; HHA, SW > 1 Hz	E1EB23AB1	190	200	110	1310	1379	3822
190–210 UTS; Plt & Forg; L-T, T-L; −65°F	E1EB23AC7	200	210	110	1379	1448	3822
190–210 UTS; GTA Weld + SR; LA, HHA, SW > 1 Hz	E1ECB2WA1	185	195	110	1276	1344	3822
190–210 UTS; GTA Weld + SR; −65°F	E1ECB2AC7	195	205	110	1344	1413	3822
HP–9–4–30							
220–240 UTS; Plt & Forg; L-T, T-L; LA, HHA, SW > 1Hz	E1GC23AB1	205	230	90	1413	1586	3127
220–240 UTS; Plt & Forg; L-T, T-L; −65°F	E1GC23AA7	215	240	90	1482	1655	3127
220–240 UTS; Plt & Forg; L-T, T-L; 600°F	E1GC23AA14	165	195	85	1138	1344	2953
HY–180(10Ni)							
Plt & Forg	E1IB13AB1	180	200	150	1241	1379	5212
Plt & Forg; DW, ASW > 0.1 Hz	E1IB13WA1	180	200	150	1241	1379	5212

*Unless noted, assume Lab Air (LA) environment and any orientation except S-T, S-L, C-R, C-L, and R-L.
† NASA NASGRO 3.0 material identifier code.

Material; Condition; Environment*	Code†	U. S. Customary Units (ksi, ksi√in)			SI Units (MPa, MPa√mm)		
		YS	UTS	K_{Ic}	YS	UTS	K_{Ic}
[E] Trade/Common-Name Steel (*continued*)							
Plt & Forg; SW > 0.1 Hz	E1IB13WB1	195	200	150	1344	1379	5212
HY-TUF							
220–240 UTS; VAR Forg	E1JB23AB1	200	230	110	1379	1586	3822
H–11 MOD							
240–260 UTS; Plt & Forg	E1LE23AB1	215	250	50	1482	1724	1737
Pressure Vessel/Piping							
HY 80							
Plt	E2AA13AB1	90	105	200	621	724	6949
Plt; 3.5% NaCl / SW > 0.1 Hz	E2AA13WB1	90	105	200	621	724	6949
HY 130							
Plt	E2CA13AB1	140	150	200	965	1034	6949
Plt; 3.5% NaCl / SW > 0.1 Hz	E2CA13WB1	140	150	200	965	1034	6949
GMA Weld	E2CAC1AB1	130	140	200	896	965	6949
SMA Weld	E2CAF1AB1	130	140	200	896	965	6949
Construction Grade							
HT–80							
Plt	E3BA13AB1	110	120	150	758	827	5212
SA Weld	E3BAH1AB1	95	115	150	655	793	5212
[F] AISI-type Stainless Steel							
AISI 300 Series							
AISI 301/302							
Ann Plt & Sht	F3AA13AB1	40	90	200	276	621	6949
1/2 Hard sht	F3AC13AB1	125	165	100	862	1138	3475
Full Hard sht	F3AE13AB1	190	205	80	1310	1413	2780
AISI 304/304L							
Ann Plt & Sht, Cast; 550°F Air	F3DA13AA13	24	64	150	165	441	5212
Ann Plt & Sht, Cast; 800°F Air, > 1 Hz	F3DA13AA16	20	63	100	138	434	3475
Ann Plt & Sht, Cast	F3DA13AB1	40	90	200	276	621	6949
Ann Plt & Sht, Cast; −320°F LN2	F3DA13LA4	100	205	200	689	1413	6949
SA weld (308 filler) + SR; 800°F Air, > 1 Hz	F3DAH2AA16	49	69	100	338	476	3475
SA weld (308 filler) + SR	F3DAH2AB1	66	107	150	455	738	5212
AISI 316/316L							
Ann Plt & Sht, Cast; 600°F Air	F3KA13AA14	32	60	150	221	414	5212
Ann Plt & Sht, Cast; 800°F Air	F3KA13AA16	20	60	100	138	414	3475
Ann Plt & Sht, Cast	F3KA13AB1	36	90	200	248	621	6949
Ann Plt & Sht, Cast; −452°F Lhe	F3KA13LA2	80	215	200	552	1482	6949
Ann Plt & Sht, Cast; −320°F LN2	F3KA13LA4	70	185	200	483	1276	6949
Cast; 600°F Air	F3KA50AA14	32	75	151	221	517	5212
Cast	F3KA50AB1	36	90	200	248	621	6949
Cast; −453°F Lhe	F3KA50LA2	79	210	200	545	1448	6949
Cast; −320°F LN2	F3KA50LA4	72	180	200	496	1241	6949
SMA weld (316 filler) + SR; 800°F Air, > 1 Hz	F3KAH2AA16	49	69	80	338	476	2780
20% CW Plt & Sht	F3KB13AB1	100	110	100	689	758	3475
AISI 400 Series							
AISI 430 VAR							
Ann Rnd, C-R	F4LA16AB1	34	59	80	234	407	2780
AISI 440C Steel							
Single temper 450°F/2 hr, T-L	F4SE12AB1	270	290	17	1862	1999	591

*Unless noted, assume Lab Air (LA) environment and any orientation except S-T, S-L, C-R, C-L, and R-L.
† NASA NASGRO 3.0 material identifier code.

Material; Condition; Environment*	Code†	U. S. Customary Units (ksi, ksi $\sqrt{in}$)			SI Units (MPa, MPa$\sqrt{mm}$)		
		YS	UTS	K_{Ic}	YS	UTS	K_{Ic}
[G] Misc. CRES/Heat Resistant Steel							
PHxx-x Alloys							
PH13–8Mo							
H1000; Plt, Forg, Extr	G1AD13AB1	200	208	100	1379	1434	3475
H1000; Plt & Forg; DW & SW, > 1 Hz	G1AD13WD1	200	208	100	1379	1434	3475
H1050; Plt & Forg	G1AF13AB1	185	190	115	1276	1310	3996
H1050; Plt & Forg; DW & SW, > 0.1 Hz	G1AF13WD1	185	190	115	1276	1310	3996
xx-xPH Alloys							
15–5PH							
H900; Rnd, C-R	G2AB16AB1	170	190	50	1172	1310	1737
H1025; Rnd, C-R	G2AD16AB1	165	175	60	1138	1207	2085
H1025; Forg	G2AD23AB1	155	165	110	1069	1138	3822
H1100; Rnd, C-R	G2AF16AB1	145	155	80	1000	1069	2780
17–4PH							
H900; Plt, L-T	G2CB11AB1	170	195	50	1172	1344	1737
H900; Plt, T-L	G2CB12AB1	170	195	45	1172	1344	1564
H1050; Plt	G2CE13AB1	155	160	60	1069	1103	2085
H1025; Rnd, C-L	G2CE19AB1	160	163	55	1103	1124	1911
H1025; Cast; HHA	G2CE50AD1	160	163	55	1103	1124	1911
H1100; Plt; HHA	G2CH13AD1	145	150	90	1000	1034	3127
17–7PH							
TH1050; Plt	G2EH13AB1	170	190	50	1172	1310	1737
AMxxx Alloys							
AM 350							
CRT; Sht, L-T	G4AH11AB1	185	205	90	1276	1413	3127
AM 367							
SCT(850); Sht	G4FC11AB1	240	243	65	1655	1675	2259
Custom xxx Alloys							
Custom 455							
H1000; Plt & Forg	G5BD23AB1	195	205	100	1344	1413	3475
H1025; Forg, C-R	G5BE26AB1	185	195	110	1276	1344	3822
Nitronic xx Alloys							
Nitronic 33							
Ann; Plt	G7AA13AB1	64	115	200	441	793	6949
Ann; Plt; −452°F Lhe	G7AA13LA2	220	260	65	1517	1793	2259
Ann; Plt; −320°F LN2	G7AA13LA4	165	220	110	1138	1517	3822
Nitronic 50							
Ann; Plt	G7CA13AB1	77	120	180	531	827	6254
Ann; Plt; −452°F Lhe	G7CA13LA2	210	275	90	1448	1896	3127
Ann; Plt; −320°F LN2	G7CA13LA4	170	230	110	1172	1586	3822
Nitronic 60							
HR, CR; Rnd Rod	G7DC18AB1	137	192	60	945	1324	2085
[H] High-Temperature Steel							
Nickel Chromium							
A286 (140 ksi)							
Plt & Sht; 600°F to 800°F	H1AB13AA15	95	138	80	655	951	2780
Plt & Sht	H1AB13AB1	100	140	100	689	965	3475
Forg, L-T, T-L, L-R	H1AB23AB1	100	140	100	689	965	3475
A286 (160 ksi)							
Plt & Sht; 600°F to 800°F	H1AC13AA15	95	138	80	655	951	2780
Plt & Sht	H1AC13AB1	105	160	100	724	1103	3475

*Unless noted, assume Lab Air (LA) environment and any orientation except S-T, S-L, C-R, C-L, and R-L.

† NASA NASGRO 3.0 material identifier code.

Material; Condition; Environment*	Code†	U. S. Customary Units (ksi, ksi√in)			SI Units (MPa, MPa√mm)		
		YS	UTS	K_{Ic}	YS	UTS	K_{Ic}
[H] High-Temperature Steel (*continued*)							
Forg. rod, L-R	H1AC28AB1	120	160	100	827	1103	3475
A286 (200-ksi bolt material)							
Forg. rod, L-R	H1AD28AB1	190	200	100	1310	1379	3475
JBK–75							
ST-CR-A; Plt, T-L	H1CB12AB1	150	180	90	1034	1241	3127
[J] Tool Steel							
AISI Tool Steel							
M–50							
61–63 Rc; Plt	J1IK10AB1	325	375	13	2241	2586	452
T1(18–4–1)							
60–63 Rc; Plt	J1MA10AB1	325	350	15	2241	2413	521
[M] 1000–9000 Series Aluminum							
2000 Series							
2014-T6							
Plt & Sht; L-T	M2AD11AB1	65	74	27	448	510	938
Plt & Sht; T-L	M2AD12AB1	63	71	18	434	490	625
2014-T651							
Plt & Sht; L-T	M2AF11AB1	64	71	22	441	490	764
Plt & Sht; T-L	M2AF12AB1	64	71	20	441	490	695
Plt & Sht; GTA Weld	M2AFB1AB1	24	47	16	165	324	556
Plt & Sht; GTA Weld, SR	M2AFB2AB1	14	27	16	97	186	556
2020-T651							
Plt & Sht; L-T	M2CB11AB1	77	82	22.5	531	565	782
Plt & Sht; T-L	M2CB12AB1	78	82	17	538	565	591
2024-T3							
Clad, Plt, & Sht; L-T; LA & HHA	M2EA11AB1	53	66	33	365	455	1147
Clad, Plt, & Sht; T-L; LA & HHA	M2EA12AB1	48	65	29	331	448	1008
Clad, Plt, & Sht; L-T; DW	M2EA11WA1	53	66	33	365	455	1147
Clad, Plt, & Sht; T-L; DW	M2EA12WA1	48	65	29	331	448	1008
2024-T351							
Plt & Sht; L-T; 300°F to 400°F Air	M2EB11AA11	52	66	33	359	455	1147
Plt & Sht; L-T; LA & HHA	M2EB11AB1	54	68	34	372	469	1181
Plt & Sht; T-L; LA & HHA	M2EB12AB1	52	68	29	359	469	1008
2024-T3511							
Extr; L-T; LA & HHA	M2EC31AB1	55	77	25	379	531	869
2024-T62							
Plt & Sht; L-T; LA, HHA & ASW	M2EG11AB1	58	66	36	400	455	1251
Plt & Sht; T-L; LA, HHA & ASW	M2EG12AB1	57	66	30	393	455	1042
2024-T81							
Plt & Sht; L-T; 350°F Air	M2EI11AA11	52	61	25	359	421	869
Plt & Sht; L-T	M2EI11AB1	63	75	22	434	517	764
Plt & Sht; L-T; DA	M2EI11AC1	63	70	22	434	483	764
Plt & Sht; L-T; HHA	M2EI11AD1	63	70	22	434	483	764
Plt & Sht; T-L; 350°F Air	M2EI12AA11	52	65	23	359	448	799
Plt & Sht; T-L	M2EI12AB1	62	68	21	427	469	730
2024-T851							
Plt & Sht; T-L; 300°F to 350°F Air	M2EJ12AA11	56	61	24	386	421	834
Plt & Sht; L-T & T-L; LA, DA, JP–4	M2EJ13AB1	64	74	23	441	510	799
Plt & Sht; L-T & T-L 3.5% NaCl	M2EJ13WB1	64	70	23	441	483	799

*Unless noted, assume Lab Air (LA) environment and any orientation except S-T, S-L, C-R, C-L, and R-L.
† NASA NASGRO 3.0 material identifier code.

Material; Condition; Environment*	Code†	U. S. Customary Units (ksi, ksi$\sqrt{in}$)			SI Units (MPa, MPa$\sqrt{mm}$)		
		YS	UTS	K_{Ic}	YS	UTS	K_{Ic}
[M] 1000–9000 Series Aluminum (*continued*)							
2000 Series							
2024-T852							
Forg; L-T & T-L , LA & DA	M2EK23AB1	57	70	28	393	483	973
2024-T861							
Plt & Sht; L-T; 300°F to 400°F Air	M2EL11AA11	56	58	36	386	400	1251
Plt & Sht; L-T; LA & HHA	M2EL11AB1	73	76	23	503	524	799
Plt & Sht; T-L	M2EL12AB1	72	76	19	496	524	660
2048-T851							
Plt & Sht; L-T; LA, DA	M2FC11AB1	63	69	35	434	476	1216
Plt & Sht; T-L; LA, DA	M2FC12AB1	62	69	30	427	476	1042
2124-T851							
Plt & Sht; L-T; 120°F to 350°F Air	M2GC11AA10	53	55	31	365	379	1077
Plt & Sht; L-T; LA, DA, HHA	M2GC11AB1	63	71	30	434	490	1042
Plt & Sht; T-L; 300°F to 400°F Air	M2GC12AA11	53	55	26	365	379	903
Plt & Sht; T-L; LA, HHA	M2GC12AB1	64	72	23	441	496	799
Plt & Sht; T-L; −200°F to −150°F GN2	M2GC12GB6	72	80	26	496	552	903
Plt & Sht; S-T, S-L; LA, HHA	M2GC15AB1	60	69	21	414	476	730
2219-T62							
Plt & Sht; L-T	M2IA11AB1	43	61	31	296	421	1077
Plt & Sht; L-T & T-L; 350°F Air	M2IA13AA11	37	46	30	255	317	1042
Plt & Sht; T-L	M2IA12AB1	43	61	29	296	421	1008
Plt & Sht; L-T & T-L; −320°F LN2	M2IA13LA4	51	76	30	352	524	1042
2219-T851							
Plt & Sht; L-T, LA, DA	M2IC11AB1	53	65	33	365	448	1147
Plt & Sht; T-L; LA, DA	M2IC12AB1	50	66	31	345	455	1077
2219-T87							
Plt & Sht; L-T; 300°F to 350°F Air	M2IF11AA11	43	49	28	296	338	973
Plt & Sht; L-T	M2IF11AB1	57	68	30	393	469	1042
Plt & Sht; L-T; −320°F LN2	M2IF11LA4	68	83	41	469	572	1425
Plt & Sht; T-L; 300°F to 350°F Air	M2IF12AA11	44	49	27	303	338	938
Plt & Sht; T-L	M2IF12AB1	58	69	27	400	476	938
Plt & Sht; T-L; −320°F LN2	M2IF12LA4	66	85	33	455	586	1147
Plt & Sht; GTA weld, PAR	M2IFB1AB1	20	42	21	138	290	730
Plt & Sht; GTA weld, PAR; −320°F LN2	M2IFB1LA4	27	46	23	186	317	799
2324-T39							
Plt & Sht; L-T	M2JA11AB1	65	72	39	448	496	1355
2090-T8E41							
Plt & Sht; L-T	M2PA11AB1	80	85	33	552	586	1147
5000 Series							
5083-O							
Plt; T-L	M5BA12AB1	20	43	45	138	296	1564
6000 Series							
6061-T6							
Plt; T-L	M6AB13AB1	41	45	26	283	310	903
Plt; GTA weld, PAR	M6ABA1AB1	23	26	26	159	179	903
6061-T651							
Plt; L-T & T-L 300°F Air	M6AC13AA10	36	37	27	248	255	938
Plt; L-T & T-L	M6AC13AB1	44	47	27	303	324	938
6063-T5							
Plt & Sht; T-L; LA	M6BA12AB1	21	27	24	145	186	834

*Unless noted, assume Lab Air (LA) environment and any orientation except S-T, S-L, C-R, C-L, and R-L.
† NASA NASGRO 3.0 material identifier code.

Material; Condition; Environment*	Code†	U. S. Customary Units (ksi, ksi$\sqrt{\text{in}}$)			SI Units (MPa, MPa$\sqrt{\text{mm}}$)		
		YS	UTS	K_{Ic}	YS	UTS	K_{Ic}
[M] 1000–9000 Series Aluminum (*continued*)							
7000 Series							
7005-T6 & T63							
Plt & Sht; L-T	M7BA11AB1	48	53	46	331	365	1598
Plt & Sht; T-L	M7BA12AB1	48	53	40	331	365	1390
7010-T73651							
Plt & Sht; L-T & L-S	M7DA11AB1	64	73	31	441	503	1077
7050-T73511							
Extr; L-T; LA, HHA, DA	M7GE31AB1	72	80	35	496	552	1216
7050-T736 & T74							
Forg; L-T	M7GI21AB1	65	72	33	448	496	1147
Forg; T-L	M7GI22AB1	62	72	24	427	496	834
7050-T73651 & T7451							
Plt & Sht; L-T; LA & HHA	M7GJ11AB1	66	77	31	455	531	1077
Plt & Sht; L-T; DA	M7GJ11AC1	66	77	31	455	531	1077
Plt & Sht; T-L; LA & HHA	M7GJ12AB1	65	77	25	448	531	869
Plt & Sht; T-L; DA	M7GJ12AC1	65	77	25	448	531	869
Plt & Sht; S-T	M7GJ15AB1	61	75	24	421	517	834
7050-T74511							
Extr; L-T; LA & DW	M7GL31AB1	70	79	36	483	545	1251
7050-T73652 & T7452							
Forg; L-T	M7GM21AB1	70	79	31	483	545	1077
Forg; T-L	M7GM22AB1	70	79	21	483	545	730
7050-T7651							
Plt & Sht; L-T; LA & HHA	M7GQ11AB1	75	80	31	517	552	1077
Plt & Sht; T-L	M7GQ12AB1	75	80	28	517	552	973
7050-T76511							
Extr; L-T; LA & HHA	M7GS31AB1	79	87	30	545	600	1042
Extr; T-L	M7GS32AB1	79	87	24	545	600	834
7000 Series							
7075-T6							
Plt, Sht & Clad; L-T & T-L; LA	M7HA13AB1	75	84	27	517	579	938
Plt, Sht & Clad; L-T & T-L; HHA	M7HA13AD1	75	84	27	517	579	938
7075-T651							
Plt & Sht; L-T; LA, DA	M7HB11AB1	76	85	28	524	586	973
Plt & Sht; L-T; HHA	M7HB11AD1	76	85	28	524	586	973
Plt & Sht; L-T; 3.5% NaCl	M7HB11WB1	76	85	28	524	586	973
Plt & Sht; T-L; DW	M7HB13WA1	76	85	24	524	586	834
Plt & Sht; S-T	M7HB15AB1	66	75	18	455	517	625
7075-T6510							
Extr; L-T; LA & DA	M7HC31AB1	79	88	28	545	607	973
7075-T6511							
Extr; L-T; LA & HHA	M7HD31AB1	80	87	28	552	600	973
Extr; T-L	M7HD32AB1	80	87	24	552	600	834
7075-T73							
Plt & Sht; L-T; LA, DA, HHA	M7HG11AB1	60	74	28	414	510	973
Plt & Sht; T-L	M7HG12AB1	61	71	23	421	490	799
7075-T7351							
Plt & Sht; L-T	M7HH11AB1	62	71	29	427	490	1008
Plt & Sht; L-T; DA	M7HH11AC1	62	71	29	427	490	1008
Plt & Sht; L-T; HHA	M7HH11AD1	62	71	29	427	490	1008

*Unless noted, assume Lab Air (LA) environment and any orientation except S-T, S-L, C-R, C-L, and R-L.
† NASA NASGRO 3.0 material identifier code

Material; Condition; Environment*	Code†	U. S. Customary Units (ksi, ksi√in)			SI Units (MPa, MPa√mm)		
		YS	UTS	K_{Ic}	YS	UTS	K_{Ic}
[M] 1000–9000 Series Aluminum (*continued*)							
Plt & Sht; T-L; LA & DA	M7HH12AB1	60	71	25	414	490	869
Plt & Sht; S-T; LA	M7HH15AB1	58	65	19	400	448	660
7075-T73510							
Extr; L-T; LA	M7HI31AB1	64	75	31	441	517	1042
7075-T73511							
Extr; L-T; LA, DA, HHA	M7HJ31AB1	65	74	33	448	510	1147
7075-T7352							
Plt, Sht, & Forg; L-T; LA & DA	M7HK11AB1	59	68	33	407	469	1147
Plt, Sht, & Forg; T-L; LA & DA	M7HK12AB1	53	68	25	365	469	869
7075-T7651							
Plt & Sht; L-T; LA & DA	M7HM11AB1	66	76	32	455	524	1112
Plt & Sht; T-L; LA & DA	M7HM12AB1	66	75	23	455	517	799
7079-T651							
Plt & Sht; L-T	M7IC11AB1	75	83	26	517	572	903
7000 Series							
7149-T73511							
Extr; L-T; LA	M7NA31AB1	66	76	31	455	524	1077
Extr; T-L; LA	M7NA32AB1	63	74	24	434	510	834
7178-T6 & T651							
Plt & Sht; L-T; LA & HHA	M7RA11AB1	84	89	24	579	614	834
Plt & Sht; T-L; LA & HHA	M7RA12AB1	79	89	21	545	614	730
7178-T7651							
Plt & Sht; L-T	M7RF11AB1	72	80	28	496	552	973
Plt & Sht; T-L	M7RF12AB1	70	79	23	483	545	799
7475-T61							
Plt, Sht, Clad; L-T; LA, DA, HHA	M7TB11AB1	74	78	31	510	538	1077
Plt, Sht, Clad; T-L; LA, DA, HHA	M7TB12AB1	71	78	26	490	538	903
7475-T651							
Plt & Sht; L-T; LA, DA, HHA	M7TD11AB1	75	85	37	517	586	1286
7475-T7351							
Plt & Sht; L-T; LA, DA, HHA, DW	M7TF11AB1	63	73	43	434	503	1494
Plt & Sht; T-L; LA, DA, HHA	M7TF12AB1	57	70	35	393	483	1216
7475-T7651							
Plt & Sht; L-T; LA, DA, HHA, DW, 3.5% NaCl	M7TJ11AB1	69	78	40	476	538	1390
[O] Misc. and Cast Aluminum							
300 Series cast							
A356-T60							
Cast	O3FB50AB1	31	40	16	214	276	556
[P] Titanium Alloys							
Ti Unalloyed							
Ti–55							
Plt & Sht	P1AA13AB1	55	65	50	379	448	1737
Plt & Sht; DW & SW	P1AA13WA1	55	65	50	379	448	1737
Ti–70							
Plt & Sht	P1CA13AB1	70	80	50	483	552	1737
Plt & Sht; DW & SW	P1CA13WA1	70	80	50	483	552	1737
Binary Alloys							
Ti–2.5 Cu; STA							
Sht; LA, HHA, DW	P2AA13AB1	97	110	50	669	758	1737

*Unless noted, assume Lab Air (LA) environment and any orientation except S-T, S-L, C-R, C-L, and R-L.
† NASA NASGRO 3.0 material identifier code.

Material; Condition; Environment*	Code†	U. S. Customary Units (ksi, ksi √in)			SI Units (MPa, MPa√mm)		
		YS	UTS	K_{Ic}	YS	UTS	K_{Ic}
[P] Titanium alloys (continued)							
Ternary Alloys							
Ti–5Al–2.5Sn; Annealed							
Sht; LA, HHA, DW	P3CA13AB1	120	130	65	827	896	2259
Ti–5Al–2.5Sn (ELI); Annealed							
Forg	P3CB23AB1	115	120	65	793	827	2259
Forg; --423°F LH2	P3CB23LA3	200	210	60	1379	1448	2085
Ti–3Al–2.5V; CW, SR (750°F)							
Extr	P3DB33AB1	105	125	50	724	862	1737
Ti–6Al–4V (MA)							
Plt & Sht, −100°F	P3EA13AA7	165	170	50	1138	1172	1737
Plt & Sht	P3EA13AB1	138	146	50	951	1007	1737
Forg	P3EA23AB1	135	145	50	931	1000	1737
Extr	P3EA33AB1	125	140	60	862	965	2085
Ti–6Al–4V; BA (1900°F/0.5h + 1325°F/2h)							
Plt & Sht; LA, DA, 3.5% NaCl	P3EB12AB1	120	135	80	827	931	2780
Forg; LA, DA, HHA, 3.5% NaCl	P3EB23AB1	110	125	80	758	862	2780
Ti–6Al–4V; RA							
Sht; L-T; LA, DA, HHA, DW, 3.5% NaCl	P3EC11AB1	140	150	60	965	1034	2085
Sht; T-L; LA, DA, HHA, DW, 3.5% NaCl	P3EC12AB1	140	150	60	965	1034	2085
Plt; −100°F	P3EC13AA7	160	175	65	1103	1207	2259
Plt; LA, DA, HHA, DW	P3EC13AB1	125	135	75	862	931	2606
Forg; LA, DA, HHA, 3.5% NaCl	P3EC23AB1	115	130	75	793	896	2606
Ti–6Al–4V; ST (1750°F) + A (1000°F/4h)							
Plt & Sht; SR (1000°F/4h)	P3ED13AA1	155	167	45	1069	1151	1564
Plt & Sht; SR (1000°F/8h)	P3ED13AB1	140	150	45	965	1034	1564
Plt & Sht; SR (1000°F/4h); −320°F LN2	P3ED13LA4	230	240	38	1586	1655	1320
Forg; SR (1000°F/4h)	P3ED20AB1	150	163	42	1034	1124	1459
Forg; SR (1000°F/4h); −320°F LN2	P3ED20LA4	225	235	40	1551	1620	1390
GTA Weld; SR; thk < 0.2"	P3EDB2AB1A	125	135	40	862	931	1390
GTA Weld; SR; thk >= 0.2"	P3EDB2AB1B	125	135	55	862	931	1911
Ti–6Al–4V; ELI; BA (1900°F/0.5h) + 1325°F/2h)							
Plt & sht; LA, 3.5% NaCl	P3EL12AB1	115	127	80	793	876	2780
Ti–6Al–4V (ELI) RA							
Plt	P3EM13AB1	120	130	75	827	896	2606
Forg; −100°F	P3EM23AA7	145	155	75	1000	1069	2606
Forg	P3EM23AB1	120	130	75	827	896	2606
Forg; −452°F LHe	P3EM23LA2	240	248	50	1655	1710	1737
Forg; −320°F LN2	P3EM23LA4	200	213	55	1379	1469	1911
Forg; EB welded, SR; weldline	P3EMD2AB1	120	130	65	827	896	2259
Forg; EB welded, SR; weldline; −320°F LN2	P3EMD2LA4	200	213	55	1379	1469	1911
Forg; EB welded, SR; HAZ	P3EMD8AB1	120	130	65	827	896	2259
Forg; EB welded, SR; HAZ; −320°F LN2	P3EMD8LA4	200	213	55	1379	1469	1911
Quaternary Alloys							
Ti–4.5Al–5Mo–1.5Cr							
Plt; LA, 3.5% NaCl	P4BA10AB1	170	182	45	1172	1255	1564
Ti–8Al–1Mo–1V							
Sht	P4CB11AB1	138	150	55	951	1034	1911
Ti–6Al–6V–2Sn MA							
Plt, Forg, Extr; LA, DA, HHA, DW	P4DA33AB1	150	160	50	1034	1103	1737

*Unless noted, assume Lab Air (LA) environment and any orientation except S-T, S-L, C-R, C-L, and R-L.
† NASA NASGRO 3.0 material identifier code.

Material; Condition; Environment*	Code†	U. S. Customary Units (ksi, ksi$\sqrt{in}$)			SI Units (MPa, MPa$\sqrt{mm}$)		
		YS	UTS	K_{Ic}	YS	UTS	K_{Ic}
[P] Titanium alloys (*continued*)							
Ti–6Al–6V–2Sn RA							
Plt	P4DB12AB1	150	160	65	1034	1103	2259
Ti–6Al–6V–2Sn BA							
Plt	P4DC12AB1	140	155	60	965	1069	2085
Ti–6Al–6V–2Sn ST (1600°F); A (1000°F/6h)							
Forg; −65°F	P4DD21AA7	205	212	28	1413	1462	973
Forg; 300°F	P4DD21AA10	165	172	53	1138	1186	1842
Forg; LA, DA, HHA	P4DD21AB1	180	190	30	1241	1310	1042
Ti–10V–2Fe–3Al							
STA(140–160 UTS, 70K_{Ic}) Plt & Forg	P4MD23AB1	140	150	70	965	1034	2432
STA(160–180 UTS, 60K_{Ic}) Plt & Forg	P4MF23AB1	153	170	60	1055	1172	2085
STA(180–200 UTS, 40K_{Ic}) Plt & Forg	P4MG13AB1	178	190	40	1227	1310	1390
STA(180–200 UTS, 30K_{Ic}) Plt & Forg	P4MG20AB1	178	190	30	1227	1310	1042
STA(180–200 UTS, 25K_{Ic}) Forg	P4MG23AB1	178	190	25	1227	1310	869
Ti–6A;-2Zn–2Sn–2Mo–2Cr (ST or STA)							
Plt; HHA	P5FB11AD1	155	165	55	1069	1138	1911
[Q] Ni Alloys/Superalloys							
Hastelloy Alloys							
Hastelloy B							
Rnd Rod	Q1AA16AB1	60	127	100	414	876	3475
Hastelloy X–280; ST (2150°F)							
Plt; 600° to 800°F Air	Q1QA10AA15	39	91	110	269	627	3822
Plt; 1000° to 1200°F Air; > 0.67 Hz	Q1QA10AA19	33	84	100	228	579	3475
Plt	Q1QA10AB1	53	109	120	365	752	4170
Inconel Alloys							
Inconel 600							
Plt & Sht; 1000°F	Q3AB10AA18	28	83	100	193	572	3475
Plt & Sht; 75° to 800°F	Q3AB10AB1	35	94	100	241	648	3475
Inconel 625							
Plt & Sht; 600°F	Q3EA10AA14	50	122	90	345	841	3127
Plt & Sht; 800°F	Q3EA10AA16	48	119	90	331	820	3127
Plt & Sht; 1000°F	Q3EA10AA18	48	119	90	331	820	3127
Plt & Sht;	Q3EA10AB1	66	133	100	455	917	3475
Inconel 706; ST (1800° to 1950°F); A(1375°F/8h; 1150°F/5–8h)							
Forg & Extr	Q3JB33AB1	145	177	85	1000	1220	2953
Forg & Extr; −452°F Lhe	Q3JB33LA2	177	230	105	1220	1586	3648
ST Plt-GTA weld-STA	Q3JBB3AB1	145	164	45	1000	1131	1564
ST Plt-GTA weld-STA; −452°F Lhe	Q3JBB3LA2	177	213	50	1220	1469	1737
Inconel 718; ST (1700° to 1850°F)+A(1325°F/8h+1150°F/10h)							
Plt; 600°F air, > 0.3 Hz	Q3LB11AA14	150	190	90	1034	1310	3127
Plt; 800°F air, > 0.3 Hz	Q3LB11AA16	142	180	85	979	1241	2953
Plt; 1000°F air, > 0.3 Hz	Q3LB13AA18	135	178	70	931	1227	2432
Sht ($t < 0.25"$)	Q3LB13AB1A	175	210	85	1207	1448	2953
Plt	Q3LB13AB1B	170	200	90	1172	1379	3127
Forg	Q3LB23AB1	165	190	90	1138	1310	3127
Forg; 300°F air, > 0.3 Hz	Q3LB26AA10	160	187	90	1103	1289	3127
Forg; 600°F air, > 0.3 Hz	Q3LB26AA14	155	185	90	1069	1276	3127
Forg; 800°F air, > 0.3 Hz	Q3LB26AA16	147	173	85	1014	1193	2953
Forg; 1000°F air, > 0.3 Hz	Q3LB26AA18	140	167	70	965	1151	2432
GTA weld-STA; 600°F air, > 0.6 Hz	Q3LBB3AA14	140	170	60	965	1172	2085

*Unless noted, assume Lab Air (LA) environment and any orientation except S-T, S-L, C-R, C-L, and R-L.

† NASA NASGRO 3.0 material identifier code.

Material; Condition; Environment*	Code†	U. S. Customary Units (ksi, ksi $\sqrt{in}$)			SI Units (MPa, MPa $\sqrt{mm}$)		
		YS	UTS	K_{Ic}	YS	UTS	K_{Ic}
[Q] Ni Alloys/Superalloys (*continued*)							
GTA weld-STA; 800°F air, > 0.6 Hz	Q3LBB3AA16	135	160	55	931	1103	1911
GTA weld-STA; 1000°F air, > 0.6 Hz	Q3LBB3AA18	132	160	45	910	1103	1564
ST plt-GTA weld-aged	Q3LBB3AB1	160	192	55	1103	1324	1911
ST plt-GTA weld-aged; −320°F LN2	Q3LBB3LA4	178	227	55	1227	1565	1911
ST plt-EB weld-aged	Q3LBD3AB1	163	205	45	1124	1413	1564
ST plt-EB weld-aged; −320°F LN2	Q3LBD3LA4	193	224	40	1331	1544	1390
Inconel 718; ST (1900°F) + A (1400°F/10h + 1200°F/10h)							
Plt; −320°F LN2	Q3LC10LA4	191	225	90	1317	1551	3127
Plt; 600°F air; > 0.6 Hz	Q3LC11AA14	158	180	90	1089	1241	3127
Plt	Q3LC11AB1	165	205	90	1138	1413	3127
Plt; 1000°F air; > 0.6 Hz	Q3LC12AA18	145	170	70	1000	1172	2432
Inconel 718; ST (2000°F) + A (1325°F/4h + 1150°F/16h)							
Plt; 800°F air; > 0.2 Hz	Q3LE11AA16	133	170	85	917	1172	2953
Plt; 1000°F air; > 0.6 Hz	Q3LE13AA18	125	157	70	862	1082	2432
Plt	Q3LE13AB1	145	187	90	1000	1289	3127
GTA weld-STA; 600°F air, > 0.6 Hz	Q3LEB3AA14	140	180	60	965	1241	2085
GTA weld-STA; 800°F air, > 0.6 Hz	Q3LEB3AA16	132	175	55	910	1207	1911
GTA weld-STA; 1000°F air, > 0.6 Hz	Q3LEB3AA18	129	167	50	889	1151	1737
GTA weld-STA	Q3LEB3AB1	147	193	60	1014	1331	2085
Inconel 718, Bolt Material							
185 ksi UTS Bolts	Q3LP18AB1	180	205	70	1241	1413	2432
225 ksi UTS Bolts	Q3LQ18AB1	205	225	55	1413	1551	1911
Inconel X–750; ST (2100°F) + A (1550°F/24h + 1300°F/20h)							
Plt; 600°F air, > 0.6 Hz	Q3SD10AA14	87	147	50	600	1014	1737
Plt; 800°F air, > 0.6 Hz	Q3SD10AA16	83	143	50	572	986	1737
Plt; 1000°F air, > 0.6 Hz	Q3SD10AA18	81	134	50	558	924	1737
Plt & Forg	Q3SD26AB1	100	150	60	689	1034	2085
Forg; −452°F Lhe	Q3SD26LA2	125	220	70	862	1517	2432
Rene and Udimet Alloys							
Rene 41; ST (1950°F) + A (1400°F/16h)							
Plt & Forg	Q7AD13AB1	138	184	75	951	1269	2606
Forg; 1100°F air	Q7AD26AA19	112	157	60	772	1082	2085
Forg; 1200°F air	Q7AD26AA20	110	155	55	758	1069	1911
[R] Misc. Superalloys							
Multiphase Alloys							
MP35N Rnd Rod	R3AB18AB1	254	274	80	1751	1889	2780
[S] Copper/Bronze Alloys							
Be-Cu Alloys							
CDA 172							
Rnd Rod	S0BA13AB1	159	179	26	1096	1234	903
C17510							
Peak-aged Plt	S1LB11AB1	110	125	75	758	862	2606
Peak-aged Plt; −320°F LN2	S1LB11LA4	120	141	110	827	972	3822
Overaged Plt	S1LC11AB1	86	99	70	593	683	2432
Al-Bronze Alloys							
CDA 630 Al-Bronze Extr	S6JB36AB1	82	117	53	565	807	1842
[T] Magnesium Alloys							
AM 503 Plt	T1AA11AB1	16	29	12	110	200	417
AZ–31B-H24 Plt	T1DA12AB1	26	39	20	179	269	695
ZK–60A-T5 Plt	T1MA12AB1	39	50	20	269	345	695

*Unless noted, assume Lab Air (LA) environment and any orientation except S-T, S-L, C-R, C-L, and R-L.
† NASA NASGRO 3.0 material identifier code

Material; Condition; Environment*	Code†	U. S. Customary Units (ksi, ksi$\sqrt{in}$)			SI Units (MPa, MPa$\sqrt{mm}$)		
		YS	UTS	K_{Ic}	YS	UTS	K_{Ic}
[T] Magnesium Alloys (*continued*)							
ZW1 Plt	T1NA11AB1	24	36	13	165	248	452
QE22A-T6 Plt	T2LB13AB1	26	33	13	179	228	452
[U] Misc. Nonferrous Alloys							
Beryllium							
Cross-rolled sht	U1CA90AB1	55	70	9.5	379	483	313
Hot-pressed blk	U1CA93AB1	35	60	10	241	414	347
Columbium Alloys							
C–103 Plt	U2CA10AB1	43	50	30	296	345	1042
Zinc Alloys							
Zn–4Al–0.04Mg Die cast alloy No. 3							
As cast; 135° to 200°F air0	U4BA50AA9	25	34	14	172	234	486
As cast	U4BA50AB1	30	41	14	207	283	486

*Unless noted, assume Lab Air (LA) environment and any orientation except S-T, S-L, C-R, C-L, and R-L.
† NASA NASGRO 3.0 material identifier code.

NASGRO 3.0 Material Constants for Selected Materials

Adapted from NASA Report JSC–22267B, March 2000

Disclaimer

The data in the following tables have been adapted from information compiled by the National Aeronautics and Space Administration (NASA). The data are believed to be representative of the material described, but, due to the wide variations possible, the values should not be relied upon to accurately represent the properties of a specific batch of material. Critical design decisions should be based on results taken from the actual lot of material to be used for fabrication, tested under conditions that as accurately as possible mirror service conditions.

The data are provided "as is" without warranty of any kind, either expressed or implied. Neither the author nor the publisher shall be liable for any damages arising from the use of these illustrative data.

Notation

Code: the identifier code used in NASMAT to identify the complete data set from which the the fatigue constants were determined.

YS: yield stress σ_{ys}.

UTS: ultimate tensile strength.

K_{Ie}: critical plane-strain stress intensity factor for elliptical cracks determined from part-through fracture test specimens. For surface-breaking cracks, a critical value of $1.1 K_{Ie}$ is employed in NASGRO 3.0.

K_{Ic}: critical plane-strain stress intensity factor for through cracks.

A_k, B_k: constants used to construct K_c from K_{Ic} values according to the formula

$$\frac{K_c}{K_{Ic}} = 1 + B_k e^{-(A_k t/t_0)^2}, \quad \text{where} \quad t_0 = 2.5 \left(K_{Ic}/\sigma_{ys}\right)^2$$

C, n, p, q: best-fit values to the four-parameter fatigue crack growth model used in NASGRO 3.0 [from Eqs. (9.14) and (9.15)],

$$\frac{da}{dN} = \frac{C(1 - f)^n \Delta K^n \left(1 - \frac{\Delta K_{th}}{\Delta K}\right)^p}{(1 - R)^n \left(1 - \frac{\Delta K}{(1-R)K_c}\right)^q}$$

where the plasticity induced crack closure function f is defined as

$$f = \frac{K_{op}}{K_{max}}$$

The NASGRO Manual observes that setting p and q to 0 reduces the foregoing equation to a closure-corrected variant of the Paris law for positive stress ratios and implies that the constants C and n from the table are still suitable.

ΔK_{th}: threshold stress intensity factor for $R = 0$. For $R \neq 0$, the NASGRO 3.0 program introduces a correction factor.

Table E.1 NASGRO 3.0 MATERIALS CONSTANTS FOR SELECTED MATERIALS U. S. Customary Units (ksi, ksi$\sqrt{in}$)

Material; Condition; Environment*	Code	YS	UTS	K_{1e}	K_{1c}	A_k	B_k	C	n	p	q	ΔK_{th} R = 0
[B] ASTM Spec.-Grd. Steel												
A10 Series												
A36												
Plt (Dyn K_{1c}, < 500 Hz): LA, HHA, 3% NaCl	B0CB10AB1	44	78	100	70	0.75	0.5	0.100E-8	3.000	0.5	0.5	7.0
A500 Series												
A508 Cl2 & Cl3												
Forg	B5AC21AB1	65	100	140	100	0.75	0.5	0.100E-8	2.800	0.5	0.5	6.0
A514 Typ F												
Plt	B5BF10AB1	105	120	115	85	0.75	0.5	0.200E-8	2.570	0.25	0.25	4.0
A533-B, Cl1 & Cl2												
Plt	B5HD10AB1	70	100	200	150	0.75	0.5	0.100E-8	2.700	0.5	0.5	6.5
SMA Weld	B5HDF1AB1	60	90	140	100	0.75	0.5	0.428E-9	3.138	0.5	0.25	9.82
[C] AISI - SAE Steel												
AISI 10xx–12xx Steel												
Low-Carbon 1005–1012												
Hot-rolled plt	C1AB11AB1	25	45	100	70	0.75	0.5	0.800E-10	3.600	0.5	0.5	8.0
Low-Carbon 1015–1025												
Hot-rolled plt	C1BB11AB1	30	58	100	70	0.75	0.5	0.800E-10	3.600	0.5	0.5	8.0
AISI 43xx–48xx Steel												
4340												
160–180 UTS; Plt & Forg	C4DC21AB1	155	170	190	135	0.75	0.5	0.170E-8	2.700	0.25	0.25	6.0
180–200 UTS; Plt & Forg	C4DD21AB1	175	190	155	110	0.75	0.5	0.130E-8	2.700	0.25	0.25	5.5
200–220 UTS; Plt & Forg	C4DE11AB1	195	210	110	80	0.75	0.5	0.130E-8	2.700	0.25	0.25	4.5
220–240 UTS; Plt & Forg	C4DF11AB1	215	230	85	65	0.75	0.75	0.130E-8	2.700	0.25	0.25	4.0
240–280 UTS; Plt & Forg	C4DG11AB1	240	260	65	55	0.75	1.0	0.130E-8	2.700	0.25	0.25	3.5

Table E.1 *continued*

Material; Condition; Environment*	Code	YS	UTS	K_{Ie}	K_{Ic}	A_k	B_k	C	n	p	q	ΔK_{th} $R=0$
[E] Trade/Common-Name Steel												
Ultra-High-Strength Steel												
18 Ni Maraging												
250 Grd; Plt & Forg	E1AD10AB1	240	260	90	75	0.75	0.75	0.350E-8	2.600	0.25	0.25	3.5
300 Grd; Plt & Forg	E1AE10AB1	280	290	85	70	0.75	0.75	0.300E-8	2.600	0.25	0.25	3.0
300M												
270–300 UTS; Plt & Forg	E1BF21AB1	240	285	65	55	0.75	1.0	0.500E-8	2.460	0.25	0.25	3.0
HY–180 (10Ni)												
Plt & Forg	E1IB13AB1	180	200	200	150	0.75	0.50	0.600E-8	2.300	0.25	0.25	4.0
HY-TUF												
220–240 UTS; VAR Forg	E1IB23AB1	200	230	150	110	0.75	0.75	0.350E-8	2.500	0.25	0.25	4.0
Pressure Vessel / Piping												
HY 80												
Plt	E2AA13AB1	90	105	250	200	0.75	0.50	0.150E-8	2.500	0.25	0.25	5.5
HY 130												
Plt	E2CA13AB1	140	150	250	200	0.75	0.50	0.300E-8	2.500	0.25	0.25	5.0
GMA Weld	E2CAC1AB1	130	140	250	200	0.75	0.50	0.150E-8	2.500	0.25	0.25	5.0
SMA Weld	E2CAF1AB1	130	140	250	200	0.75	0.50	0.120E-8	2.500	0.25	0.25	5.0
Construction Grade												
HT-80												
Plt	E3BA13AB1	110	120	200	150	0.75	0.50	0.700E-9	3.000	0.25	0.25	8.0
SA Weld	E3BAH1AB1	95	115	200	150	0.75	0.50	0.250E-8	2.700	0.25	0.25	2.5
[F] AISI-Type Stainless Steel												
AISI 300 Series												
AISI 304/304L												
Ann Plt & Sht, Cast	F3DA13AB1	40	90	280	200	1.0	0.50	.600E-9	3.0	0.25	0.25	3.5

Table E.1 *continued*

Material; Condition; Environment*	Code	YS	UTS	K_{1e}	K_{1c}	A_k	B_k	C	n	p	q	ΔK_{th} $R=0$
AISI 316/316L												
Ann Plt & Sht, Cast	F3KA13AB1	36	90	280	200	1.0	0.50	.800E-9	3.0	0.25	0.25	3.5
[M] 1000–9000 Series Aluminum												
2000 Series												
2024-T3												
Clad, Plt & Sht; L-T; LA	M2EA11AB1	53	66	46	33	1.0	1.0	0.829E-8	3.284	0.5	1.0	2.9
Clad, Plt & Sht; T-L; LA	M2EA12AB1	48	65	41	29	1.0	1.0	0.244E-7	2.601	0.5	1.0	2.9
2024-T351												
Plt & Sht; L-T; LA	M2EB11AB1	54	68	48	34	1.0	1.0	0.922E-8	3.353	0.5	1.0	2.6
Plt & Sht; T-L; LA	M2EB12AB1	52	68	41	29	1.0	1.0	0.922E-8	3.353	0.5	1.0	2.6
2024-T3511												
Extr; L-T; LA	M2EC31AB1	55	77	35	25	1.0	1.0	0.200E-7	2.700	0.5	1.0	2.9
7000 Series 7075-T651												
Plt & Sht; L-T; LA	M7HB11AB1	76	85	38	28	1.0	1.0	0.233E-7	2.885	0.5	1.0	3.0
Plt & Sht; L-T; 3.5% NaCl	M7HB11WB1	76	85	38	28	1.0	1.0	0.339E-6	2.135	0.5	1.0	3.0
Plt & Sht; T-L; DW	M7HB13WA1	76	85	32	24	1.0	1.0	0.191E-6	1.917	0.5	1.0	3.0
Plt & Sht; S-T	M7HB15AB1	66	75	23	18	1.0	1.0	0.578E-7	2.435	0.5	1.0	3.0
7075-T6510												
Extr; L-T; LA	M7HC31AB1	79	88	38	28	1.0	1.0	0.184E-6	1.869	0.5	1.0	3.0
[P] Titanium Alloys												
Ti-6Al-4V (MA)												
Plt & Sht	P3EA13AB1	138	146	65	50	1.0	0.5	0.252E-8	3.010	0.25	0.75	3.5
Forg	P3EA23AB1	135	145	65	50	1.0	0.5	0.311E-9	3.667	0.25	0.75	3.5
Extr	P3EA33AB1	125	140	75	60	1.0	0.5	0.147E-9	3.834	0.25	0.75	5.0

*Unless noted, assume a Lab Air (LA) environment and any orientation except S-T, S-L, C-R, C-L, and R-L.

Table E.2 NASGRO 3.0 MATERIALS CONSTANTS FOR SELECTED MATERIALS SI Units (MPa, MPa$\sqrt{\text{mm}}$)

Material; Condition; Environment*	Code	YS	UTS	K_{Ie}	K_{Ic}	A_k	B_k	C	n	p	q	ΔK_{th} $R=0$
[B] ASTM Spec.-Grd. Steel												
A10 Series												
A36												
Plt (Dyn K_{Ic}, < 500 Hz); LA, HHA, 3% NaCl	B0CB10AB1	303	538	3475	2432	0.75	0.5	0.605E-12	3.000	0.5	0.5	243
A500 Series												
A508 Cl2 & Cl3												
Forg	B5AC21AB1	448	689	4865	3475	0.75	0.5	0.778E-10	2.800	0.5	0.5	208
A514 Typ F												
Plt	B5BF10AB1	724	827	3996	2953	0.75	0.5	0.557E-11	2.570	0.25	0.25	139
A533-B, Cl1 & Cl2												
Plt	B5HD10AB1	483	689	6949	5212	0.75	0.5	0.176E-11	2.700	0.5	0.5	226
SMA Weld	B5HDF1AB1	414	621	4865	3475	0.75	0.5	0.159E-12	3.138	0.5	0.25	341
[C] AISI - SAE Steel												
AISI 10xx–12xx Steel												
Low-Carbon 1005–1012												
Hot-rolled plt	C1AB11AB1	172	310	3475	2432	0.75	0.5	0.576E-14	3.600	0.5	0.5	278
Low-Carbon 1015–1025												
Hot-rolled plt	C1BB11AB1	207	400	3475	2432	0.75	0.5	0.576E-14	3.600	0.5	0.5	278
AISI 43xx–48xx Steel												
4340												
1100–1240 UTS; Plt & Forg	C4DC21AB1	1069	1172	6602	4691	0.75	0.5	0.298E-11	2.700	0.25	0.25	208
1240–1380 UTS; Plt & Forg	C4DD21AB1	1207	1310	5386	3822	0.75	0.5	0.228E-11	2.700	0.25	0.25	191
1380–1520 UTS; Plt & Forg	C4DE11AB1	1344	1448	3822	2780	0.75	0.5	0.228E-11	2.700	0.25	0.25	156
1520–1660 UTS; Plt & Forg	C4DF11AB1	1482	1586	2953	2259	0.75	0.75	0.228E-11	2.700	0.25	0.25	139
1660–1930 UTS; Plt & Forg	C4DG11AB1	1655	1793	2259	1911	0.75	1.0	0.228E-11	2.700	0.25	0.25	122

Table E.2 *continued*

Material; Condition; Environment*	Code	YS	UTS	K_{Ie}	K_{Ic}	A_k	B_k	C	n	p	q	ΔK_{th} $R=0$
[E] Trade/Common Name Steel												
Ultra-High-Strength Steel												
18 Ni Maraging												
250 Grd; Plt & Forg	E1AD10AB1	1655	1793	3127	2606	0.75	0.75	0.876E-11	2.600	0.25	0.25	104
300 Grd; Plt & Forg	E1AE10AB1	1931	1999	2953	2432	0.75	0.75	0.751E-11	2.600	0.25	0.25	104
300M												
1860–2070 UTS; Plt & Forg	E1BF21AB1	1655	1965	2259	1911	0.75	1.0	0.206E-10	2.460	0.25	0.25	104
HY–180 (10Ni)												
Plt & Forg	E1IB13AB1	1241	1379	6949	5212	0.75	0.50	0.435E-10	2.300	0.25	0.25	139
HY-TUF												
1520–1660 UTS; VAR Forg	E1JB23AB1	1379	1586	5212	3822	0.75	0.75	0.125E-10	2.500	0.25	0.25	139
Pressure Vessel / Piping												
HY 80												
Plt	E2AA13AB1	621	724	8687	6949	0.75	0.50	0.536E-11	2.500	0.25	0.25	191
HY 130												
Plt	E2CA13AB1	965	1034	8687	6949	0.75	0.50	0.107E-10	2.500	0.25	0.25	174
GMA Weld	E2CAC1AB1	896	965	8687	6949	0.75	0.50	0.535E-11	2.500	0.25	0.25	174
SMA Weld	E2CAF1AB1	896	965	8687	6949	0.75	0.50	0.428E-11	2.500	0.25	0.25	174
Construction Grade												
HT–80												
Plt	E3BA13AB1	758	827	6949	5212	0.75	0.50	0.424E-12	3.000	0.25	0.25	278
SA Weld	E3BAH1AB1	655	793	6949	5212	0.75	0.50	0.439E-11	2.700	0.25	0.25	87
[F] AISI-Type Stainless Steel												
AISI 300 Series												
AISI 304/304L												
Ann Plt & Sht, Cast	F3DA13AB1	276	621	9729	6949	1.0	0.50	0.363E-12	3.0	0.25	0.25	104

Table E.2 *continued*

Material; Condition; Environment*	Code	YS	UTS	K_{Ie}	K_{Ic}	A_k	B_k	C	n	p	q	ΔK_{th} $R=0$
AISI 316/316L												
Ann Plt & Sht, Cast	F3KA13AB1	248	621	9729	6949	1.0	0.50	0.485E–12	3.0	0.25	0.25	121
[M] 1000–9000 Series Aluminum												
2000 Series												
2024-T3												
Clad, Plt & Sht; L-T; LA	M2EA11AB1	365	455	1598	1147	1.0	1.0	0.183E–11	3.284	0.5	1.0	101
Clad, Plt & Sht; T-L; LA	M2EA12AB1	331	448	1425	1008	1.0	1.0	0.609E–10	2.601	0.5	1.0	101
2024-T351												
Plt & Sht; L-T; LA	M2EB11AB1	372	469	1668	1181	1.0	1.0	0.160E–11	3.353	0.5	1.0	90
Plt & Sht; T-L; LA	M2EB12AB1	359	469	1425	1008	1.0	1.0	0.160E–11	3.353	0.5	1.0	90
2024-T3511												
Extr; L-T; LA	M2EC31AB1	379	531	1216	869	1.0	1.0	0.351E–10	2.700	0.5	1.0	101
7000 Series												
7075-T651												
Plt & Sht; L-T; LA	M7HB11AB1	524	586	1320	973	1.0	1.0	0.212E–10	2.885	0.5	1.0	104
Plt & Sht; L-T; 3.5% NaCl	M7HB11WB1	524	586	1320	973	1.0	1.0	0.442E–8	2.135	0.5	1.0	104
Plt & Sht; T-L; DW	M7HB13WA1	524	586	1112	834	1.0	1.0	0.539E–8	1.917	0.5	1.0	104
Plt & Sht; S-T	M7HB15AB1	455	517	799	625	1.0	1.0	0.260E–9	2.435	0.5	1.0	104
7075-T6510												
Extr; L-T; LA	M7HC31AB1	545	607	1320	973	1.0	1.0	0.616E–8	1.869	0.5	1.0	104
[P] Titanium Alloys												
Ti-6Al-4V (MA) Plt & Sht	P3EA13AB1	951	1007	2259	1737	1.0	0.5	0.147E–11	3.010	0.25	0.75	122
Forg	P3EA23AB1	931	1000	2259	1737	1.0	0.5	0.177E–13	3.667	0.25	0.75	122
Extr	P3EA33AB1	862	965	2606	2085	1.0	0.5	0.462E–14	3.834	0.25	0.75	174

*Unless noted, assume a Lab Air (LA) environment and any orientation except S-T, S-L, C-R, C-L, and R-L.

Author Index

Subject Index